Optimized Computational Intelligence Driven Decision-Making

Scrivener Publishing
100 Cummings Center, Suite 541J
Beverly, MA 01915-6106

Industry 5.0 Transformation Applications

Series Editors: Dr. S. Balamurugan and Dr. Sheng-Lung Peng

Scope: The increase in technological advancements in the areas of artificial intelligence (AI), machine learning (ML) and data analytics has led to the next industrial revolution "Industry 5.0". The transformation to Industry 5.0 collaborates human intelligence with machines to customize efficient solutions. This book series aims to cover various subjects under promising application areas of Industry 5.0 such as smart manufacturing, green ecology, digital medicine, supply chain management, smart textiles, intelligent traffic, innovation ecosystem, cloud manufacturing, digital marketing, real-time productivity optimization, augmented reality and virtual reality, smart energy consumption, predictive maintenance, smart additive manufacturing, hyper customization and cyber physical cognitive systems. The book series will also cover titles supporting technologies for promoting potential applications of Industry 5.0, such as collaborative robots (Cobots), edge computing, Internet of Everything, big data analytics, digital twins, 6G and beyond, blockchain, quantum computing and hyper intelligent networks.

Publishers at Scrivener
Martin Scrivener (martin@scrivenerpublishing.com)
Phillip Carmical (pcarmical@scrivenerpublishing.com)

Optimized Computational Intelligence Driven Decision-Making

Theory, Application and Challenges

Edited by
Hrudaya Kumar Tripathy
Sushruta Mishra
Minakhi Rout
S. Balamurugan
and
Samaresh Mishra

Scrivener Publishing

WILEY

This edition first published 2024 by John Wiley & Sons, Inc., 111 River Street, Hoboken, NJ 07030, USA and Scrivener Publishing LLC, 100 Cummings Center, Suite 541J, Beverly, MA 01915, USA

For more information about Scrivener publications please visit www.scrivenerpublishing.com.

Wiley Global Headquarters
111 River Street, Hoboken, NJ 07030, USA

For details of our global editorial offices, customer services, and more information about Wiley products visit us at www.wiley.com.

Library of Congress Cataloging-in-Publication Data

ISBN 978-1-394-24253-5

Cover image: Pixabay.Com
Cover design by Russell Richardson

Set in size of 11pt and Minion Pro by Manila Typesetting Company, Makati, Philippines

Printed in the USA

10 9 8 7 6 5 4 3 2 1

Contents

Preface

Optimized Computational Intelligence (OCI) is a new, cutting-edge, and multidisciplinary research area that tackles the fundamental problems shared by modern informatics, biologically-inspired computation, software engineering, AI, cybernetics, cognitive science, medical science, systems science, philosophy, linguistics, economics, management science, and life sciences. OCI aims to apply modern computationally intelligent methods to generate optimum outcomes in various application domains. This book presents the latest technologies-driven material to explore optimized various computational intelligence domains.

To begin, the first chapter discusses the emergence of computational intelligence in smart sensory settings. Chapter 2 deals with the capabilities of advanced machine intelligence in educational domains. The third chapter addresses the issue of recognizing false data injection attacks with a machine learning approach in the industrial IoT sector.

Chapter 4 discusses the analysis of fake news by using modern intelligence-based approaches to prevent misinformation propagation. The fifth chapter addresses the challenges and issues of telemedicine by applying computational intelligence techniques. Chapter 6 demonstrates how to detect and predict crop suitability by deploying machine intelligence in smart sensory settings. The seventh chapter explains the significance of using advanced intelligence methods for smart, IoT-based regulation of road traffic. The eighth chapter presents a succinct analysis of text-based corpora using social network analysis.

Chapter 9 highlights the growing role of autonomous intelligent vehicles, the challenging issues related to them, and their futuristic trends. The tenth chapter discusses the impact of smart predictive analytics in healthcare within a modern urban society. Chapter 11 show how to use advanced predictive analytics to assess depression in the modern world. The twelfth chapter discusses current scenarios that demonstrate IoT-enabled healthcare standards with revolutionized guidelines.

Chapter 13 presents a detailed analysis of Parkinson's disease risk factors and explains how to apply machine learning in detection and treatment. Chapter 14 discusses the capability of computational intelligence to monitor climatic variations that are taking place in today's world. The final chapter presents a deep analysis on the relevance of using computational intelligence to address weather fluctuation.

We are deeply grateful to everyone who helped with this book and greatly appreciate the dedicated support and valuable assistance rendered by Martin Scrivener and the Scrivener Publishing team during its publication.

1

Emergence of Advanced Computational Intelligence Coupled with Smart Environment

Risha Rani* **and Tirtha Deb**

Kalinga Institute of Industrial Technology, Deemed to be University, Bhubaneswar, Odisha, India

Abstract

In this paper we have tried to work out various models which may elevate the lifestyle of us humans by using the technology of present day IoT system concept making it possible to make things or our daily requirement smart or very easily available. As for the models, we have thoroughly studied and discussed how we would build smart homes using various sub disciplines such as building garbage monitoring systems and a system in which we detect any accident. We also went through the whole idea of the present day healthcare system and we built a model for a smart healthcare system and how we can build a weather monitoring system, an Air Pollution monitoring system. Considering the extra need for the noise surrounding the environment, we have proposed a model for Noise Pollution monitoring system. We have proposed a forest fire detection system model. We also have tried to bring the knowledge of advanced computational intelligence and artificial intelligence to our work as we believe the huge significance of advanced computational intelligence and AI inside the development of smart green terrain is a manner to attain clever green frugality.

Keywords**:** Advanced computational intelligence, scalability, security, connectivity, analyzing and integration

**Corresponding author*: 2006083@kiit.ac.in

Hrudaya Kumar Tripathy, Sushruta Mishra, Minakhi Rout, S. Balamurugan and Samaresh Mishra (eds.) Optimized Computational Intelligence Driven Decision-Making: Theory, Application and Challenges, (1–28)

1.1 Introduction

The things which we can hardly imagine are encouraged and supported by new opportunities that the IoTs is reinventing. The Internet of Things is changing our physical world to grow. Through the Internet of Things, the devices that are connected through the internet are designed to become specific, customized, and intelligent to fully fill our unique and day-to-day necessity and requirements. The word smart stands for making Specific, Measurable, Achievable, Relevant and Time-bound objects and the word environment means the surroundings. A smart environment is therefore said to be the ability to acquire knowledge and apply it to modify the needs of its residents to improve their occurrence with that environment. By using various wireless technologies we can enhance the functional capabilities of smart devices. According to reports from Cisco, 50 billion objects and devices will be connected to the Internet 2020. More than 99% of things available in the world today still remain unconnected. According to a Navigant research report, the number of smart meters installed worldwide will increase to 1.1 billion by 2023. [3] Automotive News reports that worldwide the number of Internet-connected cars will increase from 23 million in 2013 to 152 million in 2024. [5] The significant growth such as forecasting shows that the Internet of Things will become a modern society to conceive a concept of a smart environment. For the integration of IoT devices along with the smart environments, several research efforts have been made with the smart environments. The possibilities of smart objects from the combination of IoT with a smart environment are expanded by allowing remote locations of the environment to be monitored by the users. Primarily based on the software requirements, IoT can be included into various smart environments. Work on intelligent IoT-based environments can generally be divided into the areas: smart cities, smart houses, and smart health. An IoT-based system consists of objects, sensor devices, and a computing and processing unit that can be located in the cloud, a decision-making and action-invoking system. IoT things and devices play a vital role in interaction and communication through data exchange. They respond to actual events in the physical world and also have the effect of triggering processes that trigger various actions and services with or without human intervention. With the enabling current era technologies like advanced computational intelligence, the world is becoming a supremely computerized environment. These enabling technologies are planned for a smart environment focusing on restful life to live in. The digital transformations experts collaborating with the application tests providers lead in the progressed conclusion of the new innovative automation and the Internet of Things. The world in the near future will utilize the knowledge of such specialists for the remodeling of the various cities. The smart environment is characterized by complex systems which require balance between transparency and context awareness. The architecture of such a system responds to the demand space and the incorporation of modular and the design of such architecture are flexible and responsible for the right time production of appropriate services.

Main contribution of the paper:

- How energy and resource management techniques are building up the nation and how more swiftly it can be utilized
- Human computer interaction and building more sophisticated software and algorithms for a smart environment
- Predictive maintenance of the society with the proper system using new technologies

1.2 Background Works

In 2017, Kanishka Majumdar Devices for Integrated Circuit (DevIC), 23–24 March 2017, Kalyani, India. "Development Board", has exact attributes like the microcontrollers in the trade but has some additional features i.e., twice the figures of input and output pins adaptable with Arduino IDE also it is very cost efficient. [1] Soil & Surrounding Testing Module by K. P. Keyur and M. P. Sunil, "Internet of Things-IOT". A farmer can easily be benefited by this model as it is an automated analysis producer of soil moisture, pH, temperature, humidity, etc. It can advise regarding the sprout growth and the amount of further fertilizers needed. LCD screens will be of great use here. Sensors like moisture sensor, pH sensors, humidity sensors, etc., are used. [12] Water Pump ON/OFF via Phone Call by M. Fahim and A. Sillitti, "Anomaly detection, analysis and prediction techniques in IoT environment: A systematic literature review," IEEE Access. It provides an important feature of turning the water pump on or off by using a mere phone call option. It reduces a great deal of labor. [13] Solar Tracking System (Renewable Power Supply) by K. P. Keyur and M. P. Sunil, "Internet of Things-IOT". Solar tracker is built on the idea that solar can provide an alternative of renewable energy fit for the respective farmland. It acts as an automatic single axis solar tracker. Sensors such as the LDR are used here. [6] Electronic Scarecrow by H. Haddad Pajouh, A. Dehghantanha,

R. Khayami, and K.-K.-R. Choo. This model basically acts as a scarecrow which can be used to keep the harmful pests off the land. Sensors such as PIRs are being used here [11].

1.3 Integrated Smart Environment

The Internet of Things (IoT) is coming with new technologies to improve human capabilities in this modern world. These means or new technologies promise a high quality of life and professional efficiency; however, with each new advancement in IoT synthesis and human augmentation technologies coming, the challenges of the IoT go far beyond that. The integrated intelligent environment is designed with various applications such as intelligent home systems, intelligent health care, intelligent transportation, intelligent agricultural systems, intelligent electronic management system, intelligent weather monitoring system, intelligent education system, etc. IoT objects and things are connected with RFID tags and sensors that are already in various business applications of the smart environment. RFID tags that utilize intelligent barcodes to identify any item use high-frequency technology in which radio waves transmit data from the tag to a reader that acts as a translator to a computer program. In Figure 1.1, the data of an integrated smart environment are stored in cloud based applications consisting of sensors and connecting IoT devices. It makes it much easier to send and analyze the given data and predict outcomes [8].

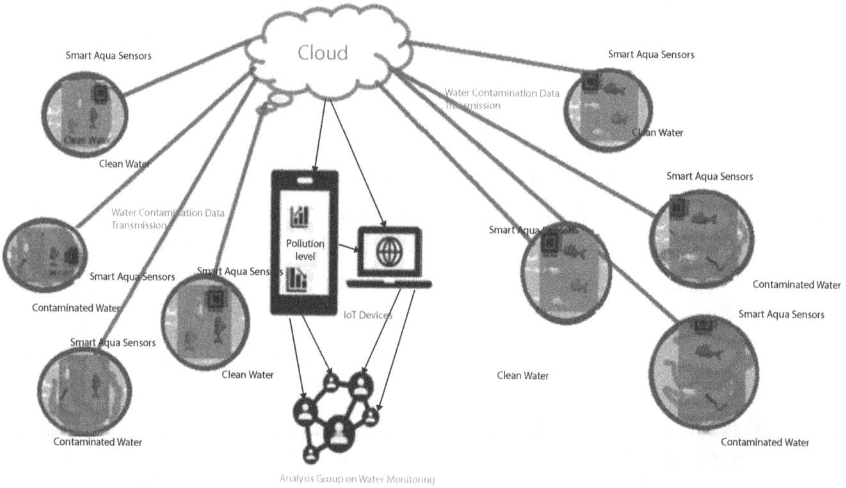

Figure 1.1 Integrated smart environment using cloud based application.

1.4 Proposed Models for Smart Intelligent Environment

Figure 1.2 denotes the sensors enabled smart environment for smart cities and homes along with smart healthcare service.

1.4.1 Smart Cities

Smart cities are technologically upgraded urban areas which use smart things, sensors, and electronic methods to collect various data and make life easier. The various functions performed by smart things can be traffic management, multi-city connectivity, pollution control, and smart lighting. The main purpose of smart cities is to make things easier and make us look at things with a new perspective. The Internet of Things makes an

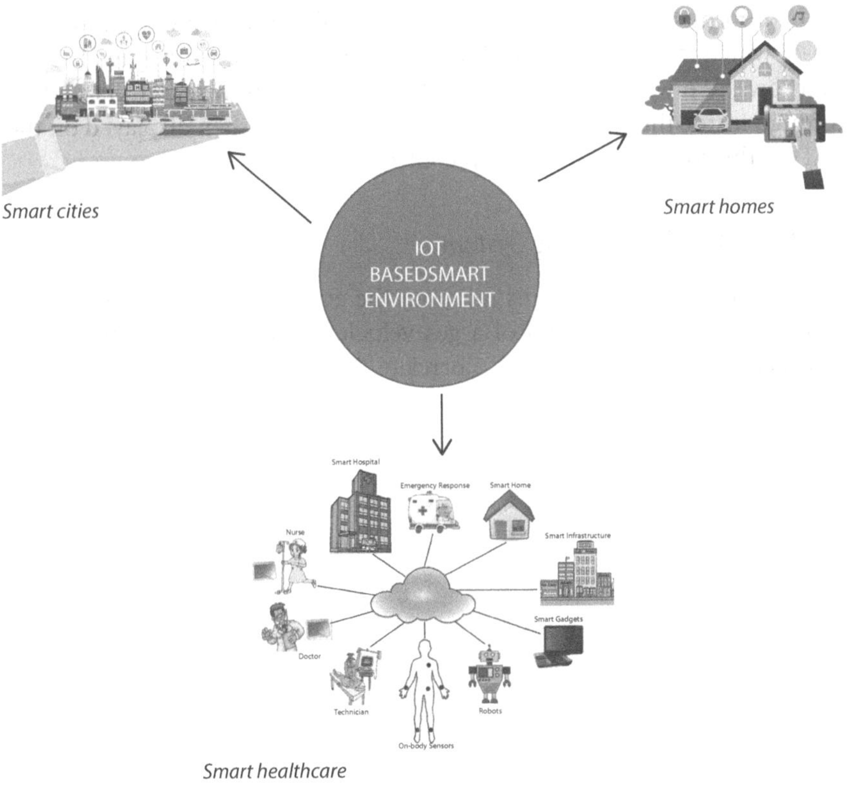

Figure 1.2 Smart environment based IoT system in smart cities, smart homes, and smart healthcare.

impact on various things such as every daily reliable life activity and also to any human's complex emotions. The Internet of things makes benefits for each day to us and the environment. Smart city can device itself as an assistant for anyone's daily schedule telling him/her to get up, make coffee or have dinner and go up to remind the person for dinner. It can detect the health condition if there is a problem or any underlying disease.

1.4.1.1 Garbage Monitoring System

Our proposed model gives an answer to the various hygiene problems we face in our daily life. Our model basically includes three subjects: 1. Smart trash, 2. Correlate, 3. Notify.

In Figure 1.3, the basic structure of our proposed model is shown. We have used different types of sensors with connecting them together to an LCD display for a better understanding of the situation. The basket contains a sensor and an ultrasonic sensor. The sensor we connected will bring in the information about the level of waste. With the help of Arduino the Wi-Fi which is fitted inside the bin will inform the authority when the bin overflows. We will receive a notification from the web server via coding through inside the Arduino. The collected details are then displayed on the LCD in the corresponding area in control [10].

1.4.1.2 Accident Sensing System

This proposed model informs us about the events or possibilities of accidents caused by concussion of a gas vehicle. This project contains three subjects: 1. Accident sensing, 2. Correlate, 3. Notify.

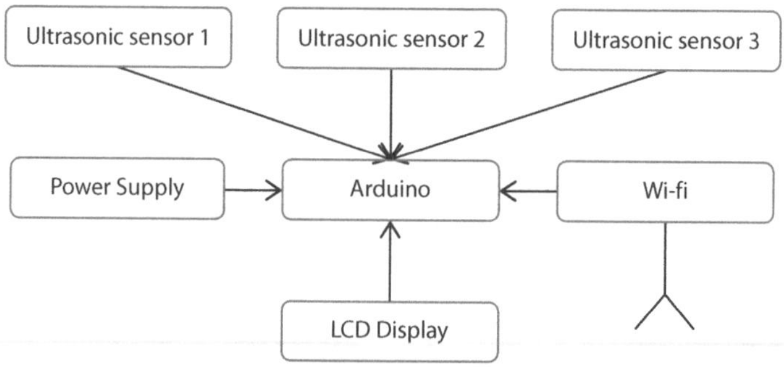

Figure 1.3 Block diagram of garbage monitoring system.

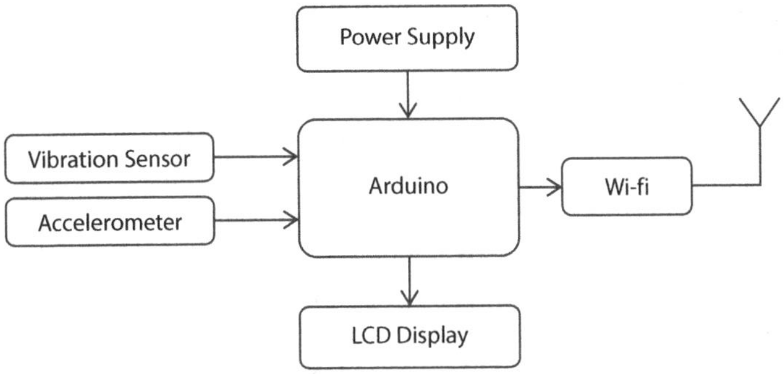

Figure 1.4 Proposed accident detection system model.

In Figure 1.4, the basic structure of the proposed Accident Sensing System is shown. The model will require power supply for the different utilities such as the sensors and we can have a clear visualization of the situation through the LCD display. This model contains two sensors, an accelerometer, and a vibration sensor. These sensors have some ability to alert a person to a possible occurrence. Wi-Fi located inside the module informs the appropriate person about the accident via Arduino. The coding programmed inside the Arduino helps the web server in sending the respective notification. The individual in charge will take action on the accident. If the casualty is not solved, the Arduino located on the module will generate constant information about the possibilities of an accident to the associated individual until the problem is solved. A crash sensing system is considered to be an economical and better way to maintain a safe environment without vehicle collision accidents [16].

1.4.2 Smart Healthcare

The Smart Healthcare system is mainly focused on the vision to provide the best healthcare to all the people across the globe. This system is done in a more than gentle and economical way. Henceforth, if you want to boost the efficiency of the healthcare system and patient care system one should focus on improving the healthcare monitoring equipment. In monitoring patients, medical fields are facing generally two problems; firstly the need that is to be present at the patient's bedside for providing the health care and caregiving, and secondly, the patients are attached to large machines and confined to a bed. The problem of providing flexible and friendly patient care, the solution was given to develop bio instrumentation and telecommunication technologies. With the help of these technologies it has become possible to design home vital signs monitoring systems to display, collect, record and transfer physical data from the body of humans to any further location. There are many reasons that motivate doing work like making healthcare accessible to all public who do not have ingress to healthcare providers and for going hospitals there is no availability of public transportations; giving care to those patients who require more time to heal and more care; avoiding in the delay of delivering the medical kit to patients for health care providers, specifically in the event of accidents or emergencies; and reducing manual patient data entry, allowing healthcare staff to effectively monitor their patients [18].

In Figure 1.5, we show how in a smart healthcare system we can use many different types of sensors and using new technologies we can infer data and collectively enhance the diagnosis of many people at the same time. Our proposed model is an automated system which senses the patient's heart rate, blood pressure or the body temperature. The features can be expanded to predict the patient's possible chronic disease or other health parameters and other various symptoms. In Level 1, the various data we collect from the IoT devices will be gathered, organized, and stored on a server. Various sensors are acquired here, for example, sensors for BP, heart rate, body motion, etc. Since the output maximum times are given as an analog output, we first have to convert the analog values into digital form using a converter IC so that raspberry pi can use it. Further, the raspberry pi with the help of its Linux OS installed, converts the data to a python code which will update the database at specific or required time periods. At Level 2, with the help of filtering, classifying, and categorizing the helpful data is obtained from the stored data. This relevant data is solely about the

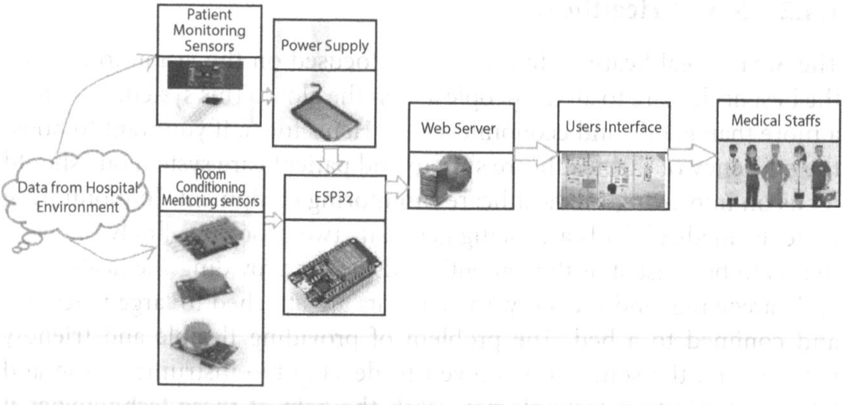

Figure 1.5 Block diagram from proposed model.

patient's real-time health information and his/her symptoms. This information will further help in predicting or diagnosing the patient. This helps the system be more self-sufficient and efficient. In phase 3, analysis/prediction is performed, techniques of data mining are utilized to understand the problem, its nature and type. i.e., the disease characteristics. Artificial Intelligence can be integrated to make the system more proficient [9].

1.4.3 Smart Homes

Nowadays, smart homes are considered some of the important applications for IoT based environments. A crucial feature involving smart homes is automation. What the goal we try to achieve here is to reduce human efforts as much as possible. At the present time, remote control systems are of great importance. The important advantage we get for using IoT in smart homes is the remote control of every device in the home. Home automation architecture can differ depending on the protocols and hardware utilized by the very system. Further following, the main services fetched out in the field of home automation are analyzed and, based on the analysis, a comparison of IoT architectures is made. There are certain advantages of using wireless technologies which cannot be established using traditional wired networks. We often call a smart home a home automation system, which uses the brand new technology to make household activities easier.

In Figure 1.6, we see how different sensors which are very justifiable to a home being secure and efficient connected to each other with the help of present technologies and thus providing a means of smart home. The proposed home automation system model includes a server, actuators,

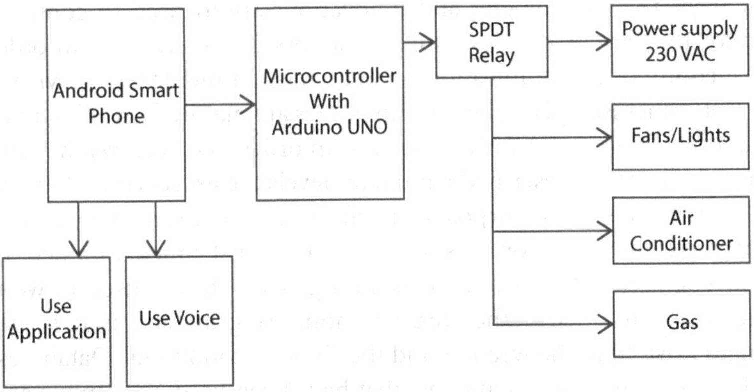

Figure 1.6 Block diagram from the proposed IoT based smart home system.

sensors and microcontrollers. The back end server will be set up to control and monitor the sensor devices. The proposed smart home system is going to be remotely controlled by wireless technology communication devices such as smartphones, cards, and other wireless devices remotely over the Internet. The room temperature could be remotely controlled, controlled, automatic fan on and off, automatic lights on and off, automatic gas leakage detected by sensors, air conditioning system, etc. are automatically controlled and controlled by the home automation system. Designed without the help of any human interaction with the home automation system monitor as well as gas leak control, fan on/off system, lights on/off system, room temperature, and humidity level control and monitoring through IoT communication device. The Node MCU is the primary need of this system and performs numerous procedures for the home appliance system. The Node MCU secures, interfaces with numerous sensors and collects real-time information for a home automation system. These contain two node MCUs. Node MCU (Node Micro Controller Unit) is an open source containing software and hardware that built a much cheaper system designed on a chip known as ESP8266. In particular, the home automation system remotely manages home appliances to make them convenient for people. This system includes for warning of any violation of safety assurance and violation of harmful events certainly will not happen in the home. A system linked to the Buzzer Alarm system can alert a person in the home with an acoustic signal to signal any problem. And there is also an alert SMS to the user's mobile phone or an email that can be sent to the affected user for home security alerts.

1.4.3.1 Weather Monitoring IoT-Based System

Nowadays, the technologies and innovations are focused to control and monitor the various devices wirelessly across the Internet, so in order to be a medium of communication between the communicating devices, the Internet comes into play. These technologies are mainly focused in managing and monitoring the various objects. In order to detect weather conditions, whether the prescribed parameter levels are exceeded and to collect the data for research purposes an effective monitoring system comes namely, weather monitoring system. Multiple instruments namely thermometers, barometers, wind vanes, rain gauges, etc. are used in weather stations inserted inside the weather monitoring system used to detect changes existing in the weather and the climatic conditions. Databases are used to store and the instruments that have been used, use simple analog calculations that are recorded later on physically. The radio stations and

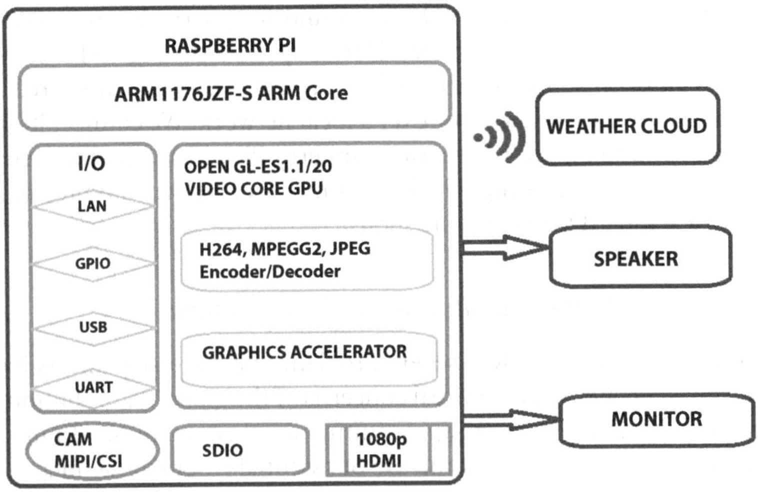

Figure 1.7 Block diagram of weather monitoring system using Raspberry Pi.

news stations collect this information lately and thus a weather report has been made. The data has been collected using a number of different connected sensors including humidity, pressure, temperature, and so forth and the statistics has been dispatched to cloud applications in order to supply them. The cloud applications then analyze and visualize the data that has been collected. These apps then send the weather alert to the users that have been logged in. The device named AirPi is used to detect air quality and weather and is able to record and upload the information on humidity, air pressure, light levels, temperature, UV levels, carbon monoxide, nitrogen dioxide, and smoke level on the Internet [2]. The proposed system is an advanced weather monitor solution that uses IoT to easily make actual time data available on a very wide scale. The changes made by the system deals with weather and climate are monitored as follows:

1. DHT11 sensors are used in monitoring humidity and temperature.
2. An anemometer measures wind speed and directions with LDR which keeps track of Light intensity.
3. GY8511 solar sensor measures UV radiation and MQ7 measures the Carbon monoxide in the air.
4. Hygrometer measures soil moisture.
5. Level sensors measure ultrasonic rainwater.
6. Raindrop sensor to detect rain or snow.

Mainly two devices namely, Dark Sky and Raspberry Pi are also used, which is an open source IoT source. Dark Sky is for storing and retrieving data. It is an open source Internet of Things (IoT) is an open source API using HTTP over the Internet or over a local network. We connect it using a Raspberry pi. In Figure 1.7 we use Raspberry Pi, which is an inexpensive mini sized credit card computer that fits into a computer monitor or TV and uses a standard keyboard and mouse.

1.4.3.2 Air Pollution Monitoring IoT-Based System

Air pollution contributes to a severe problem that adversely affects living organisms. It creates the major real concerns of the globe. Air pollution is a major global concern consisting of multinational companies, administration and broadcasting. Even some use of essential resources at an amount faster than nature's capacity to regenerate can cause pollution of plants, water and air. In addition to human activities, there are several irregular characteristic cycles that further lead to the release of dangerous things. In addition to artificial activities, nature's calamities can lead to air contamination. The Internet of Things (IoT) has become a primary conveying trends of recent times. Using this idea, it is foreseeable to secure innumerable intelligent embedded objects with low consumption among everyone and to the Internet. The ubiquitous existence of numerous wireless technologies for example, tags, RFID (Radio Frequency Identification), actuators, sensors, mobile phones form the foundation of the IoT concept. An IoT-based air pollution sensing and system design can detect dangerous gas discharge from industries and vehicles utilizing gas and weather sensors. Collected information can conceivably be analyzed to create informed conclusions regarding the pollution control application.

In Figure 1.8, we showcase what type of sensors our proposed system will use to detect the level of noise and pollution in the air. It has features of data transmission and perception so as to act to the benefit of the user.

The air pollution detection and forecasting design we proposed in this paper proposed a decent quick fix to the complication of air pollution. Utilization of some sensors makes sure that the monitoring accuracy is appropriate, also lessens monitoring costs, and fabricates the monitoring data in the monitoring area more organized and clean. The huge aggregate of field data given by the front-end sensor network builds big data analysis in the background application layer more directly and efficiently, and gives a factual and good decision-making foundation for emergency response after a pollution accident occurs.

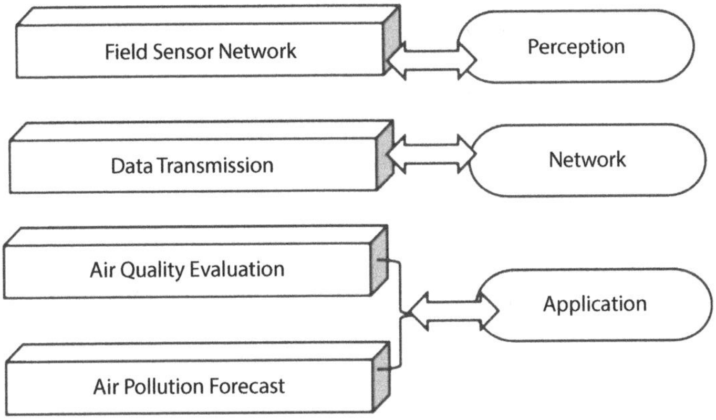

Figure 1.8 System's integral design architecture.

1.4.3.3 Noise Pollution Monitoring IoT-Based System

The pollution's growth is expanding nowadays with a certain number of factors which affects the environment and results in the dropping of bio degradation. The rapid increases of industrialization and urbanization may have led this to happen. This leads to population degradation and also affects the people's health by directly affecting people's health in one way or another. To manage a positive future and healthy lifestyle for everyone, examining air quality and noise levels are required. A great decrease is seen in the industrial and infrastructure plants and their dispatch, which has caused various ecological problems such as air pollution, water pollution, noise pollution, climate change, atmospheric dissimilarities, defects that has an environmental effect on the requirement of anatomically adaptable, efficient, an affordable and smart monitoring system. The air quality and sound pollution monitoring enables us to monitor and control the actual time air peculiarities as well as sound pollution in specialized areas through our latest IoT device automations. Noise maps for cities are created by monitoring the noise pollution of the cities. To manipulate noise stages near residential areas, schools and playgrounds, it helps policy makers by formulating policies. Various noise monitoring systems are used which are located in various parts of the cities. The cloud or servers used the collected noise level data from stations for further data aggregation and to create noise maps.

In Figure 1.9, the proposed system manages different types of sensors for detecting noise pollution in the environment. Power supply is the main key to generate these sensors. We also use buzzers for any hazard

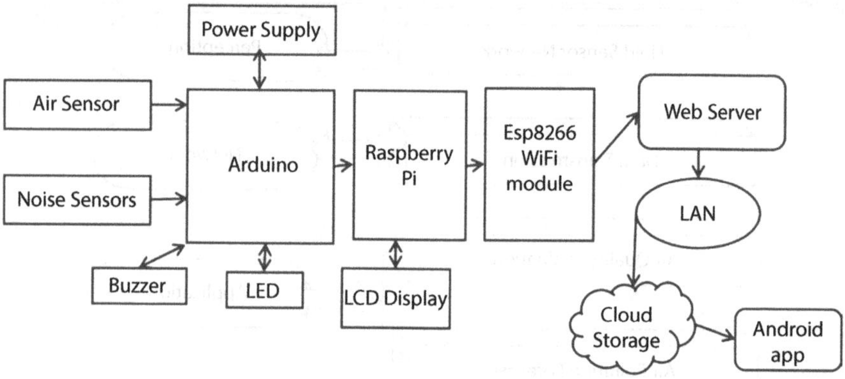

Figure 1.9 System architecture for noise pollution monitoring system.

that might occur. This model controls the problem of the greatly polluted field, which is the main problem, and aids the newest automation and sufficiently supports the idea of the good life. The contamination level used to be checked by the community through their cell phones using the applications. In this, we have focused on the efficient way to standardize the atmosphere and an efficient, cost-effective equipped model to be presented. The proposed model effectively describes the functionality of the other modules. The proposed model describes the functionality of the other modules. The IoT idea has proven itself in practice for managing certain criteria. This also gives the criteria and actuator data to the cloud. For later reviews these information are useful and also shared between various operators. These automations intensify to control various aspects of nature, such as aerial properties and the problem of sound properties is highlighted. By the use of different technological trends the plan of IoT is to spread happiness for the various communities.

1.4.3.4 Forest Fire Detection IoT-Based System

The use of detectors serving as a data collection center Temperature detectors and bank detectors are used, which should be set up in specific chambers in order to maintain a view of the entire timber area, with the ultimate goal of distinguishing the original disturbing temperature and carbon dioxide gas (CO_2) position [4]. These detectors will shoot a point to the microcontroller. Each describes transformations on Earth and naturally replies in the event of an extremity. New technologies in programmed launch widgets use cameras and PC computations to explore apparent honey impacts and development in a way that other discovery widgets can't. A number of end detectors are to be used on a case-by-case basis, which should be set up on specific wards in order to maintain a view of the entire timber region. Information gathering using IC installed in Arduino transmitter circuit – The IC ATMega 328-p (micro regulator) bedded in the Arduino stage shown in the transmitter circuit acquires the data found and collected by the detectors [7]. At this point, the regulator plays back the exertion to it and passes it to the transmitter to transmit the data to the receiving station. Transmission of information by transmitter – after entering the data from the regulator, the transmitter sends the data to a certain scope where the receiving station is extended to be used. The micro regulator is the central element of the device circuit; it governs and certifies the operation of the entire circuit, then for this situation the transmitter circuit event of data by the receiving station. After entering the data from the transmitter circuit, the receiver transmits the data to the control IC of the connected Arduino unto installed in the entering circuit in the computer frame, which makes it possible that the regulator performs altered conduct to control the temperature position and CO_2 position for honey identification. Display of temperature situations and CO_2 position on point runners is available through a private system. At the moment the temperature and CO_2 position information is ready in the IC circuits of the receiver node Mcu, which is altered with colorful library rudiments of the Ethernet guard interface, so that it's possible to produce a runner. Fire Security System using the node Mcu Ethernet guard R3(assembled) which allows the NodeMcu board to connect to the web. It relies on the Wiznet W5100 Ethernet chip(datasheet). Wiznet W5100 provides a system(IP) mound equipped for both TCP and UDP. Arduino Ethernet Shield 2 will connect your knot Mcu to the web in negligible twinkles [4]. Simply plug this module into the NodeMcu board, connect it to your system using RJ45 connection with benefits like:

- Association speed 10/100Mb
- Ethernet Controller W5500 with internal 32K support
- Association with esp8266 on SPI harborage
- Working voltage 5V (handed from the NodeMcu board)

In Figure 1.10 shows addition, a warning circuit was created to encourage the Fire security platoon to find the vulnerable section at the foremost occasion. This fire protection circuit will sound an alarm just when the temperature situations exceed the preset value.

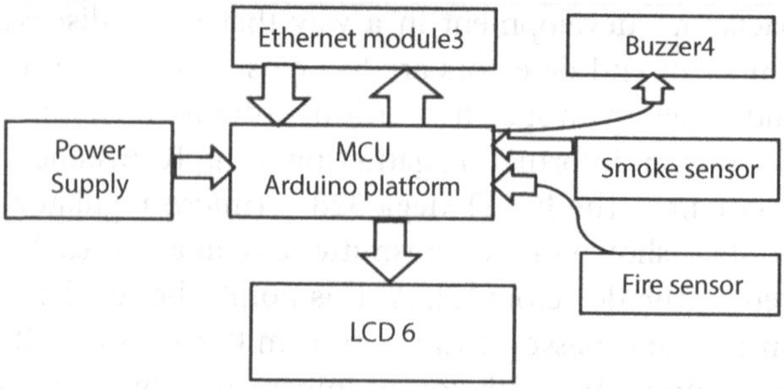

Figure 1.10 System architecture for forest fire detection.

1.5 IoT Architecture

IoT Architecture is the combination of different types of sensors, protocols, actuators and many layers that form the basic foundation of IoT networking systems. The layers in the IoT architecture allows us to properly administer, which is by good evaluation, monitoring the quality of the system. The four layers are namely : Perception Layer, Network layer, Session layer, Application layer. In Figure 1.11, As we see that each layer has different functionalities and functions which are required in the IoT architecture. Every layer differs in functionalities in terms of different services and technologies. Any possible threat to any layer in the IoT architecture may completely redirect the functionality of the layer.

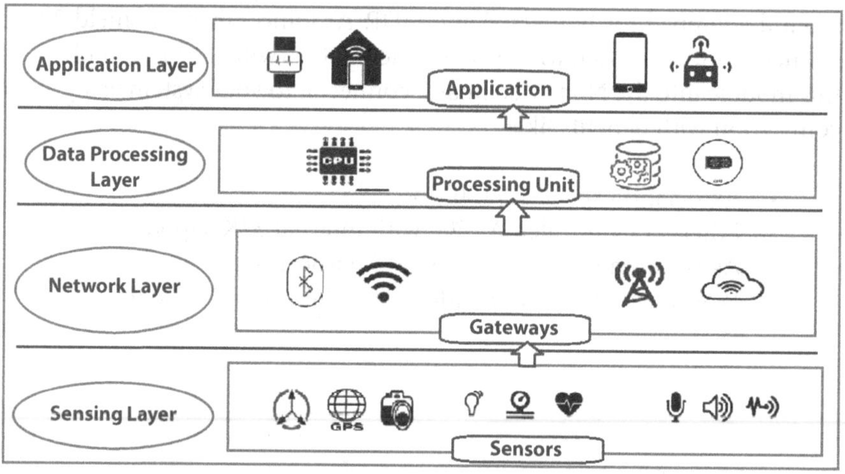

Figure 1.11 IoT architecture.

1.5.1 Perception Layer

In IoT architecture, the sensor layer is named as the perception layer. The sensor provides some special technical support and receives the data from the environment. The sensing layer uses RFID sensors and readers that have definite memory, ineffective less power consumption, and definite computing capabilities, making them unprotected. The involved devices used to capture the information through the attached sensors into it. In order to track the locations for structural applications over the network GPS is inserted inside the layer. For short distance and local communication, the combination of IoT nodes is done using the perception layer. The data is first collected, observed then processed and then transmits the information stored in the network layer. Intimidation in the perception layer focuses primarily on information gathering activities via sensors. Various security risks are associated in this layer. RFID, sensors and smart implanted modulations are liable to various harms at this layer [14]. Here are some secured procedures and the available answers to mitigate the risks:

1.5.1.1 Privacy and Verification

Aggression may include replay attack or spoofing or even eavesdropping. These attacks pose some serious security risks in the sensor network authentication and secrecy may be disturbed. Solution – Making reliable cryptographic solutions which are built to be robust and efficient in many terms.

1.5.1.2 Network Availability

Attacks include Denial of Services (DoS), which is caused when there is a threat to the sensor networks. The reason is because the sensor network is divided into layers and all sensors are vulnerable to this kind of attack. Solution – Path-based DoS (PDoS) is a solution which reduces the attack.

1.5.1.3 Service Integrity

There may be instances when the attacker may trick the network into accepting incorrect data. Solution – Good reliable and robust cryptographic solution is sure to minimize the misleading of the networks by the attackers.

1.5.1.4 *Jamming*

Jamming is also a Denial of Service attack in which the communication operations related to the reader and the tag is paralyzed thus affecting the air interface. For example, communication is disturbed when the nodes in between are occupied in the communication channel and this is called Signal jamming. Solution - Primary sensing of jamming gadgets is an optimal solution to steer clear of such attacks.

1.5.1.5 *Eavesdropping*

The attacker secretly watches the communication between the tag and the reader in the search for personal details. The attacker can use the RFID feature to get the information about the user's passwords and confidential information. Solution – A possible solution to such attacks is data encryption or restricting the space between the tag and the reader [15].

1.5.1.6 *Replay Attack*

In this attack, the invader or the attacker compares the authentication sequences by breaking the transmission connecting the tag and the reader and a duplicate tag. This type of attack affects RFID tags and air interfaces. Solution - Data encryption and token security can lessen this attack.

1.5.1.7 *Man-in-the-Middle (MITM) Attack*

MITM refers to the man in the middle attack that occurs during the communication of the data between two users communicating directly. MITM attacks are usually known as Interception and decryption. In Interception the attackers get the full visibility access of the online data transmission. Through malicious Wi-Fi hotspots which are available publicly, free attackers intercept the user .In decryption there are several methods involved namely HTTPS spoofing, SSL beast, SSL hijacking, SSL stripping used to decrypt the two way SSL traffic. By avoiding Wi-Fi connections which are not secured by the passwords we can prevent MITM attacks. By practicing activities such as immediately turning out from secured applications and during sensitive data transactions avoiding the use of public networks we can prevent the Man in the middle attack.

1.5.1.8 *Denial of Service (DoS)*

To prevent end users from accessing other networking devices by using a blocking tag to stimulate the multiple tags and resulting in the cause of DoS. The user tries to validate the non-existing tags which causes Denial of Service (DoS). The authenticated users can minimize the use of blocking attacks by detecting before the blocking device.

1.5.1.9 *Tag Cloning*

Through an illegal approach the attacker creates a duplicate tag basically on the real tag. RFID tags and air interfaces of the tag cloning are mostly affected which causes the problem related to financially in the applications. In order to prevent the user from this attack, users can access token authentication. The methods namely Synchronized Secrets Methods which is used to detect tag cloning attacks and provides secured manner of cloning. It is used to detect various tags which have their own identical IDs [19].

1.5.1.10 *Take Off i.e. Spoofing*

The spoofing attacks enables attackers to access the user's system exchanges of the data or spread malware. It makes changes in the original appearing sources and falsifies it which pertains to cybersecurity. Some of its examples are Email spoofing, Website or/URL spoofing.

1.5.1.11 *Device Tampering*

It also refers to Node captain because it enables the attacker to change the sensor node to its malevolent node. It gains complete access over the captured node. It makes the device initiate inappropriate actions.

1.5.1.12 *Outage of Nodes*

In this attack, the attacker damages the network by physically or logically enabling the stoppage of functionality. Node outage attacks are one of the hazardous attacks in WSNs.

1.5.1.13 *Leakage of Information*

This is a type of passive evasion. When there is random behavior the attacker gains access to some private important information to which authorization is not allowed.

1.5.2 Network Layer

The main role of the network layer is data routing and transmission between different IoT hubs and various other IoT devices. It basically acts as the central nervous system in the architecture of IoT. Network layer works for the data initial processing and transmission of information. Modern wireless technologies are used here, for example, Bluetooth, Wi-Fi, LTE. The attacks involved in this layer can practically damage the network's capability of communication between different IoT devices. Some of the threats are mentioned below:

1.5.2.1 *Selective Forwarding*

In this attack, the intruder selectively advances some of the personal information and discards them. The malicious node forwards the leftover traffic to show its malfeasance. Neglect and Greed is an alike type of attack, where an overthrown node jumps over routing some data haphazardly. A similar type of attack is one called Neglect and Greed, where a subverted node skips routing some messages randomly.

1.5.2.2 *Sybil Attack*

It is when the striker has numerous malicious places to attack. The malicious node or the device illegitimately holds several identities on the network. The attacks affect the fault-tolerant schemes. Solution - Individual node cites and the location of nodes can be tracked using distributed hash tables.

1.5.2.3 *Sinkhole/Black Hole Attack*

A Sinkhole or Black hole attack is described by strong competition for resources between nodes which are near to the of the attacker's node due to limited channel access and limited bandwidth. What happens after is that the network becomes congested and the energy which is consumed by the nodes increases.

1.5.2.4 *Wormhole*

It is a very acute and challenging attack that can cause failure of location dependent protocols. A kind of DoS attack where information bits in the network are moved from their previous spots over a low-latency link. Security - The authentication protection process is proposed and is known as the "Markle tree authentication" method.

1.5.2.5 *Attacks of Hello Flood*

Instigation of lofty congestion in a detection network by flooding the path with tremendous amounts of unwanted requests is known as Hello-Flood attack. An attack in which a dangerous node sends an unnecessary message to which the attacker responds, producing congestion.

1.5.3 Support Layer

Action of gathering important information or data, and data processing and identification of the physical world is done here. In bulk data processing, smart tackling of dangerous data is very crucial. Intelligent detection of malicious information is a very difficult job. The support layer is occupied with storing of the data or the information, access to cloud services for efficient use of network advancements, and data analysis to indicate accurate facts. The attackers primarily focus on attacks in the support layer in order to target the data storage technologies. A number of the maximum common attacks assaults of this IoT software layer are described underneath:

1.5.3.1 *Data Tampering*

The data tampering takes place when the benefits of an insider are taken away by manipulating the data either for their own benefits or for the benefit of any third party. Protection – Data authentication must be provided so that insiders cannot easily make changes to the data.

1.5.3.2 *Unauthorized Access*

The system data can be corrupted as an attacker sneaks into the system. The denial of access to related IoT devices and its services are considered under such attacks. For the prevention of such attacks there must be a proper security mechanism.

1.5.3.3 DoS Attack

In this layer of the IoT architecture, the possible effects of such DoS attacks are attacks such as system shutdown, which may damage the system or lead to system unavailability.

1.5.4 Application Layer

Application layer plays its role in conditioning services to various industries like Smart e-health, smart government, smart city, etc. The security requirements in the application layer vary from application to application. The one most main feature of Application layer is data sharing, which also gives rise to several problems like the need for data privacy, authentication, integrity, confidentiality, etc. The main threats related to this layer are underneath:

1.5.4.1 Sniffer

Attackers introduce logger programs inside the system in the motive of stealing from the network from the network traffic. The attacker can steal from the passwords, email bodies or also email files.

1.5.4.2 Injection

The attackers sneak a code inside an application running on the server and it results in data corruption or data loss.

1.5.4.3 Session Hijacking

In this type of attack, the personal data is exposed. Security flaws are done to the session management and authentication.

1.5.4.4 Distributed Denial of Service (DDoS)

DDoS is performed concurrently by numerous attackers, resulting in the denial of service in the network.

1.5.4.5 Social Engineering

Social media chats are used by the attackers to obtain the information of the user. Solution - To avoid the exploitation of such sensitive data, the authorization is being given only to the sole users of the data.

1.6 Smart Environment and Advanced Computational Intelligence

For developing and integrating software, AI and literacy of machines have transfigures the way to sustain smart metropolises work in its various ways for the AI and computational grounded smart cities, for growing and integrating software program and tackle systems, theoretically making plans and also fine computing models of ICT structure which is also essential. For the creation of smart terrain some of the special issues motivate experimenters and the academicians to consume their work which includes AI and advanced computational intelligence. For the advancement of the smart megacity idea, the intelligence or information via Artificial Intelligence and advanced computational intelligence has been essential. With the help of devices such as environmental detectors, business detectors, smart measures, videotape cameras and also by merging mortal intelligence within the sharp computing programs a cyber-physical space can be created.[12] As there are multiclass organizations required for the numerous people which requires a lot of time and operations because of the nature of the data. Similarly for the various unknown circumstances the detector data is constantly sensitive and noisy, also as there are imperfect detectors and missing data. In order to attain excessive delicacy bracket, it's far essential to perceive and recognize spatio-temporal correlations among detector studying smart green and smart terrain (e.g., smart home, smart plant, smart megacity, smart structure, smart parking, smart husbandry, intelligent business systems, smart auto, etc.) with the use of general computing, advanced computational literacy, artificial intelligence, cognitive detector systems, wireless and detector networks to understand cyber and bodily factors includes seeing, logic, and be in charge of the terrain are pertained for the smart surroundings. In a smart terrain the independent agents play an important role of interceding function between mortal druggies and the terrain. This is in particular genuine whilst excessive position analytic chops and computational intelligence are employed to manage with the Misgivings of the complex terrain, permitting dealers to respond meekly in a variety of situations. Machine integration, device conservation, geographical and practical extensibility, social networking, cell computing, context aware operations and offerings, mortal- in- the- circle modeling and simulation, huge records analysis, pall and part- grounded IoT fabric and environment, discipline trials and check beds are all issues that must be addressed in the introduction of a clever terrain. The number one ideal of this unique problem is to present and debate current tendencies in the subject of detector systems, in particular in terms of technology, infrastructures, algorithms, and protocols for a smart environment, with a focal point on real-world operations [20]. Motifs which might be applicable encompass, but are not restricted to Wi-Fi detector networks(WSN) in clever metropolises, clever megacity sewage, water and electricity operation, clever megacity healthcare service tracking, clever megacity education, education and social assistances, clever transit architecture, estimation, and improvements, smart domestic, smart structure and social media networks architectures, big facts analytics and system literacy pall calculating for net of effects, superior computational intelligence.

1.7 Advanced Computational Intelligences: Possible Uses in Smart Environment

1.7.1 Smart Infrastructure and Green Economy

In our 21st century the most serious challenges like city congestion, increase of chemicals, the depletion of the conventional energy resources, increase of physical and biological pollution, and also an increase in global warming are affecting the quality of life. Due to the increase of high density of population and large settlements of buildings affects this regard and later the planners think of the need for a sustainable urban living. There is a chance by intelligent green growth in order to rethink the plan for profitable and smart environment growth modules or design for a new prototype which bears sustainable growth and also leads to decreasing environmental degradation and protects natural resources. This efficient and conventional model gives the border idea of civil growth which consists of fairness, effectiveness, and economic development. For a certain fixed goal for the economy and sustainable development, growth economics has been viewed as efficient. On a global scale where the massive social and environmental changes occurred, lighting up the cities in maximum numbers has been shifted in developing the smart policies to gain sustainable mobility, there is an upgrade in building energy, renewable sources increases the energy production, improvement in the waste management and implementation in the ICT infrastructure. The main objective of transformation of Smart Cities in Smart Communities is to focus on lifestyles of the humans or the civil people for providing good cultural, economical, and advancements, safe, exciting and reactive environment. The modern techniques like artificial intelligence and Advanced computational intelligence lead in modifying sustainability of urban development in transiting to a green economy. Based on the research, 50% of the population will be in metropolitan areas so there will be 60% to 80% in energy consumption as well as 75% emissions of carbon. There is a unique chance for every city in order to green the global economy [17]. To improve ecosystems and reduce environmental hazards, the national and local authorities are responsible for decreasing carbon emissions and pollution. The regional planning, integrated urban and the technology used design strategies are available in improvement of the urban transportation, construction of the building and in the growth of the metropolitan potential, water, receiving further potential from metropolitan waste design in such a way that they do not rely on natural potential utilization of these resources in the middle of long time.

1.7.2 Resolving the Issue of Sustainability

The employment of sustainable technologies is a necessity to develop a sustainable smart city. The issue of sustainability heightened the demand for green IT. The use of fewer resources in the development of hardware and software applications is focused in Green IT. The issue of roaring demand for IoT devices can be tackled by the products of Green IT. There will be less requirements of resources in manufacturing a new RAM rather than manufacturing a new computer. If the computer equipment gets recycled rather than discarding then there is reduction of generation of hazardous elements, namely, lead and mercury; 80% of CO_2 emissions are caused by urban regions. By utilizing ICTs and various optimization techniques the cities are trying to lessen emission of CO_2 and other greenhouse gasses [13]. The potential solution for this problem can be better city layout and the adjustments in the human habits by generating and facilitating the new emerging technologies. Variety of IT services are enabled by the technology of cloud computing. With the increasing number of cloud based IT services and growth in its applications there are thousands of web servers, storages, and network devices. For end users there are a variety of offers made by Cloud Data Centres (CDC) which range from high performance computing to huge scaled data analytics. There are a total 25% of IT electricity consumption in more than one geographical location to serve disbursed users because of the massive scale of cloud statistics facilities.

1.8 Conclusion

Ground breaking technologies have evolved in order to showcase what's the best potential for any smart environment idea. In this paper, we have proposed models that we think will be best for the selected smart environment idea. We dived into the different IoT architecture and have seen what can be the effects, use, and threats to the system. We now understand what challenges may be there for any IoT based smart environment system. We have studied different case studies and seen what opportunities we might have in the future. The world is on its way to becoming a supremely motorized terrain, with current period technologies like IoT, advanced computational intelligence, and artificial intelligence guiding the way. As preliminarily stated, we've bandied about how these technologies can be used to plan a smart green terrain that is both relaxing and comfortable to live in.

References

1. Ahn, B.G., Noh, Y.H., Jeong, D.U., Smart chair based on a multi heart rate detection system, in: *2015 IEEE Sensors*, pp. 1–5, Nov 2015.
2. Banka, S., Madan, I., Saranya, S.S., Smart healthcare monitoring using IoT. *Int. J. Appl. Eng. Res.*, 13, 15, 11984–11989, 2018, ISSN 0973-4562.
3. Dwivedi, A., Bali, R.K., Belsis, M.A., Naguib, R.N.G., Every, P., Nassar, N.S., Towards a practical healthcare information security model for healthcare institutions, in: *4th International IEEE EMBS Special Topic Conference on Information Technology Applications in Biomedicine*, April 2003, pp. 114–117, 2003.
4. Ali Al-Garadi, M., Mohamed, A., Al-Ali, A., Du, X., Guizani, M., A survey of machine and deep learning methods for Internet of Things (IoT) security by NPRP Grant #8-408-2-172 from the Qatar National Research Fund (a member of Qatar Foundation).
5. Hassija, V., Chamola, V., Saxena, V., Jain, D., Goyal, P., Sikdar, B., A survey on IoT security: Application areas, security threats, and solution architectures. *IEEE Access*, 7, 82721–82743, 2019.
6. Fahim, M. and Sillitti, A., Anomaly detection, analysis and prediction techniques in IoT environment: A systematic literature review. *IEEE Access*, 7, 81664–81681, 2019.
7. Biswas, A. and Maurya, K., Governance and institutional framework for smart cities in India, in: *Urban Governance and Management in the Developing World*, J. Mugambwa (Ed.), Web. 15 May. 2018, IGI Global, pp. 1–491, 2018, Doi: 10.4018/978-1-5225- 4165-3.

8. Gupta, Agrawal, D., Chhabra, J., Dhir, P.K., IoT based smart healthcare kit, in: *2016 International Conference on Computational Techniques in Information and Communication Technologies (ICCTICT)*, March 2016, pp. 237–242.
9. Lopes, N.V., Pinto, F., Furtado, P., Silva, J., IoT architecture proposal for disabled people, in: *2014 IEEE 10th International Conference on Wireless and Mobile Computing, Networking and Communications (WiMob)*, pp. 152–158, Oct 2014.
10. Tyagi, S., Agarwal, A., Maheshwari, P., A conceptual framework for IoT-based healthcare systems using cloud computing, in: *2016 6th International Conference - Cloud System and Big Data Engineering (Confluence)*, pp. 503–507, Jan 2016.
11. Samaniego, M. and Deters, R., Internet of smart things-IoST: Using blockchain and CLIPS to make things autonomous, in: *2017 IEEE International Conference on Cognitive Computing (ICCC)*, IEEE, pp. 9–16, 2017.
12. Perera, C., Zaslavsky, A., Christen, P., Georgakopoulos, D., Context aware computing for the Internet of Things: A survey. *Commun. Surv. Tutor., IEEE*, 16, 1, 414–454, 2014.
13. Narendra Kumar, G., Swamy, C., Nagadarshini, K.N., Efficient garbage disposal management in metropolitan, cities using VANETs. *J. Clean Energy Technol.*, 2, 3, July 2014.
14. Ruqsar, H., Chandana, R., Nandini, R., Surekha, T.P., Internet of Things (IOT) based real time gas leakage monitoring and controlling. *Proceedings of the 2nd International Conference on Current Trends in Engineering and Management ICCTEM -2014*, Mysore, Karnataka, India, 17–19, July 2014.
15. Jose, A.C. and Malekian, R., Improving smart home security: Integrating logical sensing into smart home. *IEEE Sens. J.*, 17, 13, 4269–4286, 2017.
16. Namboodiri, V., Aravinthan, V., Mohapatra, S.N., Karimi, B., Jewell, W., Toward a secure wireless-based home area network for metering in smart grids. *IEEE Syst. J.*, 8, 2, 509–520, 2014.
17. Ngu, A.H., Gutierrez, M., Metsis, V., Nepal, S., Sheng, Q.Z., IoT middleware: A survey on issues and enabling technologies. *IEEE Internet Things J.*, 4, 1, 1–20, Feb 2017.
18. Eckhoff, D. and Wagner, I., Privacy in the smart city—Applications, technologies, challenges, and solutions. *IEEE Commun. Surv. Tutor.*, 20, 1, 489–516, 2018.
19. Jose, A.C. and Malekian, R., Improving smart home security: Integrating logical sensing into smart home. *IEEE Sens. J.*, 17, 13, 4269–4286, 2017.
20. Salahuddin, M.A., Al-Fuqaha, A., Guizani, M., Shuaib, K., Sallabi, F., Softwarization of Internet of Things infrastructure for secure and smart healthcare, arXiv preprint arXiv:1805.11011, 2018.

2

Machine Learning-Enabled Integrated Information Platform for Educational Universities

Sai Smurti Sahu*, **Rishav Kumar, Soumya Sahoo, Balwant Kumar and Padmabati Mohanta**

Department of Computer Science and Engineering, C.V Raman Global University, BBSR, Odisha, India

Abstract

The integrated information platform for universities is a comprehensive online resource that provides information on all colleges, along with various features such as college comparison, the ability to create a favorite list, and a chatbot for user assistance. This platform aims to assist prospective students, parents, and educators in making informed decisions about college choices. By centralizing information about different colleges, the platform offers users access to detailed data including program offerings, faculty profiles, admission requirements, campus facilities, and more. The college comparison feature allows users to evaluate multiple colleges based on criteria like tuition fees, academic programs, location, student-to-faculty ratio, and extracurricular activities. This functionality empowers users to weigh the pros and cons of various colleges, facilitating the selection of the best fit for their needs and preferences.

***Keywords*:** College comparison, favorite list, chatbot, admission requirements, campus facilities, machine learning, cloud computing, students' needs and preferences

**Corresponding author*: saismrutisahu38610@gmail.com

Hrudaya Kumar Tripathy, Sushruta Mishra, Minakhi Rout, S. Balamurugan and Samaresh Mishra (eds.) Optimized Computational Intelligence Driven Decision-Making: Theory, Application and Challenges, (29–48) © 2024 Scrivener Publishing LLC

2.1 Introduction

India is home to a vast and diverse higher education system, with over 900 universities and over 40,000 colleges. These institutions offer a wide range of academic programs, from undergraduate degrees to postgraduate research opportunities, in a variety of fields such as engineering, medicine, social sciences, and humanities. The integrated information platform of Universities In an era characterized by the rapid advancement of technology and the increasing need for seamless connectivity, our website has embraced the power of integration to create a comprehensive platform that caters to the diverse needs of the students. This integrated information platform serves as a centralized hub, bringing together various aspects of university life and streamlining access to essential resources, services, and communication channels. It is designed to enhance efficiency, convenience, and a dynamic environment for students so that they can choose a better university for them. With this platform, students can easily choose which university is better for them by getting information like, University information, their fees structure, placement report, campus environment, feedback of that particular university and academic course registration, schedules, including the last year students' top result of the particular branch they want to join. They can also interact with their professors. By comparing one university to another university they can easily get a better result to choose the best college.

2.2 Cloud-Based Web Application for University

2.2.1 Overview

We currently live in a time when science and technology are advancing at a rapid pace, having a significant impact on society. Universities are today hampered by rising infrastructure and facility expenses, a lack of funding, and a lack of innovation in new teaching methods. Without the requirement for extensive IT experience in those regions, cloud computing can provide students and educators in rural parts of the state with the most cutting-edge software and educational resources and services. At the same time, it's giving the already overstretched education finances the essential relief.

2.2.2 Working Principles

Once the proposed work [1] for the Integrated platform of the university is developed, it can be deployed on the cloud server infrastructure using the

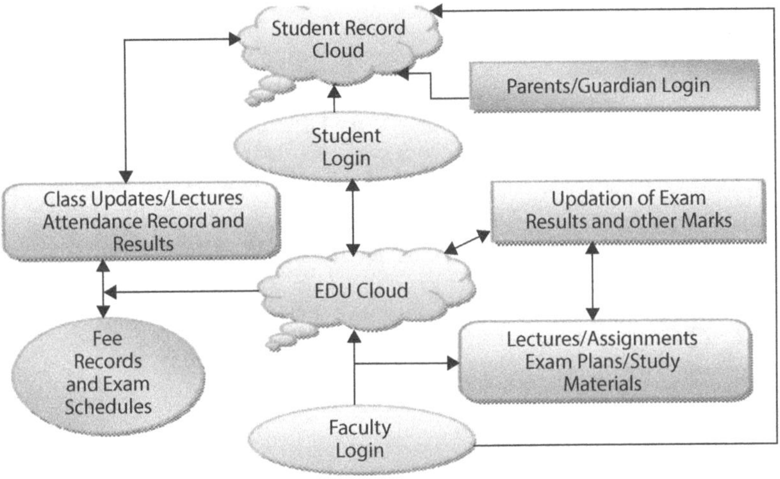

Figure 2.1 Education system using cloud computing.

Note. The proposed model shown in Figure 2.1 is describing, using two cloud the students can access the EDU Cloud record and Student Record. From "A Proposed Model for Education System Using Cloud Computing," by Engr. Ali Almed, Huma Ali Ahmed, 2018 3rd International Conference on Emerging Trends in Engineering, Sciences and Technology (ICEEST).

(https://www.researchgate.net/publication/331204461_A_Proposed_Model_for_Education_System_Using_Cloud_Computing)

cloud service provider's deployment tools. The website [2] can be accessed by users from anywhere, anytime, using a web browser or a mobile app. The website data, such as files, images, videos, and databases are stored on the cloud server infrastructure. Data of universities and students login will be managed by the cloud server and database with admin side. Both students and faculty can access the website for their uses. All the functionality and features will be performed and controlled by the cloud server and databases.

2.2.3 Cloud Computing Techniques

When it comes to building an integrated information platform for a university using cloud computing techniques, there are several key considerations. Cloud computing offers scalability, flexibility, cost-effectiveness, and accessibility, making it an ideal choice for universities to manage their diverse information systems.

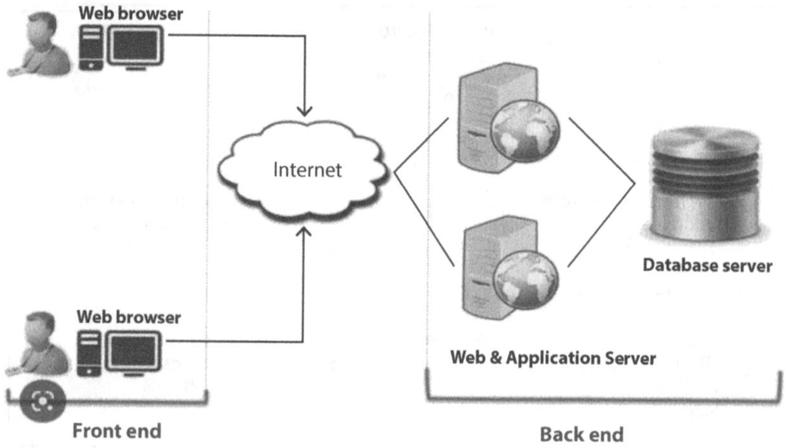

Figure 2.2 Cloud computing system working.

Note. The proposed model shown in Figure 2.2 is describing that the cloud is connected to the both Front end and Back end. From "Optimal Resource Management in the Cloud Environment- A Review" Suresh Gyan Vihar University, Jaipur International *Journal of Converging Technologies and Management (UCTM) Volume 4, Issue 1, 2018 ISSN: 2455-7528* (https://www.researchgate.net/publication/333555228_Optimal_Resource_Management_in_the_Cloud_Environment-_A_Review)

The primary elements of our proposed designed Model as [3]:

1) Cloud User Interface: Interface screens can be provided to users. Using this they can request for the Service according to their use. Every University that purchases any service from a cloud service provider (CSP) is required to accept a standard service level agreement (SLA) from the provider or negotiate such an agreement. All services can be managed with the help of Service Manager can be used to manage any Service Catalog will provide information about the details about the services that are offered.
2) Service Management Platform: The environment of Cloud is very dynamic, virtualized, and automated. Its service management will need to undergo considerable modifications in order to handle this environment.
3) Software as a Service (SaaS): SaaS refers to the capacity to obtain software via the Internet as a service.

A full application for software as a service is available as a service on demand. A single instance of the program operates on the cloud and provides support for numerous end users or client companies. The best example of a remote application service is Google Apps, which offers a number of business apps using a regular Web browser.

2.2.4 Cloud Database

At this point, cloud databases are seen as a wise choice for programmers that need to store the data for their apps in a flexible and incredibly accessible backend. Database-as-a-Service is a reference to these administrations.

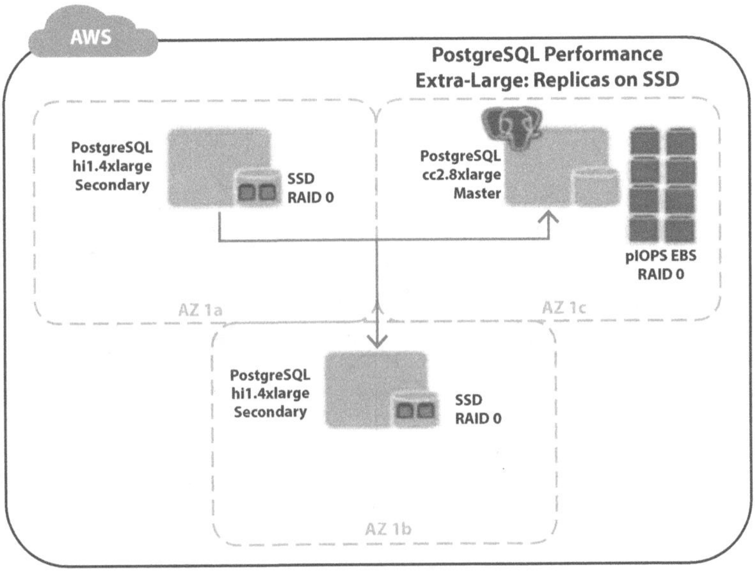

Figure 2.3 SQL performance.

Note. The proposed model shown in Figure 2.3 is describing SQL Performance in AWS. from, "A Survey on Database in Cloud Computing with Reference to the Traditional Database" Eter Basar, Ankur Pan Saikia, Dr. L. P. Saikia International *Journal of Advanced Research in Computer Science and Software Engineering, vol. 7, no. 11, Advance Academic Publisher, Nov. 2017, p. 150. Crossref.* (https://www.researchgate.net/publication/326074652_A_Survey_on_Database_in_Cloud_Computing_with_Reference_to_the_Traditional_Database)

The information [4] on various server farms located in various locations is stored in the cloud database. Because of this, the cloud database structure and the framework for objective database administration are different. Over a cloud database, there are different hubs for question administration for business farms that are server farms as well as being located in various geographic locations. To access the database on cloud administrations conveniently and completely, this connection is necessary. On the other hand [5], programs can use social database administration frameworks (DBMS) enabled by the cloud to store their data. As an illustration, clients of the base as a provider of administration, like Amazon, can transmit DBMS in virtual machines and use these to provide database administration to their applications.

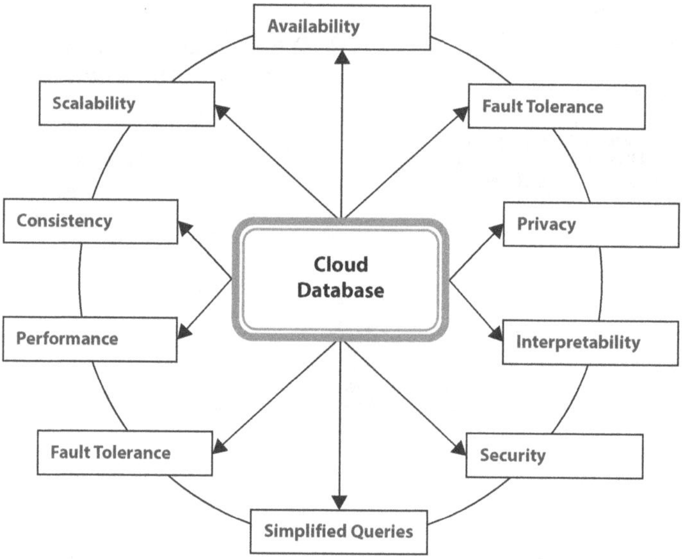

Figure 2.4 Cloud database.

Note. The proposed model shown in Figure 2.4 is describing support features of Cloud Database. From, "Cloud Databases: A Paradigm Shift in Databases" Indu Arora and Dr. Anu Gupla *LICSI International Journal of Computer Science Issues, Vol 9, Issue 4, No 3, July 2012 ISSN(Ontre)1694-0814(https://www.researchgate.net/publication/268002679_Cloud_Databases_A_Paradigm_Shift_in_Databases)*

2.2.5 Advantages of Cloud-Based Database

- Self-Service On-Demand
- Wide-ranging Network Access
- Pooling Computer Resources
- Rapid Elasticity Service
- Measured Service
- Multi-Tenacity Services

2.2.6 How Cloud-Based Website is Better than Non-Cloud Websites

- Cloud-Based Websites [6] are more reliable and user friendly as compared to Non-cloud based websites.
- Fully customizable by the users according to their needs and requirements.
- Hosting and managing the server becomes easy and user friendly.
- We can use enough of database as required on the cloud without using any hardware system.
- Fast and Accurate services of hosting and monitoring will be provided to users.
- All the API and Database will be handled by itself without any other hardware required.

Through this Cloud-based Website for Integrated platform of Indian Universities, every students and faculty of different states and colleges can easily access the information of all universities directly from the cloud service.

At the moment, a different measuring database as a service has advanced thanks to cloud databases. With the help of this service, businesses may take advantage of the resources offered by their suppliers without worrying about where to store their hardware or how to program it. They receive administration from DBaaS [7] providers and benefit from the adaptability of a constantly accessible database. As well as advantageous [8] circumstances, there are drawbacks. The popularity of cloud databases has, however, demonstrated that their benefits outweigh their drawbacks. A number of beneficial characteristics are provided by cloud database services.

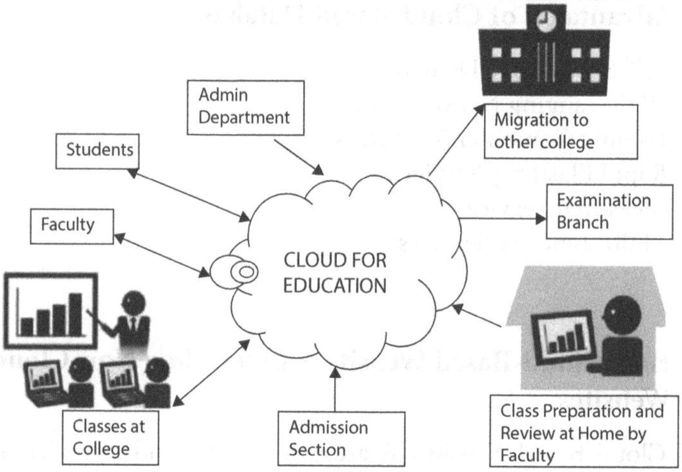

Figure 2.5 Services of education cloud.

Note. The proposed model shown in Figure 2.5 is describing Cloud requirements in education services. From, "APPLICATION OF CLOUD COMPUTING IN EDUCATION" ES Margianti, Achmad Benny Mutiara, 2015 The 1st International Joint Conference Indonesia-Malaysia-Bangladesh-Ireland (IJCIMBI)At: Banda Aceh, Indonesia (https://www.researchgate.net/publication/292607782_Application_of_Cloud_Computing_in_Education)

2.3 Integrated Information Platform of Indian Universities Using Machine Learning

2.3.1 Overview

Nowadays, Machine learning has diverse uses in different areas of technology, The first application of machine learning in education is adaptive learning, which means that rather than putting all of the students' requirements and skills in one basket, the educational approaches are adjusted to each individual student's needs and abilities. The system will detect whether the student is having difficulty with the content or whether she may find it to be too simple for her; any change can be made at an early stage to avoid performance issues. Adaptive learning can either be a stand-alone educational system or it can be incorporated into an additional electronic classroom solution. Additionally, it can assist in spotting more minute aspects in the classroom, such as an outmoded curriculum or an uneven allocation of the teacher's focus. By nearly totally automating numerous repetitive, data-intensive, but time-consuming tasks, machine learning enables both educators and educational systems.

2.3.2 Applications of Machine Learning in Integrated Information Platform

Machine learning [9] is employed by internet search engines to better online learning environments and admissions procedures. Search engines [10] can recognize spam or duplicate information due to the use of algorithms that learn pattern recognition. Specific characteristics of low-quality material frequently parallel each other, such as the proliferation of hyperlinks leading to unrelated pages or too much reliance on general terms. Teacher and instructor insights on student performance can be improved by utilizing machine learning. They gain [11] important details about the styles of learning, areas of strength and weakness, and other relevant concerns of the pupils. This information enables teachers to take the proper steps to support students in sticking to their course agendas. Admissions and enrollment both benefit from machine learning, but precise demand forecasting is especially important. Higher educational [12] institutions to forecast the number of applications they should accept in order to reach the ideal participation levels.

2.3.3 What are Uses of Machine Learning in This Platform

- Takes Data of all Indian Universities with all their features as the attributes.
- According to the data of Universities, Machine learning algorithms helps the users by giving them a good recommendation by doing prediction of colleges as per theUsers searched items [13].
- Analyze all the search items or college over the platform, algorithm can help the students to find the relevant colleges.
- All the data are kept over the cloud from different datasets of universities [14].

2.4 Applications Used to Designed This Web Platform

When it comes to the overall view/design we need a better UI so that the work we have done, it will give a better look for all these we have added the features.

2.4.1 Front-End Development

- HTML
- CSS
- JAVA SCRIPT
- BOOTSTRAP

HTML: Hyper-text markup language [15] is the markup language which is used to design the structure of webpages for the websites and it is understood by the browser and shown on the screen as per the codes given in the HTML Extension.

CSS: Cascading style sheet [16] is the styling language used for designing the color and styles of the website appearance.

JAVASCRIPT: This is the scripting language [17] used to make the website responsive and reactive according to the functions and provides all the responsive and dynamic facilities

2.4.2 Backend Development

- PHP
- MY-SQL(CLOUD-BASED)

PHP: It is the most widely used open source and general purpose server [18] side scripting language used mainly in web development to create dynamic websites and applications.

MYSQL: This is very famous database [19] used to store and manipulate the data stored and data transfer between frontend and backend is perform by the database and SQL codes.

2.5 Analysis Result

2.5.1 Home Page

In this website [20] the user will go to the website and the home page (shown in the Figure 2.6) after going to the home page the user can access the option available there, but some of them user cannot access without login to the web platform.

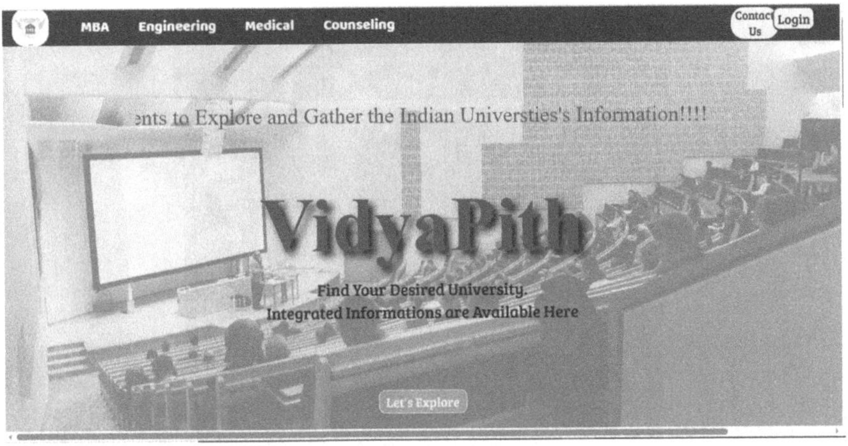

Figure 2.6 Home page.

Note. The proposed model shown in Figure 2.6 Home page, it contains the overview and guides the users to relevant information or actions they may want to take on the website. From "The proposed work".

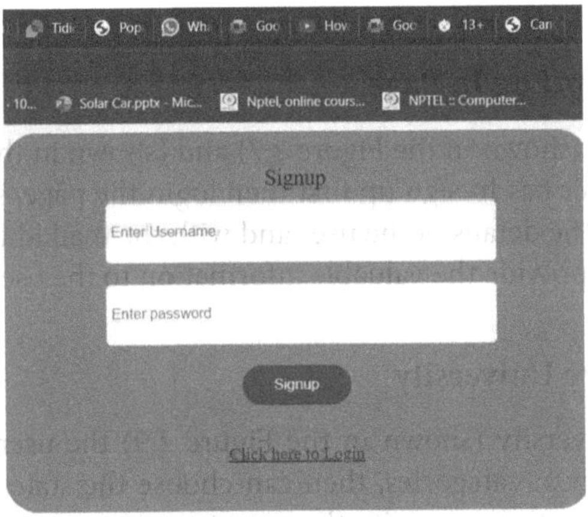

Figure 2.7 Sign up page.

Note. The proposed model shown in Figure 2.7 Sign up page contains the registration of users, user have to provide name and password so that the page can collect information and provide necessary information latter. From "The proposed work".

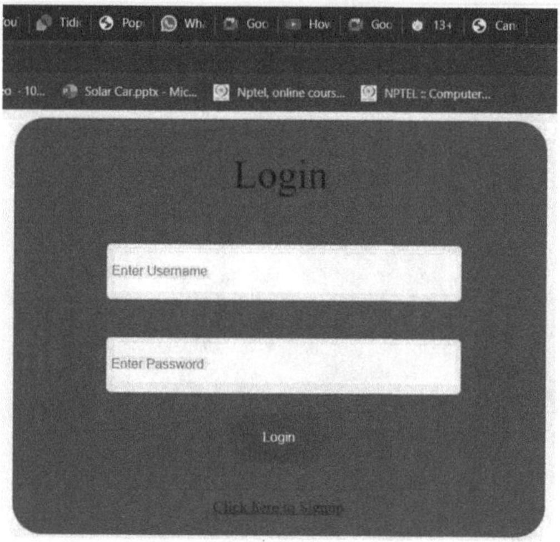

Figure 2.8 Login page.

Note. The proposed model shown in Figure 2.8 of Login page to get into the entire web page and to access all necessary features. From "The proposed work".

2.5.2 Sign up/Log In Page

In Sign up/Log (shown in the Figure 2.7) and (shown in the Figure 2.8) in the page the user has to sign up first then login the page, so the web platform will have the details of the user and with the mail id they have given, the system can provide the valuable information to the user.

2.5.3 Explore University

In Explore University (shown in the Figure 2.9) the user can check the university by given categories, then can choose the states (shown in the Figure 2.10) then the university they want to choose according to them. (shown in the Figure 2.11). After that they can see all the details of the university (shown in the Figure 2.10), (shown in the Figure 2.12).

2.5.4 Recommended Comparison

In Recommended comparison (shown in the Figure 2.14) the user will see the Recommended university comparison given by the web platform. After that the comparison (shown in the Figure 2.15).

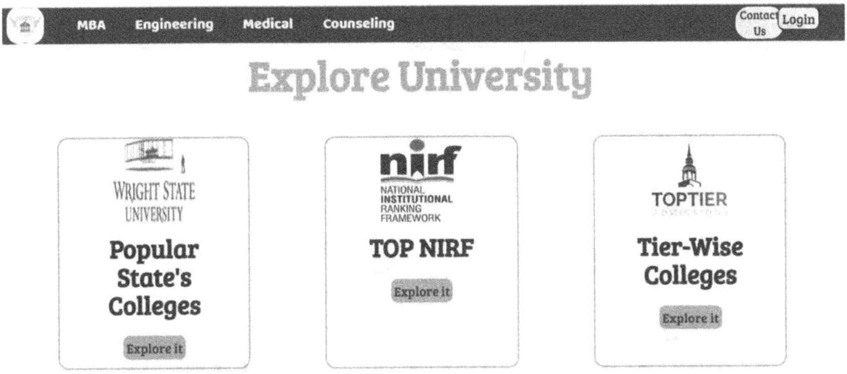

Figure 2.9 Explore university.

Note. The proposed model shown in Figure 2.9, the students can check the university from the given categories as per they want. From "The proposed work".

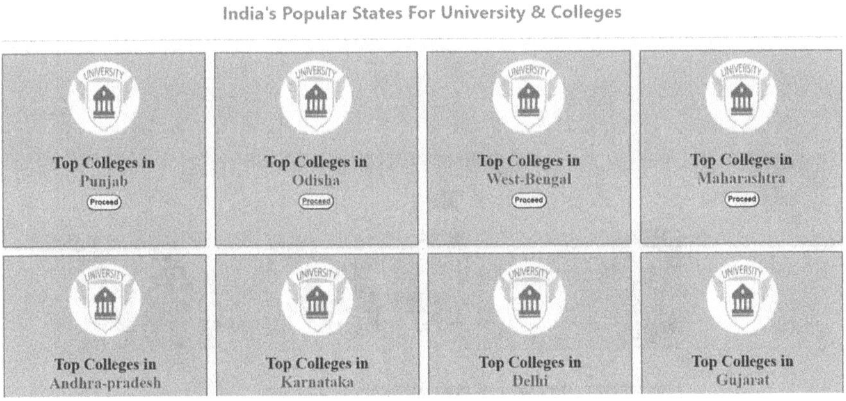

Figure 2.10 Statewise College.

Note. The proposed model shown in Figure 2.10, State wise College provides the states where students can choose the state they want to explore the colleges. From "The proposed work".

Figure 2.11 College list.

Note. The proposed model shown in Figure 2.11, after choosing the particular state users can see the college list of that state. From "The proposed work".

Figure 2.12 College details.

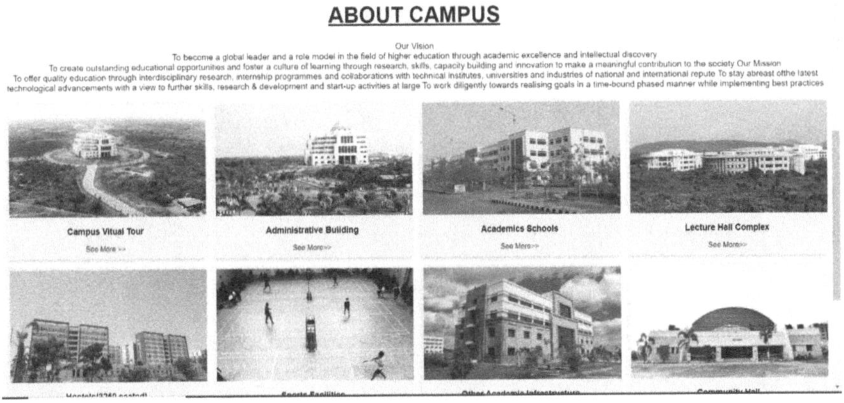

Figure 2.13 College details.

Note. The proposed model shown in Figure 2.12 and Figure 2.13, shows the details of the college that chosen by the students.

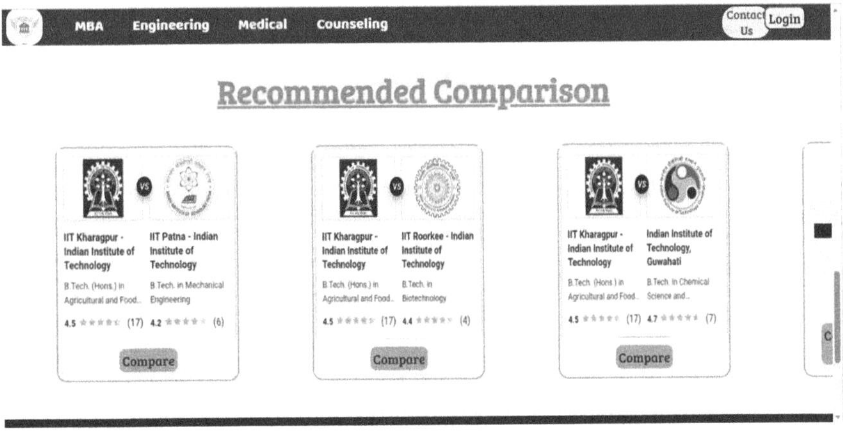

Figure 2.14 Recommended comparison.

Note. The proposed model shown in Figure 2.14, Recommended comparison shows the college comparison of the recommended college that given by the web platform. From "The proposed work".

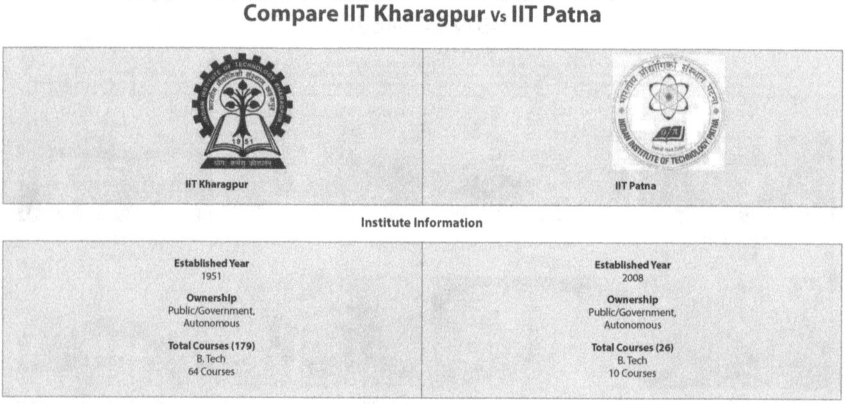

Figure 2.15 Comparison details of recommended colleges.

Note. The proposed model shown in Figure 2.15, shows the details of the college and students can see the difference and can compare the college, so that it will be helpful to choose the best college.

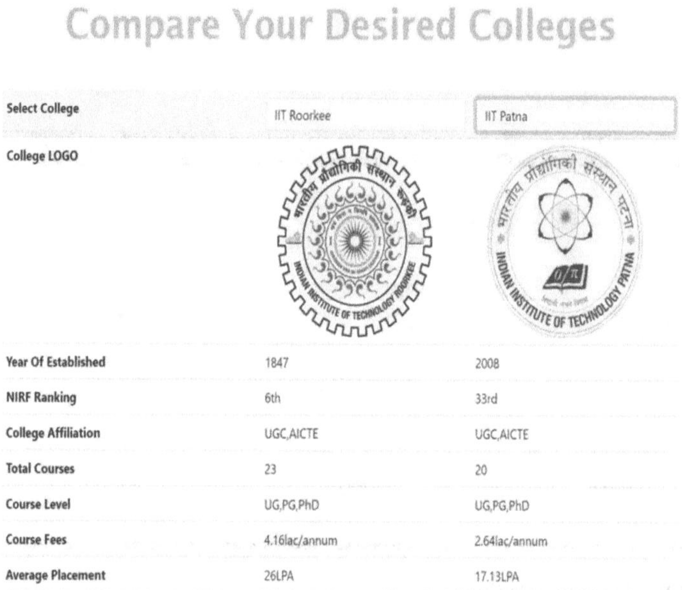

Figure 2.16 Manually comparison.

Note. The proposed model shown in Figure 2.16, Manual comparison is different from Recommended comparison. Here the students can manually add the colleges, they want to compare and it will show the comparison.

2.5.5 Manually Comparison

In manually comparison the user will compare the university by choosing according to them (shown in Figure 2.14).

Conclusion

The conclusion regarding an integrated information platform of universities that provides information of all colleges, including college comparison and other features, would depend on the specific implementation and user experience of the platform. An integrated information platform that is easily accessible and user-friendly can save time and effort for individuals seeking information about colleges. Users can access the platform from anywhere, at any time, using their preferred devices. This convenience allows for efficient research and exploration of various colleges, facilitating a smoother application process. Having a centralized platform that provides information on all colleges can be highly beneficial for prospective students, parents, and educators. It allows users to access detailed information about different colleges, including programs offered, faculty profiles, admission requirements, campus facilities, and more. This comprehensive information can assist in making informed decisions about college choices. A platform that offers college comparison features enables users to compare multiple colleges based on various criteria, such as tuition fees, academic programs, location, student-to-faculty ratio, accreditation, and extracurricular activities. This functionality can help users evaluate and weigh the pros and cons of different colleges, assisting them in finding the best fit for their needs and preferences. However, it's important to ensure that the platform's information is accurate, up-to-date, and unbiased. Regular maintenance and updates should be performed to reflect any changes in colleges' offerings, policies, or other relevant details. Additionally, user privacy and data security should be given utmost importance to protect personal information shared on the platform.

References

1. Ahmed, A. and Ahmed, H., A proposed model for education system using cloud computing, pp. 1–4, 2018, 10.1109/ICEEST.2018.8643331.
2. Rajput, R. and Pant, A., Optimal resource management in the cloud environment- a review, 2018.

3. Hogan, M., Liu, F., Sokol, A., Tong, J., *NIST cloud computing standards roadmap*, Special Publication 500–291, July 2011.
4. Chen, L., Liu, Y., Gallagher, M., Pailthorpe, B., Sadiq, S., Shen, H.T., Li, X., Introducing cloud computing topics in curricula. *J. Inf. Syst. Educ.*, 23, 3, 315–324, 2012.
5. Al Shehri, W., Cloud database database as a service. *Int. J. Database Manag. Syst.*, 5, 1–12, 2013, 10.5121/ijdms.2013.5201.
6. Bellemare, C., Bissonnette, L., Kröger, S., Statistical power of within and between-subjects designs in economic experiments. Discussion Paper No. 8583, 2014, Retrieved from http://ftp.iza.org/dp8583.pdf.
7. Bloor, R., What is a cloud database?, 2011, Retrieved 25th November 2012 fromhttp://www.algebraixdata.com/wordpress/wp-content/uploads/2010/01/AlgebraixWP2011v06.pdf.
8. Basar, E., Saikia, A., Saikia, L., A survey on database in cloud computing with reference to the traditional database. *Int. J. Adv. Res. Comput. Sci. Software Eng.*, 2017, 7.150.10.23956/ijarcsse.v7i11.477.
9. Luckin, R., Holmes, W., Griffiths, M., Focier, L.B., Intelligence unleashed: An argument for Al in education, Pearson, 2016.
10. Noorian, M., Bagheri, E., Du, W., Machine learning-based software testing: Towards a classification framework. *Proceedings of the 23rd International Conference on Software Engineering & Knowledge Engineering*, 2011.
11. Amershi, S. and Conati, C., Unsupervised and supervised machine learning in user modeling for intelligent learning environments. *Proceedings of 12th International Conference on Intelligent User Interfaces*, 2007.
12. Lou, Y., Ren, R., Zhao, Y., A machine learning approach for future career planning.
13. Liu, D. and Lin, C., Sherlock: A semi-automatic quiz generationsystem using linked data. *13th International Semantic WebConference*, 2014.
14. Greff, K., Srivastava, R.K., Koutník, J., Steunebrink, B.R., Schmidhuber, J., LSTM: A search space odyssey. *Trans. Neural Networks Learn. Syst.*, 2015.
15. Peroni, S., Osborne, F., Di Iorio, A., Nuzzolese, A., Poggi, F., Vitali, F., Motta, E., Research articles in simplified HTML: A web-first format for HTML-based scholarly articles. *PeerJ Comput. Sci.*, 3, e132, 2017, 10.7717/peerj-cs.132.
16. Genevès, P., Layaida, N., Quint, V., On the analysis of cascading style sheets, 2012, 10.1145/2187836.2187946.
17. Delcev, S. and Draskovic, D., Modern JavaScript frameworks: A survey study. *2018 Zooming Innovation in Consumer Technologies Conference (ZINC)*, Novi Sad, Serbia, pp. 106–109, 2018, doi: 10.1109/ZINC.2018.8448444.

18. Cui, W., Huang, L., Liang, L., Li, J., The research of PHP development framework based on MVC pattern. *2009 Fourth International Conference on Computer Sciences and Convergence Information Technology*, Seoul, Korea (South), pp. 947–949, 2009, doi: 10.1109/ICCIT.2009.130.
19. Letkowski, J., Doing database design with MySQL. *J. Technol. Res.*, 6, 2015.
20. Collegedunia, Top colleges and universities in India, https://collegedunia.com/.

3

False Data Injection Attack Detection Using Machine Learning in Industrial Internet of Things

Hafizunisa*, **Prerna Rai and Damini Sinha**

Kalinga Institute of Industrial Technology, Deemed to be University, Bhubaneswar, Odisha, India

Abstract

With the sudden global change and adoption in the field of Internet-of-Things (IoT), it has also made the infrastructure prone to serious cyberattacks. By allowing system autonomy and giving real-time response, the Industrial Internet of Things has aided in resolving numerous industrial issues, but the security challenges for these applications should also be thoroughly explored and taken care of for their successful implementation. There has been a rise in False Data injection (FDI) attacks in cases involving the IIoT. These attacks successfully undermine the decision-making power of the targeted industries by manipulating the critical sensor readings. Recent studies have found that Machine Learning algorithms have the capacity to handle such cyberattacks using their decision-making capabilities and are helpful in detecting a variety of threats arising in the domain of IIoT. In our research, we offer a method for identifying FDI attacks that uses a neural network method called autoencoders (AEs). The AEs helps in finding the deviation between the correlational structure of the corrupted data and the expected learned correlation structure. This is why we suggest using it as a classification strategy to identify FDI attacks. Once the presence of the corrupted data has been detected, we use denoising autoencoders (DAE) to reconstruct the originally recorded data. Finally, we compare the complexity of the AEs with that of the Support Vector Machine (SVM), which is also a Machine Learning methodology.

Keywords: Industrial internet of things (IIoT), false data injections, machine learning, Autoencoders (AE), neural networks, Denoising Autoencoders (DAE), Support Vector Machine (SVM)

**Corresponding author*: hafizunisamohammed@gmail.com

Hrudaya Kumar Tripathy, Sushruta Mishra, Minakhi Rout, S. Balamurugan and Samaresh Mishra (eds.) Optimized Computational Intelligence Driven Decision-Making: Theory, Application and Challenges, (49–68) © 2024 Scrivener Publishing LLC

3.1 Introduction

The Industrial Internet of Things (IIoT) is linked to industrial applications and is basically an online network of sensors, instruments, and self-contained machines. Using this network, we can gather data, conduct analysis, and optimize production in such a way that it boosts effectiveness and lowers production and service costs. Industrial applications are an extensive ecosystem of technology that links the equipment and personnel responsible for overseeing logistics, production lines, and massive distribution. Its features include increased flexibility and improved process optimization, which lead to lower deployment and maintenance costs while delivering new services and adapting processes to the needs of manufacturers, operators, and customers. The IIoT system works by layering various components together. A few examples of devices in this system are sensors, GPS, locators, and machines. The connectivity layer consists of the network created between these devices and servers through edge computing, cloud storage, or a network. IIoT offers services like analyzing data collected from various sensors and processing it to find some meaningful conclusions. Content, or end devices (also known as human interface), acts as a medium between the user and the system; it also includes gadgets like glasses for augmented or virtual reality. Figure 3.1 shows a detailed block diagram of IIoT. The block diagram is structured into four sections: Enterprise Operations, Cloud Platform, Edge System, and the final level of IIoT and Data Sources. Enterprise Operations is the stage in IIoT where organizations are progressively investing in linked physical objects that are integrated with computing devices that help drive operational efficiency, minimize human labor, and generally speed up business processes. It is then further subdivided into two categories: the business domain and the application domain. The enterprise applications and the dashboard are part of the business domain, whereas logic, rules, and API reports are part of the application domain.

It is then followed by the Cloud Platform, which manages the storage and resources. It is then further divided into two domains: Information and operations. The information from the business domain and the data from the application domains then flow into the information domain, which helps in the provision of data and analytics services. Whereas the operation domain manages provisioning, prognostics, deployment, assets, metadata optimization, APIs, portals, monitors, and diagnosis.

The Data from the Information domain benefits management from the operations domain and orchestration from the Application domain, which flows into the Edge system, which also includes the sensors and actuators

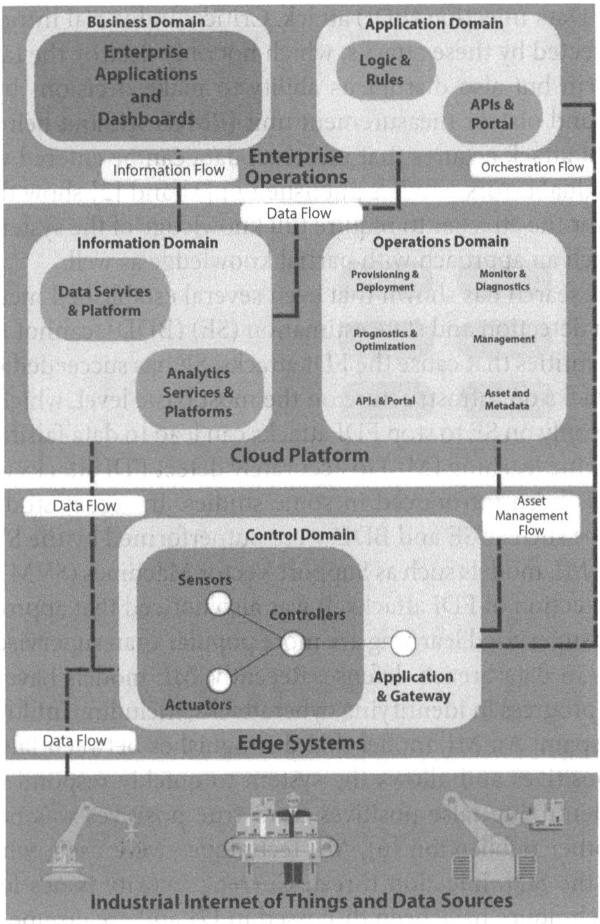

Figure 3.1 The general block diagram of IIoT.

that are linked to controllers, which are then linked to applications and gateways. The data sources are the end users, and they produce a continuous flow or stream of data. The Productivity and innovation are largely driven by IIoT applications in business. With the advancement of IIoT and the rapid adoption of this technology in global industries, critical infrastructure is prone to various malicious activities.

Now, in order to successfully deploy IIoT applications, there is a need to thoroughly address and investigate the security threats to such applications. Any loose ends in security can lead to grievous losses, sometimes at the expense of human beings. One of the recently discovered frequent attacks, which include hydraulics, the oil and gas industry, and other areas, is the False Data Injection (FDI) attack. Critical Industrial Infrastructure is severely affected by these attacks, which not only destroy the target industrial platform but also disrupt its ability to make decisions by falsifying the sensor and phasor measurement unit (PMU) without being detected. This kind of attack ensures that incorrect data can be entered while minimizing residual errors. Studies published in [1] and [2] show that it is not necessary for the attacker to require full knowledge of the system; they can perform such an approach with partial knowledge as well.

Recent research has shown that even several established methods, such as bad data detection and state estimation (SE) (BDD), cannot identify the disruptive entities that cause the FDI attacks. SE has succeeded in identifying FDI attacks on infrastructure on the most basic level, which is crucial. But relying only on SE to stop FDI attacks can lead to data falsification. The use of machine learning (ML) to accurately detect FDI attacks to solve this problem has been introduced in some studies. In recent studies [3] and [4], methods such as SE and BDD were outperformed by the SVM, which introduced ML models such as Support Vector Machines (SVM) to provide effective detection of FDI attacks. It was also noticed that approaches such as partially supervised learning are more popular than supervised learning approaches in data breach defense. Recently, ML models have also made significant progress in identifying cyberattacks, including infiltration, malware, and spam. An ML model that distinguishes between false positives and true positives and allows the system to quickly respond to cyberattacks by identifying false positives from true positives was presented in [5]. In another publication [6], ML techniques were used, which helped determine the SQL injection threat. Current security issues for machine learning techniques have been discussed in [7] and are grouped as part of the training and testing phases. Additionally, security, data security, and privacy evaluation mechanisms were also added to the subcategories of machine learning techniques. This paper [8] proposes two ways to detect false imputation, the first of which uses labeled data, which would then be involved in training a model using supervised learning, and the second of which does not require data training to detect measurement bias. In both approaches, Principal Component Analysis (PCA) is used for the mapping of the data into a low-dimensional space, which is based on the observation that normal and compromised data tend to separate after projection. Therefore, traditional methods such as SE and BDD were found a little less effective in front of SVM to detect cyberattacks. The major disadvantage of not being able to successfully adjust to changing transfer behavior patterns over time is the use of supervised anomaly detection [9]. In order to distinguish between the distribution of disturbance states and typical operating conditions, the authors of the above article proposed a Gaussian mixture model. Over the past few years, there has been a significant increase in the promise that deep learning has shown for completing challenging tasks. Now, as part of data-driven deep learning techniques, a function approximator is created using gradient descent over a predetermined set of data points.

The neural networks' capacity to learn complex operations as well as the accessibility of enormous datasets have successfully contributed to The success of deep learning. In this paper, we have looked at the use of autoencoders (AE) to detect FDI attacks. According to AE, "labeled" data cannot be used to train the model. [10] shows the use of autoencoders to train latent features that reduce the size of the feature vectors used to feed the ML classifier. In [11], to improve the classification performance of the neural network, composite AEs are used, and the AEs that are used to learn critical information will be delivered to the state of multi-layer perception (MLP) detection methods. Any change that changes the data correlation structure can be detected by an AE-based detector.

Other common detectors (classifiers) must be trained with labeled datasets for each attack they are intended to detect because they cannot be guaranteed to catch any additional attacks for which they have not been trained.

Therefore, the article's main contributions can be summed up as follows:

- In order to get the fundamentals of the proposed model, the technical background and basic principles of Autoencoders and Denoising autoencoders are discussed in the paper.
- To detect the FDIs using autoencoders, which can study the false data by accessing how far the corrupted data correlation structure is from the expected structure, unsupervised methods are used.
- In order to reconstruct the corrupted data while taking the altered data as input and applying the principles of autoencoders to recover the original data by rebuilding the original correlation structure, Denoising autoencoders were used.
- A comparative and complexity analysis of our proposed model and between Autoencoders and SVM classifiers
- The various advantages and limitations of Autoencoders and related work were also discussed.

Machine learning algorithms have advanced to the point where complex tasks can be performed using simple concepts. An enormous potential has been detected in the application of ML to problems such as forecasting and anomaly detection. In order to find false negatives and false positives for positive classification, [12] applies the collected data from Knowledge Discovery in Databases (KDD) to implement, test, and evaluate machine learning (ML) methods such as Decision Table, Random Forest, Random Tree, Naive Bayes, and Bayes Network.

This paper conducts a Systematic Literature Review (SLR) on machine learning (ML) models that detect anomalies in their applications [13]. Four angles—applications for anomaly detection, machine learning techniques, performance metrics for ML models, and classification for anomaly detection—have been used to examine the models. Between 2000 and 2020, 290 research articles that explored ML techniques for anomaly detection were published, according to this review, and many of the applications use machine learning algorithms. They are largely helpful in circumstances that involve large databases and data mining, where the subject is poorly known and thus really difficult for humans to represent. These techniques can be used to process large data sets, and data from many sources can be mixed to factor in past knowledge when doing analyses. [14] In this study, machine learning algorithms were used to predict diabetes in large datasets.

We were motivated by the performance of ML and Deep Learning algorithms in prediction and anomaly detection to apply AE, i.e., neural networks, for FDI detection in the recorded data set. Then, using DAE, the original dataset was recovered from the corrupted one.

3.2 Literature Review

In the following section, we will discuss some of the previous works done by researchers on the topic of Detection of Attacks using Machine Learning techniques in Industrial Internet of things. The authors in [1] presented a systematic but efficient approach that needs a little knowledge of the network to examine and identify the possible attacking region of a single line. The six IEEE standard systems were used for validating the proposed attacking strategy. Their study has revealed the real-time topology of a power grid's vulnerability and is very helpful in creating topological attack defense for smart grids. Taking into account the challenges of gathering network information, the suggested attack method employs the least amount of network data possible. In [2], two solutions for minimizing the risks of misleading data injection in electrical power networks were proposed. Firstly, a method of developing unnoticeable attacks on the

AC power flow equations was presented, which describes in detail how an adversary can launch a secret attack to accomplish a task. Offering a defense tactic that can find out attacks and introduce known disruption by purposefully probing the system in a predetermined, structured manner.

The method of identifying the presence of fake data injection attacks, their locations, and details about the changed data values is the next step. Using local information, they have shown how to conduct innocuous attacks on the AC flow equations. In addition, a particular probing defense strategy for identifying attacks through the use of probe Sequences and the recording of amounts that were observed was provided.

The defense is non-responsive to the sort of attack and probe types. As the adversary's view of the system changes, research into the productiveness of the attacks and the probing defense will continue.

In [3], two Strategies were used. In the first technique, labeled data and supervised learning were used to train a distributed support vector machine (SVM). High convergence rate of the alternating direction multiplier technique and established optimal method is used to create a distributed SVM. The second technique explores measurement variance without the need for training data. Principal component analysis is used to decrease the dimensionality of the data to be evaluated, which reduces processing complexity for both approaches.

In [4], machine learning techniques were applied to determine whether measurements were secure or compromised. Because of the proposed method, an attack detection framework to take advantage of any previous knowledge of the system and get over limitations brought on by the limited nature of the problem was supplied. For the representation of the detection problem, online learning approaches (supervised and semi-supervised) were combined with decision- and feature-level fusion. The exploration of relationships between statistical and geometric aspects of attack vectors utilized in attack scenarios and learning algorithms was done in order to detect unnoticeable attacks using statistical learning techniques. Later, many IEEE test systems were put to use to evaluate the proposed methodologies. According to the experiment, the suggested attack detection framework's proposed attack detection algorithms exceed machine learning techniques. The concept of attack detection was envisioned as a machine learning problem, and the effectiveness of supervised, semi-supervised, classifier, feature space fusion, and online learning methods were assessed for a range of assault situations.

The Introduction of a cyberthreat detection AI method based on artificial neural networks was done in [5]. A deep learning-based detection method to boost cyber-threat identification by breaking down a huge number of recorded security events into unique event profiles was used for the suggested technique. They developed an AI-SIEM system employing a combination of event profiling for data preprocessing and other artificial neural network techniques such as FCNN, CNN, and LSTM for this project. Thanks to the system's emphasis on differentiating between genuine and false positive signals, security analysts can now respond to cyberattacks more swiftly. Experiments were conducted using the five traditional machine-learning algorithms (SVM, k-NN, RF, NB, and DT) to evaluate their performance with current methods. The experimental results of the study showed that when applied in the real world, the suggested methods outperform traditional machine-learning techniques and can be employed as learning-based models for network intrusion detection. In addition to that, it can also help security analysts respond quickly to cyber threats scattered across a large number of security events by reducing false-positive warnings.

3.3 Technical Methodology

In this section we will discuss the technicalities of achieving the assessment of false data injection attack. Using Machine learning methods can make the task of anomaly detection simpler as discussed in the previous section. Raw sensor data from various equipment which are prone to false data injection attack is assessed using machine learning techniques. An already proven work [15] portraits how with the use of AE (Autoencoders) anomalies can be detected in such sensor readings. After discovering the anomalies in the data it is important to recover the original data, hence DAE (Denoising Autoencoders) help do so. In the following section we have discussed about the working of AE and DAE.

3.3.1 Autoencoders (AE) used for Identifying False Data

Autoencoders mostly work on two levels namely encoding and then decoding. These levels are mostly referred as network as it includes the use of network of layers.

3.3.1.1 Encoder Network

Mostly the input to an encoder network are high dimensional data, and the role of encoder network is to transform the data and convert it into a lower

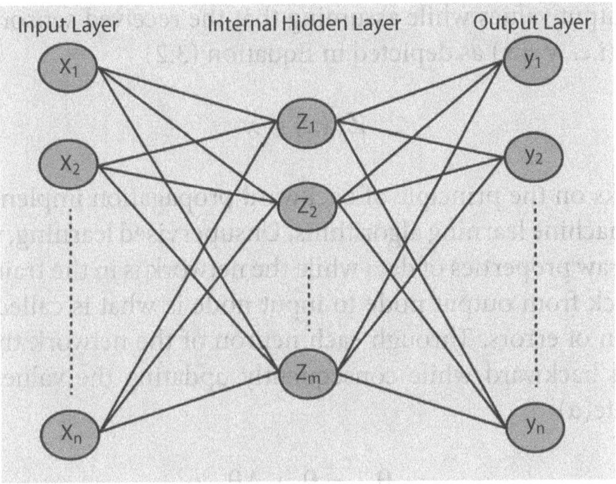

Figure 3.2 AE architecture with one hidden layer.

dimensional data that is by reducing the size of data encoder network is compressing the input data and producing it onto output data.

3.3.1.2 Decoder Network

After discovering that an attack has been made, and false values exist in the data, the original values is recovered using the decoder network, in this the opposite of the encoder network is performed that is decompressing the input data and producing it onto the output data, and all this compression and decompression of data occurs mostly in the bottleneck hidden layers of Autoencoders.

A fully connected unsupervised neural network, Autoencoders helps to assess the false data injection attack, with the help of a technique called backward propagation of errors.

Figure 3.2. adapted from [16], there is one input layer, one output layer and multiple internal hidden layers and this makes up the architecture for AE.

$$y = Z_\theta (x) \tag{3.1}$$

Over here x represents the input vector and z_θ, the fully forward propagation function and θ is the collection of biases and values required for neural network to learn while the training phase is going on. In the initial stage where the network is still in the learning phase the output value is set equal to output values while assuming that the received sensor data is not corrupted (i.e., y = x) as depicted in Equation (3.2)

$$Z_\theta (x) = x \tag{3.2}$$

AE works on the principle of backward propagation implemented by a family of machine learning algorithms, Unsupervised learning, which used unlabeled raw properties of data while the network is in the training phase. Tracing back from output node to input node is what is called backward propagation of errors. Through each neuron of the network the output is propagated backward while consequently updating the values based on learning rate(α)

$$\theta_{i+1} = \theta_i + \Delta\theta_i \tag{3.3}$$

Where i denotes the index of the training data $\Delta\theta_i$ and is the values updates, specified as

$$\Delta\theta = -\alpha \frac{\delta E}{\theta i} \tag{3.4}$$

The measure scale used to keep track of the quantity of error we are using mean-square error(MSE) function which is given by:

$$E = \frac{1}{M} \Sigma_m (y_m - x_m)^2 \tag{3.5}$$

For Equation (3.5) which represents the MSE value, the size of data is denoted by M, and at the m^{the} position x_m represents the input vector while y_m is the output vector. Compressing and decompressing of data leads to destruction of its correlation by AE. After compression the input the decoder network is used to reconstruct the correlation between data values. Strong compressible nature defines strong correlation between data values. To achieve this strong correlation the number of internal layers in the neural network should be kept smaller than that of input and output layers.

Although the output of DAE is the reconstructed version of the input data the data which the internal layers generate is the compressed version of the input data having lower dimensions. This compression of the size of data helps to determine whether a FDI attack was made or not in the raw sensor data. In case of an attack the estimated MSE value will significantly bigger than the predicted MSE value at the output of AE. The estimated MSE value is compared with specified threshold value, if the difference between the two values is too high then an FDI attack is declared in the recorded data and the raw data is passed into DAE in order reconstruct the data from damaged data.

3.3.2 Denoising Autoencoder (DAE) used for Data Recovery

Another type of AE is Denoising Autoencoder (DAE) using the same principle of backward propagation of errors to reconstruct the corrupted data. Once an FDI attack has been identified the corrupted data is fed as a input to DAE and the original data is recovered at output of DAE. Hence target output:

$$Z_\theta (x) = x$$

X denoted the corrupted set of data and the $Z_\theta(X)$ is denotes the output of DAE that is the original recovered data. We use DAE to make the function more efficient while dealing with real world data that might be of a larger size. DAE is mainly used to retrace the corrupted data without just copying the input data but by capturing the important features of the training data that is the corrupted data. During this training phase the DAE is responsible for recreating the correlation between the input data using original corrupted data.

3.4 Proposed Model for Detecting False Data and its Correction

In this study the identification of False Data Injection attack has been discussed. Figure 3.3 shows the flow of working of how the raw recorded data which has been injected by false data is recovered using a fully connect neural network AE and DAE using the procedure of backward propagation of errors. When detected whether an attack was not made, data is passed on for further processing. On detection of a variance the corrupted data is placed as an input to DAE which is responsible for recovering the original data from corrupted data.

In the currently presented model, the attacker can change the raw readings data from more than one sensor anytime anywhere in order to make

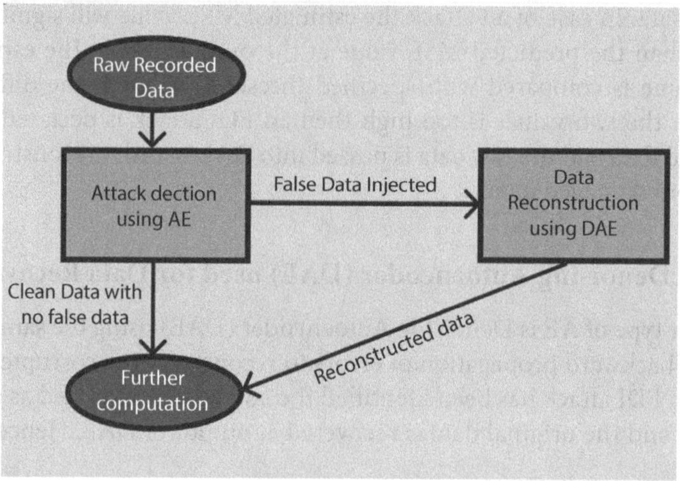

Figure 3.3 Flowchart of working of AE and DAE in detecting and correcting of data.

the false data within the accurate measurement range so that it can defy the readings and weaken the decision making capabilities of the target system having critical infrastructure.

The false data attack on measured data can be represented as follows:

$$X_a = X + p \tag{3.6}$$

Where X_a represents the readings vector from the attack sensors that may have been altered, X represents the record vector that is clean, and free from any alterations, i.e.

$X = [X_1 \ldots\ldots\ldots\ldots X_m]^T$ and $p = [p_1 \ldots\ldots\ldots\ldots p_m]^T$ denotes the information that has been added to the actual values, that is the vector where p_i denotes the attacker's capacity to change the ith reading and replace different false value for its corresponding true value.

3.4.1 Attack Detection Algorithm

As discussed earlier AE is been used for attack detection because of its proven ability to capture data's structure and explore the correlation between features and reading. Later after an attack has been detected the corrupted data is then passes on to DAE which reconstructed the correlation of data in order to recover the original data.

The initial sensor data is input into AE throughout the training phase. Only during this training phase the data recorded from various sensors are vulnerable to attacks. The recorded data is set as target values. Following the training phase, the data is sent into AE, which detects falsified data.

The neural network tries to compress and then decompress the input vector after feeding the data into the AE. The entered data is then used to calculate the MSE (Mean Square error), which is larger than the original data because the corrupted data lacks the necessary correlation between features. To guarantee that the data is correct, the MSE is matched with the previously determined threshold calculated by the mean of the MSE. An attack is made when MSE value exceeds the threshold value. Moving on a validation step is necessary in order to prevent over fitting. The reason being, as the validation error continues to rise, the training data is decreasing. By keeping an eye on the validation error, the training process is abruptly stopped if it starts to rise.

3.4.2 Denoising Autoencoders-Based Data Cleaning

As indicated in the flowchart, if the data is deemed clean by the AE, it is appropriate to move on nevertheless, after an FDI attack is discovered, the data is sent to DAE. The original data and corrupted data are used as target outputs for training DAEs. The DAEs have the capacity to reconstruct the correlation between the inputs, as was previously observed. The cleaned version of the false data is generated from the DAE after being fed into it, and it may then be utilized as the actual sensor readings that were being attacked and are suitable for use in system processing going forward.

3.4.3 Algorithm for False Data Detection and Correction

The Algorithm that is proposed performs the detecting and cleaning procedure in 4 stages which are as follows Preparing, Training, Testing and Denoising of corrupted data. The detailed steps and procedure is explained below

---

1. **Procedure to Prepare Data Set**
2. Divide the data into training, validation, and testing sets as [60:20:20].
3. training_data, validation_data, testing_data = split_data(-data, [0.6, 0.2, 0.2])
4. Create a copy of the testing set with faulty sensor readings.
5. faulty_testing_data = create_faulty_copy(testing_data)
6. **Procedure to Train the model**

7. Initialize some Random weights.
8. weights = initialize_weights()
9. Enter each sensor reading and assign them to M1.
10. for data_point in training_data:
11. M1 = data_point
12. Back Propagation Learning Technique: $m(y_m - x_m)^2$
13. $Error = (y - model(M1))^2$
14. While Validation error is decreasing do:
15. while validation_error_decreasing:
16. For each training data span perform the following:
17. for training_point in training_data:
18. Calculate MSE(Mean Square Error) considering current weights and entered data.
19. mse = calculate_mse(training_point, weights)
20. Based on calculated MSE and learning rate, update the weights value.
21. weights = update_weights(weights, mse, learning_rate)
22. In order to avoid profitless data make sure the validation error is decreasing.
23. if validation_error_not_decreasing:
24. break
25. End for
26. End while
27. End procedure
28. Procedure to test the Prepared model
29. Keep the complexity factor O(e.p) for each iteration, where e denotes the dimensions of input and p denotes the encoding dimensions.
30. complexity_factor = calculate_complexity_factor(input_dimensions, encoding_dimensions)
31. Enter testing data and faulty data.
32. for data_point in [testing_data, faulty_testing_data]:
33. While Results not achieved do:
34. while results_not_achieved:
35. Change the testing parameters for e.g. Batch size , hidden layers, etc.
36. change_testing_parameters()
37. Repeat the training procedure.
38. trained_model = train_model(training_data, validation_data)
39. **End while**

40. **End Testing Procedure**
41. **Procedure to perform reconstruction of Data**
42. Input the corrupted data and target as original data for Training.
43. training_input = corrupted_data
44. training_output = original_data
45. Using detected falsified data as inputs for Testing.
46. testing_input = detected_falsified_data
47. Corrected data as Output.
48. corrected_data = denoise_data(testing_input, trained_model)
49. **End Procedure**

3.5 Complexity Analysis of Proposed Model

Another machine learning algorithm that has also been proven effective in detection of anomalies is SVM. It is a powerful classifier that is basically a construction method. It creates a boundary between two classes in order to enable label prediction. This decision boundary also known as hyperplane, is positioned to be as far from the nearest data points from each class as is practical. Support vectors are the closest locations in the model. In [17] the authors talk about how SVM is used for false data detection. In this study, an SVM algorithm is used to distinguish between malicious and trusted packets by detecting injected false messages. Another article [18] discusses the implementation of SVM to stop false data injection attacks. However, SVM can be used for the detection of false data injection attack, but in comparison it with Autoencoders, it was proven to be more complex and time consuming. In [19] the comparative analysis between Autoencoders and SVM classification model was simulated for the case of hyper spectral images and it was found out that the testing time and training time for SVM classifier was significantly more than that of AE model. This paper also demonstrates the accuracy analysis for SVM and SSRN and Autoencoders after which it was found out that their proposed model based on Autoencoders was 98% accurate while classifying the hyper spectral images. In the paper [20] the training complexity was found out to be $O(N^3)$, N representing the size of data considered. While the training complexity of AE was found out to be O(e.p) for each iteration, where e denotes the dimensions of input and p denotes the encoding dimensions.

Hence the training time for AE is less as compared to SVM. In this paper we are proposing to recover the attacked data using a DAE which can reconstruct the correlation structure of the data recorded by various sensors. The DAE makes use of MSE (Mean Square Error) to reconstruct the correlation model. MSE is used to illustrate the disparity between the original and attacked data.

3.6 Advantages of the Model

- Autoencoders (AEs) training does not need "labeled" data as they are considered as unsupervised learning techniques.
- The feature vector is successfully trimmed using Autoencoders in an unsupervised manner.
- AS we know Autoencoders are trained on unlabeled data, therefore using Autoencoders for classification for their capacity to identify additional attacks is one of its the major advantages [21, 22].
- A detector system using an autoencoder would recognize any attack that disrupts the structure of the data correlation.
- In case of a utility company that does not have a lot of information about attacks, the unsupervised detection method should be an appropriate option for detecting False Data Injection Attack. The localization of intrusion comes as an additional advantage for detection, as no separate model needs to be trained.

For creating input data and to condense the data collection, self-supervised machine learning models popularly known as autoencoders are used. Since these models were created using supervised machine learning techniques and then used unsupervised methods for inference, they are known as self-supervised models.

Its usages are as follows:

- Dimension Reduction. The use of autoencoders trains the network to transform the data's inherent structure into a useful lower-dimensional representation, which is accomplished by using decoding and encoding methods for the reduction of the reconstruction error.

- Detecting Anomaly: While training, autoencoders attempt to lower the reconstruction error. Examining the size of the reconstruction loss allows for the detection of anomalies.
- Image Denoising: It basically means restoring a corrupted image back to its original image. The autoencoder receives the noisy input image , and the noiseless output is created by rebuilding it while minimizing the loss from the original goal output (noiseless).

3.7 Future Scope and Limitations of the Proposed Model

Additional testing on these models will be done on more complex instances, such as 118- or 300 bus systems.

- The proposed model will help testing on future emergencies or straits(contingencies).
- Application of auto-encoders to sub-grids or local pools of devices which have high correlations.

The visible constraints of the model are outlined below.

- The algorithm suggested has the ability to point out attempts to wreck the correlation between sensor values. Therefore the Autoencoder might not be able to identify the assault if it can preserve this association.
- While comparing Autoencoders, which does not have the requirement of supervised learning, the DAE must to be trained for each sort of attack. Therefore, the DAE can only be used to defend against specific types of threats for which it was designed.

3.8 Conclusion

Talking about (IOT) Internet-of-things, the main Idea behind it is combined and integrated development of industrial automation and networking of everything Technological. In the above paper we have briefly discussed about IIOT, its benefits, challenges and have tackled one of its major problems which is a great threat nowadays i.e. FDI (False Data Injection) attacks. We have proposed a practical solution for the concerned issue that is using Autoencoders(AEs).

The AEs performs better in terms of detection because the Autoencoders are simpler to train and don't require labeled training. It may also get to know about hidden complicated data sets that show correlation, which allows them to identify various attacks. The detection systems which use autoencoders can easily identify any attack which can alter these correlation patterns noticeably. Other additional classifiers like SVM, are taught to recognize "specific" assaults and as a result, there is no assurance that they will detect any additional attacks for which they have not been trained. The above study has also talked about the Applications of autoencoders in anomaly detection and the various limitations of the use of autoencoders as well as its future applications.

References

1. Liu, X. and Li, Z., Local topology attacks in smart grids. *IEEE Trans. Smart Grid*, 8, 6, 2617–2626, Nov. 2017, doi: 10.1109/TSG.2016.2532347. doi:10.1109/MCOM.2018.1700592.
2. Davis, K.R., Morrow, K.L., Bobba, R., Heine, E., Power flow cyber attacks and perturbation-based defense. *2012 IEEE Third International Conference on Smart Grid Communications (SmartGridComm)*, pp. 342–347, 2012, doi: 10.1109/SmartGridComm.2012.6486007.
3. Esmalifalak, M., Liu, L., Nguyen, N., Zheng, R., Han, Z., Detecting stealthy false data injection using machine learning in smart grid. *IEEE Syst. J.*, 11, 3, 1644–1652, Sept. 2017, doi: 10.1109/JSYST.2014.2341597.
4. Ozay, M., Esnaola, I., Yarman Vural, F.T., Kulkarni, S.R., Poor, H.V., Machine learning methods for attack detection in the smart grid. *IEEE Trans. Neural Networks Learn. Syst.*, 27, 8, 1773–1786, Aug. 2016, doi: 10.1109/TNNLS.2015.2404803.
5. Lee, J., Kim, J., Kim, I., Han, K., Cyber threat detection based on artificial neural networks using event profiles. *IEEE Access*, 7, 165607–165626, 2019, doi: 10.1109/ACCESS.2019.2953095.
6. Ross, K., SQL injection detection using machine learning techniques and multiple data sources. Master's Projects. 650, 2018.
7. Liu, Q., Li, P., Zhao, W., Cai, W., Yu, S., Leung, V.C.M., A survey on security threats and defensive techniques of machine learning: A data driven view. *IEEE Access*, 6, 12103–12117, 2018, doi: 10.1109/ACCESS.2018.2805680.

8. Esmalifalak, M., Liu, L., Nguyen, N., Zheng, R., Han, Z., Detecting stealthy false data injection using machine learning in smart grid. *IEEE Syst. J.*, 11, 3, 1644–1652, Sept. 2017, doi: 10.1109/JSYST.2014.2341597.
9. Kundu, A. *et al.*, A3D: Attention-based auto-encoder anomaly detector for false data injection attacks. *Electr. Power Syst. Res.*, 189, 106795, Dec. 2020. 10.1016/j.epsr.2020.106795. Accessed 27 Aug. 2020.
10. Yousefi-Azar, M., Varadharajan, V., Hamey, L., Tupakula, U., Autoencoder-based feature learning for cyber security applications. *2017 International Joint Conference on Neural Networks (IJCNN)*, pp. 3854–3861, 2017, doi: 10.1109/IJCNN.2017.7966342.
11. Vincent, P. *et al.*, Stacked denoising autoencoders: Learning useful representations in a deep network with a local denoising criterion Yoshua Bengio Pierre-Antoinenzagol. *J. Mach. Learn. Res.*, 11, 3371–3408, 2010, www.jmlr.org/papers/volume11/vincent10a/vincent10a.pdf?ref=githubhelp.com.
12. Almseidin, M., Alzubi, M., Kovacs, S., Alkasassbeh, M., Evaluation of machine learning algorithms for intrusion detection system. *2017 IEEE 15th International Symposium on Intelligent Systems and Informatics (SISY)*, pp. 000277–000282, 2017, doi: 0.1109/SISY.2017.8080566.
13. Lee, C.K.M., Zhang, S.Z., Ng, K.K.H., Development of an industrial Internet of things suite for smart factory towards re-industrialization. *Adv. Manuf.*, 5, 4, 335–343, nov 2017, doi: 10.1007/s40436- 017-0197-2.
14. Meyfroidt, G. *et al.*, Machine learning techniques to examine large patient databases. *Best Pract. Res. Clin. Anaesthesiol.*, 23, 1, 127–143, Mar. 2009. 10.1016/j.bpa.2008.09.003.
15. Zhou, C. and Paffenroth, R.C., Anomaly detection with robust deep autoencoders. *Proceedings of the 23rd ACM SIGKDD International Conference on Knowledge Discovery and Data Mining*, 4 Aug. 2017, 10.1145/3097983.3098052.
16. Aboelwafa, M.M.N., Seddik, K.G., Eldefrawy, M.H., Gadallah, Y., Gidlund, M., A machine-learning-based technique for false data injection attacks detection in industrial IoT. *IEEE Internet Things J.*, 7, 9, 8462–8471, Sept. 2020, doi: 10.1109/JIOT.2020.2991693.
17. Cabelin, J.D., Alpano, P.V., Pedrasa, J.R., SVM-based detection of false data injection in intelligent transportation system. *2021 International Conference on Information Networking (ICOIN)*, pp. 279–284, 2021, doi: 10.1109/ICOIN50884.2021.9333942.
18. Esmalifalak, M., Liu, L., Nguyen, N., Zheng, R., Han, Z., Detecting stealthy false data injection using machine learning in smart grid. *IEEE Syst. J.*, 11, 3, 1644–1652, Sept. 2017, doi: 10.1109/JSYST.2014.2341597.
19. Ramamurthy, M. *et al.*, Auto encoder based dimensionality reduction and classification using convolutional neural networks for hyperspectral images. *Microprocess. Microsyst.*, 103280, Sept. 2020, 10.1016/j.micpro.2020.103280. Accessed 27 Sept. 2020.

20. Ns, A., Time complexity analysis of support vector machines (SVM) in LibSVM. *Int. J. Comput. Appl.*, 128, 975–8887, 2015, 10.5120/ijca2015906480.
21. Mohapatra, R., Mishra, S., Mohapatra, T., Coverage problem in wireless sensor networks. *Comp. Cytogenet.*, 2, 1, 67–72, 2012.
22. Mishra, S., Mishra, S., Kayal, A., Chudi, S.R., Simulation in wireless sensor networks. *Int. J. Electron. Commun. Comput. Technol. (IJECCT)*, 2, 4, 176, 2012.

4

Fake News Detection: Restricting Spreading of Misinformation Using Machine Learning

Shubham Choudhary* **and Pratyush Mishra**

Kalinga Institute of Industrial Technology, Deemed to be University, Bhubaneswar, Odisha, India

Abstract

Everyone now relies on several web sources for news in our day of pervasive internet use. Rapid usage of social media has increased the circulation of bogus news among large number of people in short span of time. The propagation of incorrect information has extensive effects, the formation of slanted opinions, which influences of election results in favor of particular political party or leaders. In addition to it, people spam and utilize click-bait titles from news articles to generate income through advertisements. With the use of principles from artificial intelligence (AI), natural language processing (NLP), and machine learning (ML), we seek to perform operations to categorize various news articles binarily, that are available online. We want the user to give the option to label news as bogus or false news. Classification whether the news is bogus or not, it is very important as to stop misleading people, as news is one of the prime factor as, The newspaper or news websites are very effective medium for providing information about our country, world, sports, economy etc., it provides us a vast amount of knowledge. It is one of the most orthodox means of communication and the way how people stay connected to the whole world, Newspaper providing the information about what's happening around us help us in stay updated. It gives us proper information how government and people connected to it are working, that gives the people the opportunity to elect the most eligible candidate. Job opportunities are also available for young people.

**Corresponding author*: 2028182@kiit.ac.in

Hrudaya Kumar Tripathy, Sushruta Mishra, Minakhi Rout, S. Balamurugan and Samaresh Mishra (eds.) Optimized Computational Intelligence Driven Decision-Making: Theory, Application and Challenges, (69–84) © 2024 Scrivener Publishing LLC

Keywords: Artificial intelligence (AI), fuzzy logic, bogus news, naive-based classifier, aggregator, recommendation, Support Vector Machine (SVM)

4.1 Introduction

The modern planet is evolving quickly. Without a question, the digital age has many benefits for us, but it also has drawbacks. Several issues exist in this technological age. False rumor and news are one of them. False news is easy to circulate. Spreading fake news can damage a particular person's image, and can even damage an organization's honor. The misinformation can be inclined towards a politics outfit or any other entity. Misleading information are mostly spread using the following Social media platforms such as face-book, Whatsapp, Twitter, most of the users are generally younger generation, i.e., people less than 25 years of age, who are misled easily by some people, Algorithms are developed using ML and AI to carry out a lot of tasks in Donepudi back in 2019. Datasets are mostly trained on these models using different techniques such as supervised, unsupervised, reinforcement learning. These algorithms are frequently employed and used for various jobs post training phase. Data analytics is applied in many industrial entities to undertake several works. The majority of prediction enabled techniques are used to predict hidden objects. Users benefit from online platforms since they can readily access rumors on them. The problem is that this allows cyber criminals the chance to use these platforms to spread rumors or false information. This rumor is frequently shown to be destructive to an individual or community. Users consume news and start to believe it without any independent validation. Identifying fake news might be difficult because it is not a simple task. If fake news isn't exposed quickly, people will start spreading it and everyone will start believing it. Fake news, rumors, and political parties can all have an impact on people, businesses, or political parties. The fake news in the 2016 US election has hurt people's perceptions and judgments (Dewey, 2016). Many academics are working to spot fake news. In this field of study, machine learning is being employed, which is beneficial. Researchers are employing several models to spot ambiguous news. As per academics (Wang, 2017), spotting false news is a significant challenge. Machine learning must be used in order to detect fake news. According to researchers (Zhou *et al.*, 2019), fake news is growing more and more common over time. This makes it crucial to recognize fake news. Models for predictive intelligence are designed with this purpose. Once trained, data analytics models can automatically detect untrue information in a given data set. The numerous research issues will be addressed by this examination of the literature. This literature analysis will demonstrate the value of machine learning in identifying false information.

Figure 4.1 essentially reflects our entire motivation for conducting this research.

It shows how the perceived news from any source is verified whether it is true or false, if false it is discarded to stop misleading people and to avoid circumstances that can lead to chaos in the city or country.

The steps involved in the above model are shown in Figure 4.2 and are summarized as follows:

First the dataset are collected from different websites that will be used in our Machine learning model. Data has to go certain pre-processing to remove users name, punctuation marks, and empty columns etc. Categorical values are converted into Continuous values as our model only takes continuous values as input. Using Feature extraction process we extract the required features that are used to train our ML model, and then there are independent variables used to predict the dependent variable. They often affect reader's opinion on different topics. Their content is always fake to attract people, while few companies advertise their products; they use this as a marketing strategy. They always use attention probing words and news formats which are found to be rarely accurate. Absolutely not true many occasions as shown in Figure 4.3 given below. The need to identify incorrect information and rumors arises from the fact that many people share misleading information on social media in an effort to mislead the public. For instance, during election season in India, a lot of fake news, rumor and bogus news against a Political Party or a Leader, is used to mislead the voters, and there is a lot of confusion, etc., to prevent people or institutions from degrading their brand as a result of misleading data or rumors.

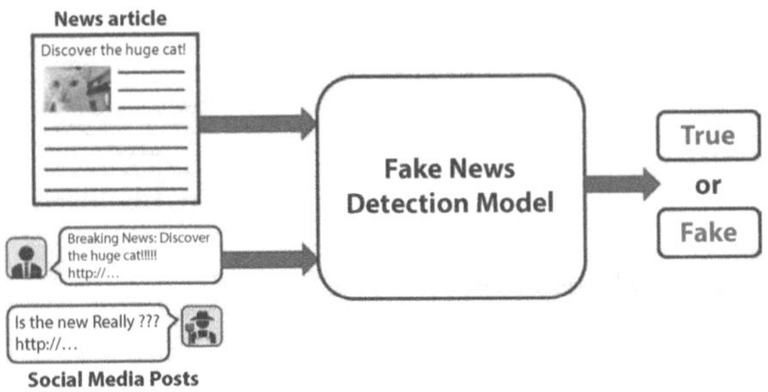

Figure 4.1 Fake news detection model.

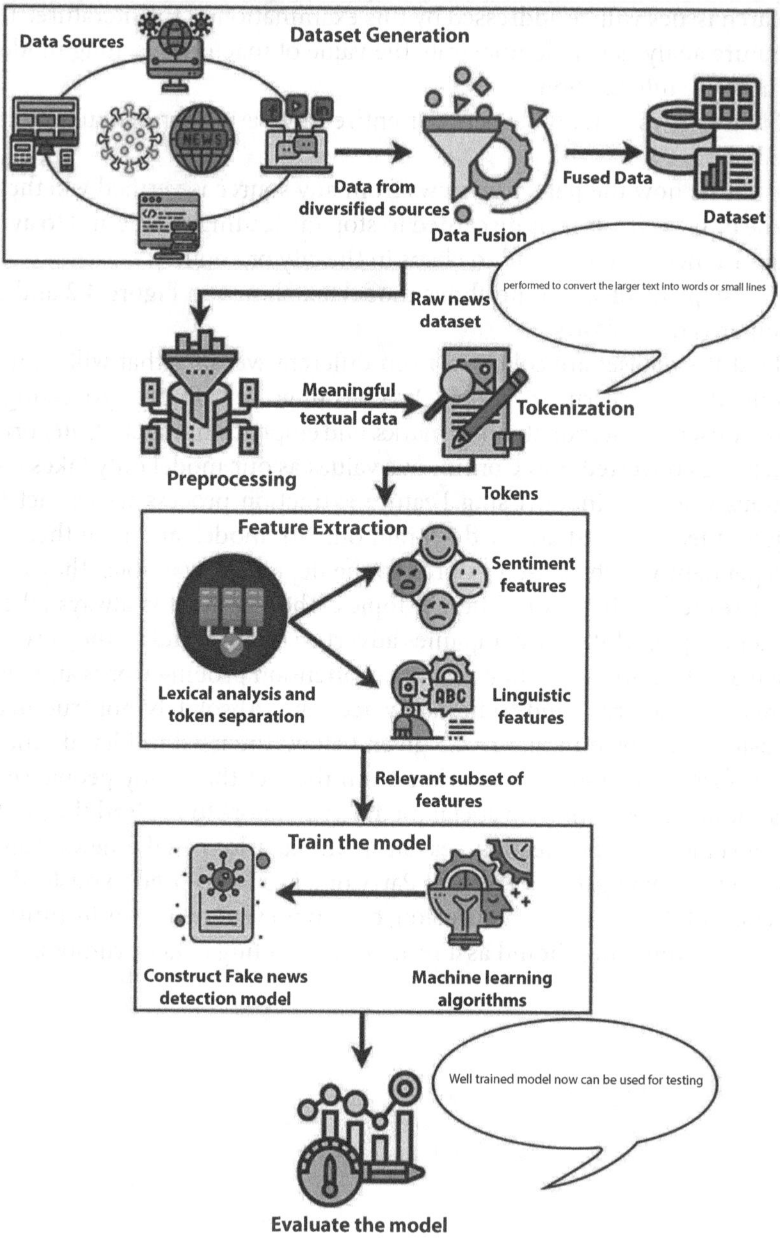

Figure 4.2 Fake news model overview.

4.2 Scope of False News Detection

In the following paper, you can investigate the efficacy of incorporating external information sources into machine learning models for false news identification, such as social network analysis, sentiment analysis, and fact-checking databases. The interoperability of machine learning models for false information recognition may be investigated in this study, as well as the extent to which these models can explain their predictions. The need for identification of incorrect information and rumors is very important from the fact that many people share misleading information on social media, to mislead the public. By Comparing several machine learning model we forecast the information to be untrue. In this study, you may examine the effectiveness of various machine learning techniques for spotting fake news, including Natural Language Processing, KNN, and SVM, logistic regression, decision trees, and neural networks. As we have shown in the following research paper.

4.3 Main Highlights of the Analysis

4.3.1 Approach

Nowadays, it is challenging to differentiate among various categories of news due to the multifaceted nature of fake news. It goes without saying that a practical approach needs to include several viewpoints in order to tackle these kinds of problems precisely. For this reason, the suggested approach combines the Naive Bayes classifier, Support Vector Machines, and semantic analyses. Using such calculations which cannot accurately find subjective abilities, they recommend a technique that is dependent on artificial intelligence and it shows results, in which it is correctly identifies between the real and the false news. The multi-phase method integrates predictive computations that are partitioned into optimized analytical operations with strategies for creating distinctive language.

4.3.2 Naive Bayes

The Naive Bayes classifier, a supervised machine learning algorithm, employs the Bayes theorem to address classification problems. The variables utilized in the model are assumed to be independent of each other. It has been observed that this classifier yields satisfactory outcomes independently, showcasing its effectiveness.

$$P(X|C_i) = P(x1|Ci)^* \, P(x2|Ci) \, ^*...P(xn|Ci) \, ... \quad (4.I)$$

The greatest posterior, which is equal to the maximal P(Ci|X) given the aforementioned equation, is determined using the Bayes theorem to carry out the classification (4.I). This assumption considerably reduces the computing cost by simply accounting for the class distribution. The well-known Naive Bayes approach employs the multinomial Naive Bayes theorem to assess the accuracy of a certain data collection or a specific news item, determining whether it is true or false. It's not only the way that these classifiers are trained; there are various algorithms that focus on recurring themes.

Figure 4.3 shows how to use naive Bayes to tell if the news is true or false, using the different steps involved in Naive Bayes Classification, and predicting the final outcome i.e. the news is bogus or not.

4.3.3 Support Vector Machine (SVM)

The SVM has the capability to retrieve the two labels on basis of the dataset provided to the framework. The objective within the presented methodology is to categorize the study into two classes: genuine and deceptive. The Support Vector Machine (SVM), a supervised classifier, can be utilized for both regression and classification tasks. As illustrated in Figure 4.5 shown below, its fundamental principle revolves around identifying the hyperplane that optimally separates a given dataset into two distinct classes. Decision boundaries referred to as hyperplane aid the machine learning model in categorizing the datasets. The diagram below depicts how the

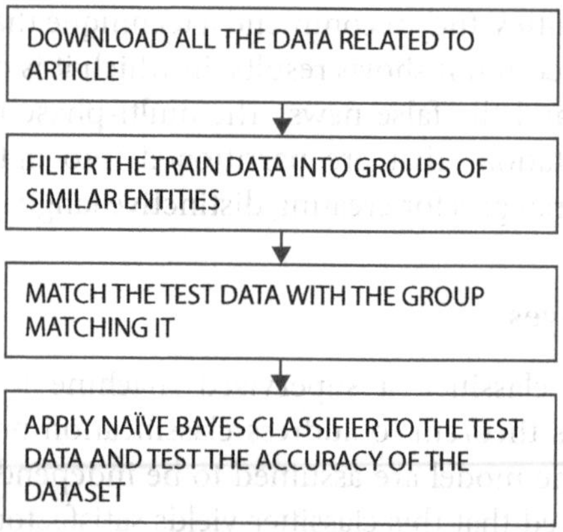

Figure 4.3 Use of Naïve Bayes classifier.

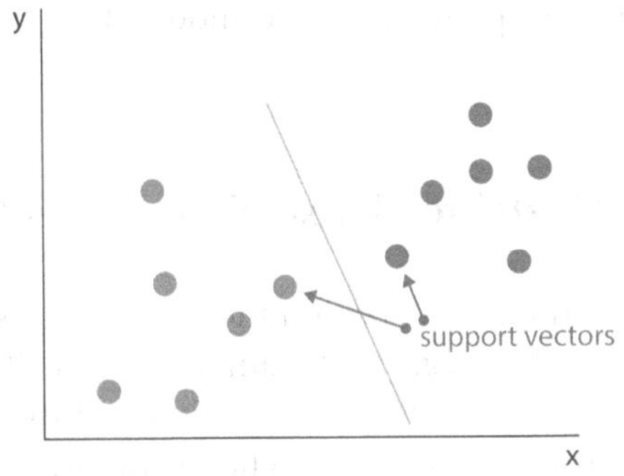

Figure 4.4 Illustration of the hyper-plane classifying the data-set into two groups.

data point's classification is conducted using hyper-planes. Additionally, employing the SVM technique offers the advantage of being generally highly accurate and exhibiting exceptional performance on partially structured datasets. Moreover, due to its frequent application in organizing or even selecting numerical values, this approach demonstrates remarkable flexibility.

Figure 4.4 show the result of using SVM that is Support Vector Machine another ML algorithm that shows the hyper-plane classifying the data-set into two groups.

As demonstrated in Figure 4.5 for system design, support vector machines can also handle high dimensional spaces and are often memory

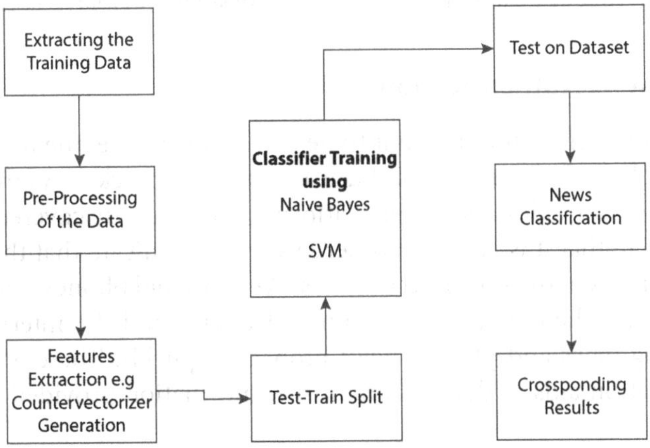

Figure 4.5 Flow chart – classifier training.

efficient; the various steps to train the model, that helps in false news detection.

4.4 A Novel Model for False News Detection

Figure 4.6 illustrates the categorization model we suggest in this study for spotting bogus news. The suggested framework constitutes 3 main units which include aggregator, news authentication and the news recommendation system. These are described in detail below as:

4.4.1 Aggregator

Customers can view news, rumors, and updates from several sources in one convenient location on news aggregator websites. They gather the data, label and categorize it, and present it in a way that makes it easier to use. As most of the popular news websites, like Google-News, Feeds, Daily-Hunt, In-shorts, and others, offer semi-structured news and rumor data. Users can make things simple and practical for themselves by using the RSS Aggregator plugins. In essence, an aggregator improves the reliability and quality of the news. Any news aggregator's principal goal is to collect datasets. Checking RSS Feeds regularly is a fundamental practice; all items are taken from different news websites and datasets are gathered from them. It is frequently used with keyword-based techniques and systems to find related content. When all the steps have been completed, vital or current news that is accurate and as true to life as possible is displayed on the page.

4.4.2 News Authentication

To find out if the news turns out to be correct or untrue, the new authentication takes a few stages. It will use a data-set to check news that is provided from our sides and other platforms; News are discovered there, it will indicate that it is accurate; otherwise, it will indicate that there hasn't been any news of the sort in recent days. We can avoid phoney news thanks to this. Since the emergence of online platforms and the internet today, misleading news and phoney news propagate quickly. Therefore, a news authentication aids in determining if the information provided is true or false.

4.4.3 News Suggestion/Recommendation System

News suggestions make reference to recent news as well as information about rumor or news that the user has provided as part of the verification process. Whether, the news is not true or phoney, this suggestion of the news provided by the pertinent information on the topic. This information recommendation are made on basis of vital words, provides whether the news is true that is verified.

Figure 4.6 shows the novel prototype that is a flowchart of the whole paper the different stages the data need to go through to obtain the desired result. The detection whether the news obtained is true or false. The aim of our work is to determine if the reported information is true or not, to stop spreading rumors and misleading people. The four current approaches NLP, Naive Bayes, KNN, and SVM—are taken into consideration for implementation. The accuracy is among the top 200 outcomes

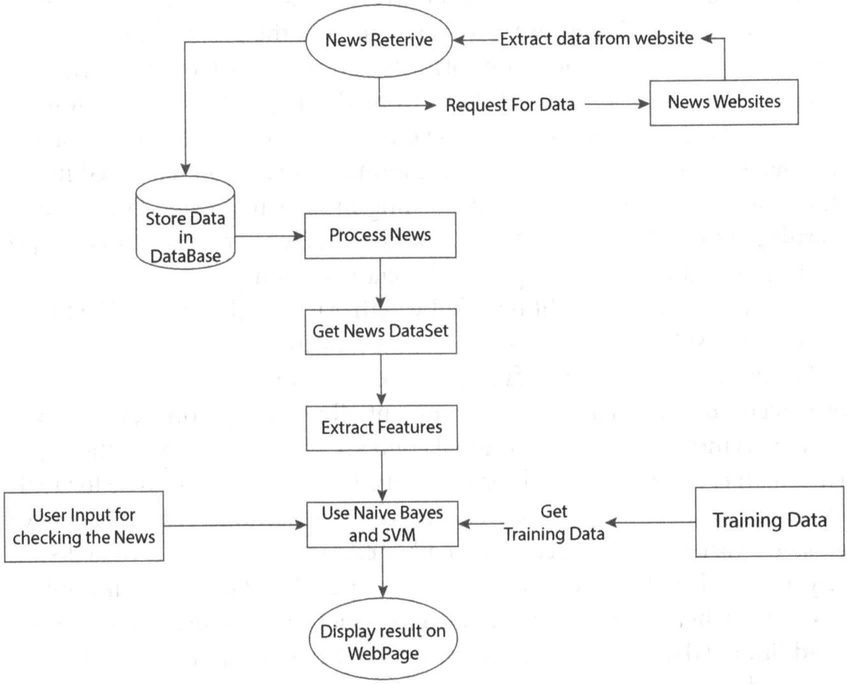

Figure 4.6 Working of the proposed model.

of a particular data set that is mentioned in the table after the results of the four Principles that are discussed and compared with the suggested Model or Principle. The trained model performs the best in case where Naive Bayes, SVM and NLP methods were implemented collectively and it gives the best accuracy of 93.50%.

4.5 Literature Review

Fake news are detected, we take reference From the paper Garnik *et al.* [1] A manually labeled news data set will be used in the study to investigate how well this strategy performs for this specific challenge and to determine whether or not artificial intelligence should be used to detect false news. From the paper Martinez-Garcia *et al.* [2] The project recognized diverse perspectives on the attributes, functionalities and degree of cases in the context of learning, drawing from existing literature and its pilot work. Instead of aiming to establish universal design criteria derived from a unified framework of usecase driven analytics, the project embraced the plurality of interpretations to inform its approach. Detection of whether news is true or false, we seek help from the Paper Humanante-Ramos *et al.* [3] Also, the availability of several technological platforms for the management and distribution of educational information has aided in the development of new non-formal learning models that cater to particular learning demands. From the paper we have taken Granskogen *et al.* [4] Each user is assigned a weight in the classification process using an LR model. A post's total weight revealed whether it is a fake or not. The likelihood of a post being fake increases with increasing weight. From the paper Rubin *et al.* [5] Fake news, false press releases, and hoaxes are examples of inaccurate information that can potentially be dangerous, especially if they are removed from their original contexts and sources. From the paper Hiramath *et al.* [6] The need for clean pixels might not hold true for real-world hyper spectral datasets. In order to solve this issue, researchers have created a variety of advanced end member extraction techniques that do not rely on the idea of a pure pixel. From the paper Shu *et al.* [7] Unlike online interfaces where extra informative contents and related data may be used as additional data to aid in the detection of false news, traditional information medium mainly is dependent upon the information domain to detect misleading data. The following outcome will outline how to extract and display important characteristics from news material and society point of view. From the paper Kumar *et al.* [8] The consequences of spreading misleading information range from skewing attitudes and voting outcomes in favor of certain parties. Moreover, spammers utilize alluring news articles to profit from click-bait advertisements and start misinformation. From the paper Shete *et al.* [9] The broadcasting of untrue information happens through online networking platforms and internet newsletters. Yet, this concern can be handled using data analytics policies and provide news which may be validated. The article use Logistic Regression to detect fake news. From the paper Rao *et al.* [10] A homogenous pool of classifiers was created, and the Random Subspace approach was employed to diversify it. The disparity between the fundamental cart and the support aggregation approaches, along with the utilization of majority voting methodologies, can yield superior outcomes compared to a mere 89.1% accuracy rate as the number of features expands. From the paper Ahmad *et al.* [11] The rationale behind employing ensemble modeling is akin to a familiar concept we encounter in our daily lives. Just as we seek viewpoints of various professionals before reaching to a crucial decision to minimize the likelihood of making a wrong choice that may result in undesirable outcomes, ensemble modeling leverages the collective wisdom of multiple models to enhance overall performance and mitigate the risk of erroneous decisions. From the paper Zhang *et al.* [12] Several validations were held to determine the novelty of the model using an open source dataset gathered through a China based online interface. The results demonstrate that the proposed method exhibits lower training time cost and achieves higher classification accuracy when compared to baseline methods. From the paper Ţucudean *et al.* [13] In order to enhance performance, an augmented approach referred to as back translation is employed in combination with the SVM classifier. From the paper Manzoor *et al.* [14] The following paper gives us a thorough examination of various ML approaches utilized for detecting fake and fabricated news. The limitations of these approaches are discussed, along with an exploration of how deep learning can be implemented to address and improve upon these limitations. From the paper Jain *et al.* [15] In this paper, a model and approach for identifying fabricated news are introduced. The researcher employs machine learning and natural language processing techniques to collect news data and subsequently assess the genuinity of the news articles using Support Vector Machine. From the paper Sami *et al.* [16] The effectiveness of algorithms is evaluated using the confusion matrix, revealing that the Decision Tree algorithm attained the highest score in terms of True Positive (7008/13470), successfully identifying fake news. This outcome highlights the superior performance of the Decision Tree algorithm compared to the other algorithms considered in this study, solidifying its efficacy in accurately detecting fake news. From the paper Awan *et al.* [17] Accuracy assessments were performed on various classifiers, and a comprehensive analysis of accuracy, as well as other performance metrics, was conducted. The model employed feature extraction techniques such as term frequency and inverted document frequency, along with a decision tree classifier. Notably, this combined approach yielded a remarkable accuracy rate of 99.59%. From the paper Moreno-Vallejo *et al.* [18] A website, employing a web service architecture, has been constructed to facilitate access to the classifier. This entails integrating the machine learning model into the web service, allowing for real-time availability and utilization. From the paper Fifita *et al.* [19] The researchers leveraged latest technical models to extract 158 BioIE-derived attributes from a pool of over 1000 COVID-19-related news articles. The selection of BioIE enabled attributes for pandemic related misleading data detection was based on their pertinence and relevance in gathering the clinical data generally seen in pandemic oriented news records. These features were extracted to encompass and portray the crucial elements of biomedical information, encompassing symptoms, treatments, and medical procedures associated with the disease. The underlying theory was that integrating these BioIE-based features into the machine learning models would enhance the accuracy of fake news detection by augmenting the models' knowledge base.. From the paper Malik *et al.* [20] Machine learning algorithms and deep learning algorithms are comparable in terms of there working. However, there is a significant distinction. Data is viewed differently by layers in deep learning systems. The system of such procedures is known as artificial neural networks. Similar works related to the domain are also undertaken in previous analysis [21, 22].

4.6 Results and Analysis

The four current approaches—NLP, Naive Bayes, CNN, and SVM—are taken into consideration for implementation. The accuracy is among the top 200 outcomes of a particular data set that is mentioned in the table after the results of the four Principles that are discussed and compared with the suggested Model or Principle. Python programming on R Studio and a few machine learning methods are used to complete the demonstration.

Table 4.1 shows the different parameters to show the implementation of different Machine Learning algorithms and to classify the data received that is in form of news into True or False/Bogus news to stop misleading the people reading it.

Table 4.1 Result comparison.

Section	Accuracy	Precision	Recall	Method of implementation
R.V.L, Yimin and C.N.J (2016)	76%	73%	71%	NLP
M. Garnik and V. Mesyura (2017)	74%	72%	70%	Naive Bayes
Y. Seo, D. Seo, and C. S Jeong (2018)	86.65%	81%	85%	KNN
Jain A., Khatter H., Shakya A. (2019)	93.50%	89%	94%	Naive Bayes, SVM, NLP

4.7 Conclusion

In the twenty-first century, most work is done online. Online news sources and apps like Google, Facebook, Twitter, and Instagram are gradually taking the place of hard-copies of news that were newspaper. Another useful resources are Whatsapp forwards. Since fake news and information aim to sway people's opinions and attitudes about using digital technologies and other things, the problem will only get worse. One of two things could take place when someone is misled by the actual news: First, individuals might start to believe that what they currently know is the real truth about a certain topic. In order to solve the problem, we created our Fake News Detection system, which analyses user input and categorizes it as real or false. This necessitates the use of several AI concepts, such as NLP and machine learning strategies. The trained model is on an appropriate dataset, and a number of performance indicators are employed to evaluate its effectiveness. The best model, or the prototype with the high accuracy, is used to categorize breaking news or headlines. Our top model ended up becoming, having Logistic Regression which had an accuracy of 93%, as is visible above for static search. We consequently used grid search parameter optimization to enhance the performance of logistic regression, which gave us an accuracy of 93.50%. We can thus say that there are 90% chances that a user would correctly categorize a particular news piece or its title to its real nature since it seems to be true owing to its correctness. Online research is available for news items, keywords, and the reliability of websites. The dynamic system's accuracy is 93.50%, and it gets better with each iteration. We want to create our own data collection that will be updated regularly with the most recent information. The authors will employ SVM as their chosen and most effective Naive Bayes technique in the article. For binary classification, SVM excels. As a result of its propensity to propagate erroneous information and spark social and political upheaval, fake news has grown to be a significant problem in today's society. Due to its capacity to handle massive volumes of data and spot patterns that humans might overlook, machine learning has emerged as a viable method for identifying fake news. It has been shown through the investigation of several machine learning algorithms that accurate, precise, and recall-high fake news identification is possible. Additionally, it has been demonstrated that the effectiveness and interoperability of false news detection models are improved by the incorporation of external knowledge sources, cross-lingual detection, transfer learning, and explain ability. Yet, the risk of hostile attacks continues to be a problem that needs more research. This research suggests that machine learning can be an effective tool for mitigating the spread of fake news, but continuous improvements and advancements in the field are necessary to stay ahead of this evolving problem.

References

1. Granik, M. and Mesyura, V., Fake news detection using naive Bayes classifier, in: *2017 IEEE First Ukraine Conference on Electrical and Computer Engineering (UKRCON)*, 2017, pp. 900–903, IEEE, 2017, May.
2. Martinez-Garcia, A., Morris, S., Tscholl, M., Tracy, F., Carmichael, P., Case-based learning, pedagogical innovation, and semantic web technologies. *IEEE Trans. Learn. Technol.*, 5, 2, 104–116, 2011.
3. Humanante-Ramos, P.R., García-Peñalvo, F.J., Conde-González, M. Á., PLEs in mobile contexts: New ways to personalize learning. *IEEE Rev. Iberoam. Tecnol. Aprendizaje*, 11, 4, 220–226, 2016.
4. Granskogen, T. and Gulla, J.A., Fake news detection: Network data from social media used to predict fakes, in: *CEUR Workshop Proc*, vol. 2041, No. 1, pp. 59–66, 2017.
5. Rubin, V.L., Chen, Y., Conroy, N.K., Deception detection for news: Three types of fakes. *Proc. Assoc. Inf. Sci. Technol.*, 52, 1, 1–4, 2015.
6. Hiramath, C.K. and Deshpande, G.C., Fake news detection using deep learning techniques, in: *2019 1st International Conference on Advances in Information Technology (ICAIT)*, IEEE, pp. 411–415, July, 2019.

7. Shu, K., Sliva, A., Wang, S., Tang, J., Liu, H., Fake news detection on social media: A data mining perspective. *ACM SIGKDD Explor. Newsl.*, 19, 1, 22–36, 2017.
8. Kumar, V., Kumar, A., Singh, A.K., Pachauri, A., Fake news detection using machine learning and natural language processing, in: *2021 International Conference on Technological Advancements and Innovations (ICTAI)*, IEEE, pp. 547–552, November, 2021.
9. Shete, A., Soni, H., Sajnani, Z., Shete, A., Fake news detection using natural language processing and logistic regression, in: *2021 2nd International Conference on Advances in Computing, Communication, Embedded and Secure Systems (ACCESS)*, IEEE, pp. 136–140, September, 2021.
10. Rao, A., Shetty, A., Uphade, A., Thawani, P., Priya, R.L., A proposal for a novel approach to analyze and detect the fake news using AI techniques, in: *2020 3rd International Conference on Intelligent Sustainable Systems (ICISS)*, IEEE, pp. 582–589, December, 2020.
11. Ahmad, I., Yousaf, M., Yousaf, S., Ahmad, M.O., Fake news detection using machine learning ensemble methods. *Complexity*, 2020, 1–11, 2020.
12. Zhang, Q., Guo, Z., Zhu, Y., Vijayakumar, P., Castiglione, A., Gupta, B.B., A deep learning-based fast fake news detection model for cyber-physical social services. *Pattern Recognit. Lett.*, 168, 31–38, 2023.
13. Ţucudean, G. and Bucos, M., The use of data augmentation as a technique for improving fake news detection in the romanian language, in: *2022 International Symposium on Electronics and Telecommunications (ISETC)*, IEEE, pp. 1–4, November, 2022.
14. Manzoor, S.I., and Singla, J., Fake news detection using machine learning approaches: A systematic review, in: *2019 3rd International Conference on Trends in Electronics and Informatics (ICOEI)*, IEEE, pp. 230–234, April, 2019.
15. Jain, A., Shakya, A., Khatter, H., Gupta, A.K., A smart system for fake news detection using machine learning, in: *2019 International Conference on Issues and Challenges in Intelligent Computing Techniques (ICICT)*, vol. 1, IEEE, pp. 1–4, September, 2019.
16. Sami, M. and Shawkat Ali, A.B.M., Machine learning algorithms performance investigation in fake news detection, in: *2023, MayBig Data Intelligence and Computing: International Conference, DataCom 2022*, Denarau Island, Fiji, December 8–10, 2022, Proceedings, Springer Nature Singapore, Singapore, pp. 95–110, May, 2023.
17. Awan, M.J., Yasin, A., Nobanee, H., Ali, A.A., Shahzad, Z., Nabeel, M., Shahzad, H.M.F., Fake news data exploration and analytics. *Electronics*, 10, 19, 2326, 2021.
18. Moreno-Vallejo, P.X., Bastidas-Guacho, G.K., Moreno-Costales, P.R., Chariguaman-Cuji, J.J., Fake news classification web service for Spanish News by using artificial neural networks. *Int. J. Adv. Comput. Sci. Appl.*, 14, 3, 2023.

19. Fifita, F., Smith, J., Hanzsek-Brill, M.B., Li, X., Zhou, M., Machine learning-based identifications of COVID-19 fake news using biomedical information extraction. *Big Data Cogn. Comput.*, 7, 1, 46, 2023.
20. Malik, A.M., Choudhary, A., Ather, D., Forge news detection: A systematic review. Available at SSRN 4160778, 2022.
21. Mohapatra, R., Mishra, S., Mohapatra, T., Coverage problem in wireless sensor networks. *Comp. Cytogenet.*, 2, 1, 67–72, 2012.
22. Mishra, S., Mishra, S., Kayal, A., Chudi, S.R., Simulation in wireless sensor networks. *Int. J. Electron. Commun. Comput. Technol. (IJECCT)*, 2, 4, 176, 2012.

5

Adaptability, Flexibility, and Accessibility Through Telemedicine

Dipti Verma*, Somyajyoti Talukdar and Kumari Alankrita Sharma

Kalinga Institute of Industrial Technology, Deemed to be University, Bhubaneswar, Odisha, India

Abstract

The term "telemedicine" describes the delivery of therapeutic services over long distances using electronic audio and visual methods in real-time. Distance no longer has any impact because of advanced technologies like medical imaging, robotics, and remote controls that have transformed the hospital and operating rooms into accessible zones throughout the globe. In this paper, We have examined this rapidly expanding sector, not only for the benefit of isolated or rural localities with minimal ingress to medical professionals but also for the comfort and simplicity of telemedicine medium. Additionally, understand projects in telemedicine that have examined the viability of using a variety of technologies. To emphasize well-designed, meticulously executed investigations, We assumed that a manuscript that has undergone peer review is guaranteed to meet at least some minimal level of adequacy. Therefore, we focused our attention on publications that were published in scholarly and professional journals. Considering the importance of the proposed model we also focused on building a wellness monitoring system that uses sensors to sense wellness parameters which lead to better monitoring of the patients and a centralized database for the delivery of real-time information. In addition, we talked about the different advantages and limitations of the suggested model, which improves its accuracy. Finally, we have observed the prospects of telemedicine in the coming years.

***Keywords*:** Telemedicine, healthcare, Internet of things, sensors, health monitoring system, multimedia systems, telehealth

**Corresponding author*: diptiverma060@gmail.com

Hrudaya Kumar Tripathy, Sushruta Mishra, Minakhi Rout, S. Balamurugan and Samaresh Mishra (eds.) Optimized Computational Intelligence Driven Decision-Making: Theory, Application and Challenges, (85–106)

5.1 Introduction

According to the definition of telemedicine "When distance is a major factor in the exchange of precise understanding for the detection, monitoring, and treatment of illness and injury, for studies and assessments, and for the maintenance of healthcare professional's education, all with the goal of improving the health of each individual and their sections, it is important that all healthcare professionals use information and communication technology to deliver healthcare services." As a result, it includes several medical procedures including diagnosis, illness prevention and treatment, ongoing health care consumers, and provider's education research and assessment public health is referred to as telehealth. Services are provided to non-locals at a distance who aspire to be healthy but who are not necessarily ill Independent. Quick access to remote and shared medical knowledge using information and telecommunications regardless of the location of the patient. Especially for initiatives involving private nonprofit and commercial sponsorship or finance, there is currently no comprehensive inventory of apps for telemedicine due to the industry's rapid expansion in terms of consumers. Figure 5.1 shows an overview of telemedicine, its working, the features and functions of IoT, and how its applications are used in the process of Telemedicine. It shows how patients communicate with doctors, make appointments for medical treatment, and consult with equipment. All of these activities require a reliable and safe internet connection, a video and audio platform, competent technical support, the ability to record video and audio, and telehealth accessories that use various IoT applications. Information about patients is entered into the database record instead of doing it manually, which enhances the security and privacy of the stored information. The figure also shows the process is conducted in a safe and secure manner.

The April 1924 cover of the journal Radio News foreshadowed telemedicine by representing a "radiologist" connected to a patient by sound and a live image. Since radio had only recently begun to permeate American homes at the time, the first conjectural television transmission didn't truly occur prior to 1927. Despite the belief that telehealth is a relatively latest use of technology for communication given the current increase in interest in the topic, Telemedicine has been in some way for more than thirty years. The origins of telemedicine can be traced back hundreds of years when the place of an individual's access to healthcare was determined by the availability of a doctor. The distinction between patient and doctor has had significant effects on the volume of healthcare delivery and its

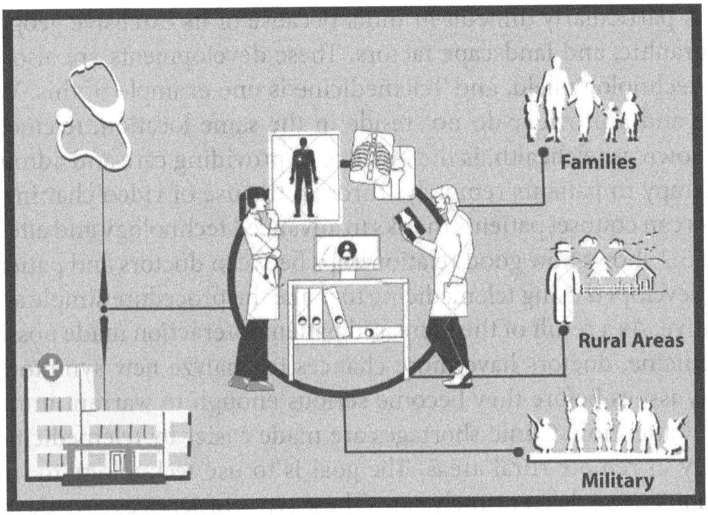

Figure 5.1 Overview of telemedicine using IoT.

associated expenses. Rural regions saw a boom in telemedicine as people with little access to healthcare could now consult specialists from a distance [1]. In the 1960s and 1970s, telemedicine study was supported by many prestigious departments. It is also synchronous and asynchronous telemedicine accessible [2]. Asynchronous and synchronous telemedicine are also available. Synchronous technology allows the medical expert to conduct consultations in real time. However, it can also occur between providers, such as when a healthcare professional consults with an expert to determine the ideal course of therapy for a patient. It usually involves a videotelephony between a long-suffering and a healthcare professional. Using remote programming for neural therapies, such as deep brain stimulation and spinal cord stimulation, is another instance of synchronous telemedicine. Asynchronous telemedicine involves gathering medical data, such as voice and picture recordings and medical scans, and sending it to a healthcare provider for offline evaluation. It allows different resources and websites for contact between patients and physicians. This can use other AI chatbots that after learning about the patient's symptoms, can direct patients in deciding whether they need consultancy. In order to deliver healthcare services, telemedicine makes use of digital and communication networks. Collaboration between healthcare professionals and patients is made possible across geographic boundaries for tasks related to patient management, promotion of good health, diagnosis, and timely referral. Healthcare advancement has always been difficult in emerging nations, but it is particularly difficult in India because of its extensive geographic, demographic, and landscape factors. These developments are also visible in the technology field, and Telemedicine is one example of this. When a patient and a provider do not reside in the same location, telemedicine, also known as telehealth, is the practice of providing care and administering therapy to patients remotely. Through the use of video chatting tools, doctors can counsel patients thanks to advanced technology and alteration. Figure 5.2 shows how good relationships between doctors and patients are being developed using telemedicine to make the procedure simple and collaborative. As a result of the increased patient interaction made possible by telemedicine, doctors have more chances to analyze new symptoms and identify issues before they become serious enough to warrant immediate medical attention. Clinic shortages are made easier by telemedicine, particularly in remote rural areas, The goal is to use technology to improve accessibility and deliver timely care. These types of services save the patient time and travel as they can consult the doctor from their home as per their comfort zone.

Telemedicine, in its broadest sense, is a word that refers to a variety of technologies used in a coordinated program for managing health-service information resources. Telemedicine can be divided into three groups: choice-support tools, cooperation arrangements for autonomous immediate-time patient care, and remote measuring. (WHO, 2001).

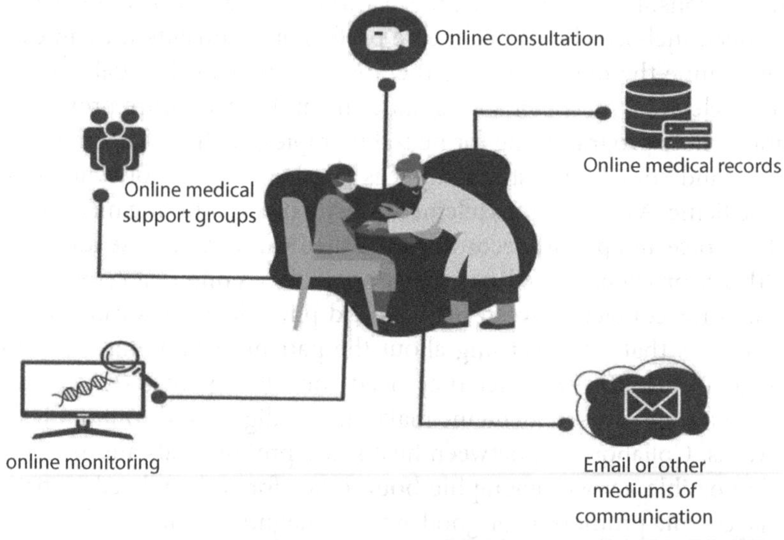

Figure 5.2 The connection between doctors and patients.

The ability of telemedicine to aid in decision-making is its most ancient feature, followed by the transfer of patient data such as x-rays, electrocardiograms, patient histories, etc. Additionally, telemedicine can help two doctors who are geographically separated from one another work together to share ideas and abilities in order to best serve the patient. The referral and payment policies, personnel credentials, liabilities, and the possibility of cross-jurisdictional licensure are all significant issues raised by this collaboration. The healthcare sector as a whole is now developing quickly. It has grown more accessible and versatile as a result of technology's extensive adoption in the medical field. Telemedicine's ability to solve problems of long-distance medical attention helps different nations to earn profit. The Telehealth sector as a whole is now developing quickly. It has become more accessible and versatile. In the year 2020, Over a billion online medical consultations will be conducted thanks to the convenience and ease of use of different telehealth platforms and resources.

The following are the key takeaways of this publication:

- We have looked at the quickly growing field of telemedicine, both for the comfort and ease of telehealth platforms and for the benefit of remote or rural locations with limited access to medical specialists.
- In the 1960s and 1970s, telemedicine research first gained traction. Over the years, we have witnessed the growth of telemedicine applications and hereby in our research, we have discussed them including a current illustration of Covid-19.
- We have proposed a sensing and automation-based wellness monitoring system that consists of a variety of sensors. This system serves as a sample working model for the process and precisely describes the complete process.
- Finally, we also discussed the various benefits and drawbacks of the suggested model, which contribute to increasing the model's accuracy.

5.2 Related Works

The viability of using a variety of technologies, including fax, radio, and telephone as well as still image transmission and real-time interactive television, has been tested through telemedicine programs. Services that are integrated with the speed of a digital network, which is nearly twice as fast as regular telephone lines are designed especially for the transfer of digital data, the majority of telemedicine applications today rely on integrated services digital networks rather than traditional telephone lines for information transmission. The necessity and function of multimedia systems and integrated services digital network technology for some telemedicine applications that can meet the technological and financial requirements for future telemedicine applications are demonstrated in the study named telemedicine needs for Multimedia [3].

Peter W. Shaughnessy, Ph.D. along with his team covered aspects in his paper named Effects and effectiveness of telemedicine [4] about the military forces that have long been interested and involved in telemedicine and mobile health services. This includes looking at technology like telepresence, telerobotic bandaid surgery, and imaginary truth, as well as the physiological monitoring of deployed soldiers. The Army has tried using telemedicine to treat people who reside on isolated islands in the Pacific Ocean. Advanced telecommunications technology was used in conjunction with mobile health units during the war in the Persian Gulf, demonstrating that these two technologies can be integrated, even under difficult geographic and climatological circumstances, with beneficial effects. In mobile modular military medical units that were placed in the Saudi desert along the borders with Iraq and Kuwait, computerized tomography (CT) scanners were installed. However, a procedural study was not performed at the time, despite the fact that both the mobile parts' operation and the CT images dispatched by satellite were of high quality [5]. Li Su analyses in his paper related to the applications of the system made for the treatment of breast diseases and examined the situation of hospital medical treatments in China, with a focus on the Web-based medical diagnostic system. It also discusses the innovation and related philosophy that the system requires, as well as the system's research trajectory. The system functions are thoroughly examined in the demand analysis of the medical diagnosis system for breast patients based on client attributes and diagnostic features. The scientific needs of the system and the conceptual structure of the mechanism function are realized via the application and analysis of the particular design process. Finally, after system analysis and design, for the existing medical system design, diversity analysis, and implementation, and finally improvement of application design and testing [6]. Telemedicine's sudden rise to fame can be accredited to the COVID-19 pandemic as, to prevent the irrelevant spread of the virus, medical professionals were compelled to search ways of treating patients in the ideal situation for telemedicine to be used: without direct physical contact. The ongoing epidemic made it easier for medical professionals to give better health services at many sites around the nation by using telecommunication methods like video conferencing [7]. To extract bibliographic information and carry out a bi-clustering study, we used gCLUTO and Bicomb. The networks of illnesses, technologies, and healthcare services were then built and displayed using R-studio and the Gephi tool. Mobile devices were mainly used to deliver healthcare, conduct remote consultations, control infection, and track down contacts. Another area where telemedicine has been used in 1972 is dermatology, Murphy along with his colleagues (1972) conducted the primary dermatologic treatment remotely utilizing a set of 75 color slides that were pondered onto a display to create a 3-by-2-foot image [8]. A television camera was used to capture the image, and both colorless and colored television displays the projected image. Diagnostic results were marginally less accurate for the broadcast images when put against the slides when compared to a "gold standard" of established diagnoses. The accuracy of colored pictures was marginally higher than that of black and white. A recent survey paper examines how far telemedicine and remote patient monitoring have progressed in terms of neurological problem detection today [9]. The primary topic of the research is Parkinson's disease, and it demonstrates that patients are increasingly accepting of using telemedicine, with a reported 100% acceptance rate in a trial conducted in 2014. The 44 participants in this study need to be explored further, although the outcome is encouraging. Additionally, the viability of using telemedicine and remote monitoring in the treatment of various neurological conditions such as multiple sclerosis, epilepsy, dementia, and depression is briefly examined. Some other recently published papers discussed different aspects of neurology such as Block sparse Bayesian learning for EEG signal detection is mentioned in a paper cited by Nadeem *et al.* (2015) [10]. An artificial neural network-based system for monitoring diabetes patients is presented by Karan *et al.* (2012) [11]. It is notoriously difficult to prove that psychiatry works. However, there have been initiatives to assess the use of telepsychiatry. In the attempt, four communication channels—in-person, telephonic, hands-free telephone, and a video conference system—were put up to the task [12]. The report's description of the methodology left some questions unanswered, and the sample was significantly biased towards the mental disorders. The use of telemedicine in ENT was revealed. Diagnosis of patients became easier for the ENT doctor utilizing the information supplied. In one instance, the specialist conducted a face-to-face assessment of the patient after examining the patient via IATV from a different room. All 17 patients' diagnoses were completely in agreement, which would have been more significant if a second physician had conducted the in-person tests [13]. The use of telemedicine during a surgical procedure is known as telesurgery. It allows a physician to deliver their skills over a distance, which could be little or practically infinite. A survey paper has discussed this in detail [14]. A robotic system reads the surgeon's instructions remotely and utilizes the capabilities of local computers and robots to realize them on the patient's side. This unique computerized intelligent surgical operation can be executed by a regional or remote surgeon. All instruments operating on the patient's end are controlled signals and data that pass through an effective broadband communications link while the surgeon is remote. As a result, telesurgery combines surgical procedures (mainly robotic) with the other three classic telemedicine components: information and communication technologies, participant distance, and medical knowledge. Several government stakeholders like the military believed that the concept should be utilized in remote surgery applications after successful localized robotic surgery studies. The goal was to allow doctors in far-off places to perform operations on injured soldiers while they were still on the battlefield [15]. Heart-related monitoring devices are the most popular category of monitoring systems, and telemedicine has also been successful in this area. Recently in the paper of Lanata *et al.* (2015) [16], it is discussed that a Number of smartphone applications used in many facets of cardiology, such as user-guiding applications for heart disease prevention and remote patient monitoring are analyzed. Utilizing contacts Hardware and software tools are used in a "telemonitoring system" to keep track of various heart-related disorders. Android runs software applications. It was noted that the use of telemedicine in fetal ultrasonography applications was successful in 1995. Different technologies are used to do ultrasound, with good results. It produces a picture of the unborn child inside the mother (uterus). It is a secure method of evaluating the health of an unborn child. A fetal ultrasound examines the baby's heart, skull, and spine in addition to other body parts. Applications in telemedicine made sure the baby wouldn't be harmed by the sound waves and imaging equipment utilized in the procedure [17]. Another important work is where the high-frequency microwave system's network configurations are described in depth. The description includes important benefits including high capacity, large propagation distances, and a wide range of frequency bands. Therefore, crucial duties like first aid, a doctor's diagnosis, and giving patients instructions may be carried out via a telemedicine system. The design of a telemedicine system built on a high linearity Ku-band transmitter module, which can boost both capacity and propagation distance, is described in this work. Engineers can save a lot of time, money, and resources by utilizing the designed transmitter module for telemedicine systems, which will ultimately lead to a reduction in the fatality rate. Therefore, a high-frequency sensor link like this can be employed for telemedicine in hostile environments [18].

5.3 Proposed Model for Remote Health Monitoring System

Firstly, researchers in the health monitoring system here planned a well-being observing framework utilizing ATmega8 or 18F452 microcontroller along with Wireless Body Area Sensor Network (WBASN). The sensors that can be used are different types of sensors that measure temperature, Blood pressure, and Heart thump. These detectors are put on the patient's body that can help with checking the medical issue without upsetting the everyday timetable of the patient and these wellbeings are sent to the specialist's server utilizing long reach remote innovation, Global System for Mobile communication (GSM). The well-being observing framework comprises sensors, a microcontroller, and a Liquid-Crystal Display (LCD) show along with a Global System for Mobile Communication (GSM) modem to communicate or get well-being-related information to or from the specialist. Additionally, at the clinic, a similar GSM modem is utilized. Thus, the Global System for Mobile Communication (GSM) modem helps form an organization between a patient's server and a specialist's server. The patient is provided an LCD display to see the immediate results. Here researchers utilized a Negative Temperature Coefficient (NTC) thermister as a temperature sensor, InfraRed (IR) Light emitting diode (LED), and a red Light-emitting Diode (LED) utilized for pulse checking and Pressure transducer, on the other hand, the sensor in light of is utilized to gauge the extra systole blood pressure and extra diastole Blood pressure. The embedded controller peruses information as given by the temperature sensor, circulatory strain detector, and Espressif Systems ESP8266 pulse sensor, and handling it gives the result as computerized and it gets straightforwardly shown on LCD (Liquid-Crystal Display) or it gets dispatched to the specialist's server through a GSM (Global System for Mobile Communication) router. The framework gives precise and momentary results with high exactness which gets straightforwardly shown on LCD (Liquid-Crystal Display). Using GSM (Global System for Mobile Communication) remote technology, it takes no more than 4-5 seconds to view the specialist's server. This framework gets some margin to know the well-being state of the sufferer and afterward conveys the detail to the physician. Secondly, Systems utilizing a similar framework, and well-being boundaries are sent by utilizing Radio frequency identification (RFID) per user, Bluetooth, a Global system for mobile communication (GSM), and a Universal Mobile Telecommunication system (UMTS). This framework allows the office to screen the pulse of patients. The well-being boundary straightforwardly sends to the control unit, embedded controller ATmega328, LCD (Liquid-Crystal Display) show. The sensor of pressure is straightforwardly associated with the sleeve, which is swelled or collapsed by means of an engine and valve. Here and there, engine switches are constrained by the chip at the correct time. Because of changes in the switches of the engine, the sleeve on the wrist gets expanded furthermore, emptied, this tension is estimated by the strain sensor. The pressure sensor creates the well-being boundary in the simple sensor. Advanced yield is achieved which is shown on the LCD (Liquid-Crystal Display) using the results obtained by the microcontroller or on the private processor or computer utilizing Recommended Standard 232 (RS232). Magneto Resistive Random Access Memory (MRAM) stores the worth of systolic and diastolic pulse and is straightforwardly associated with the microcontroller. Additionally, there is a compelling reason to siphon the sleeve the hard way, the microcontroller constrains the framework. It isn't expected to compute or notice circulatory strain physically. The aim of this research is to plan and implement an IoT-based framework for health observation, as illustrated in Figure 5.3, and to create a

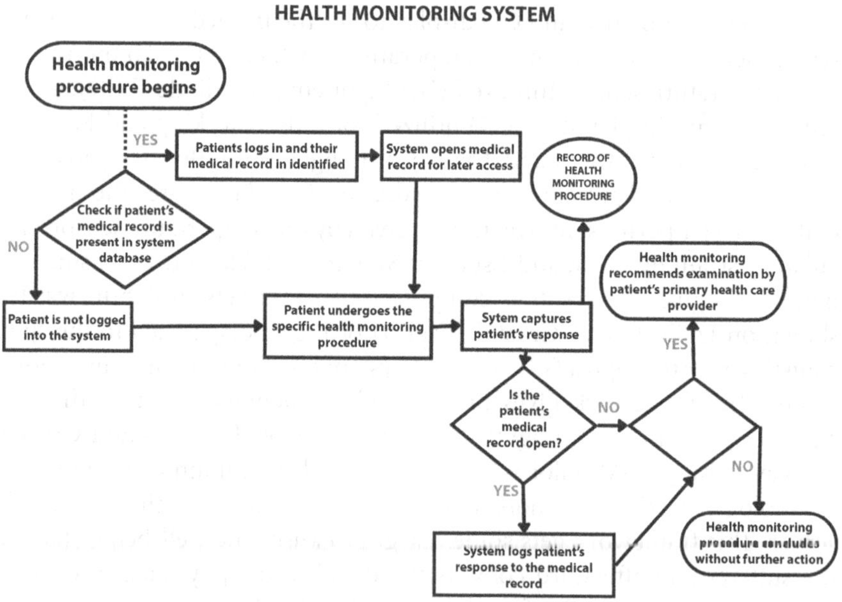

Figure 5.3 Proposed model for health monitoring system.

sample working model for the process that precisely defines the entire process. The model is created to ensure process safety and cost-effectiveness for the telehealth process. For the proposed model, various sensors and machines are used. These detectors are used to measure the sufferer's pulse rate (BPM) and oxygen saturation (SPO2). These detectors are linked to a management system that manages the benefits of the many sensors. These predetermined characteristics are discernible on the device's LED display and should be visible on the phone when using Blynk. Values decided by the app will be presented on a website that can be retrieved by using an IP address. Thingspeak can be used for knowledge sharing in general. The framework includes another website that, after entering the sensor data, forecasts the illness.

5.3.1 Microcontroller and Sensor

(a) ESP32 (Espressif Systems) - It is a minimal-expense framework. It is an enhancement for the well-known ESP8266 that is broadly utilized in different industrial situations for sensing and automation. It has many connectivity capacities including Bluetooth and Wi-Fi, that give it the capabilities to be an all-adjusted chip for the improvement of different sensing and automation projects, and implanted frameworks overall. The IoT device with touch pins that can detect changes in anything that has an electrical charge is the sensor ESP32, which is depicted in Figure 5.4 as being used in the processed model.

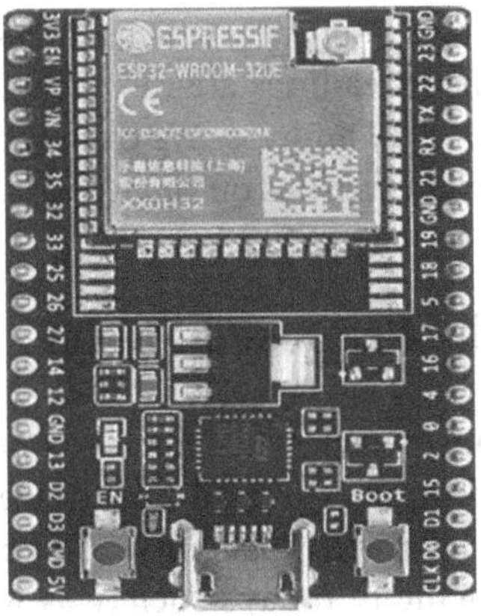

Figure 5.4 ESP32.

Figure 5.5 MAX30100.

(b) MAX30100 - It measures the patient's heart rate (BPM) and oxygen saturation (SPO2). The amount of oxygen in the digit with blood on it is measured by tiny light emissions. To accomplish this, it employs the formula: to estimate fluctuations in heart rate and the amount of light ingested by oxygenated or deoxygenated blood.

BPM is equal to 60*f, where f is the frequency of the beat.

The MAX30100, which is clearly visible in Figure 5.5 and used in the processed model as both a pulse oximeter and to track the heartbeat, is composed of Light Emitting Diodes, a photodetector, and a number of low noise signal processing components. The sensor can estimate the percentage of oxygenated hemoglobin in the blood as a result of these distinctive reactions. The sensor element consists of a pair of LEDs that emit both red and infrared light that can pass through the finger. Majorly the realistic finding for distance-capable tracking, in the sphere of well-being observation, is the Internet of Things. This paper developed an IoT-based framework for monitoring well-being. The framework checked body beat through immersion utilizing sensors, which are additionally depicted on an LCD. Then these sensor readings are dispatched from the application, along with, if necessary, an IP address. We can predict coronary illness by putting these specific characteristics on the website.

5.4 Benefits of the Proposed Model

By implementing the most recent telemedicine concepts, we can get many benefits. Telemedicine improves overall healthcare facilities for patients, reduces expenses, and increases capability and revenue [19]. There are many facets of telemedicine in today's society as we have witnessed with its development, A wellness monitoring system that takes into account

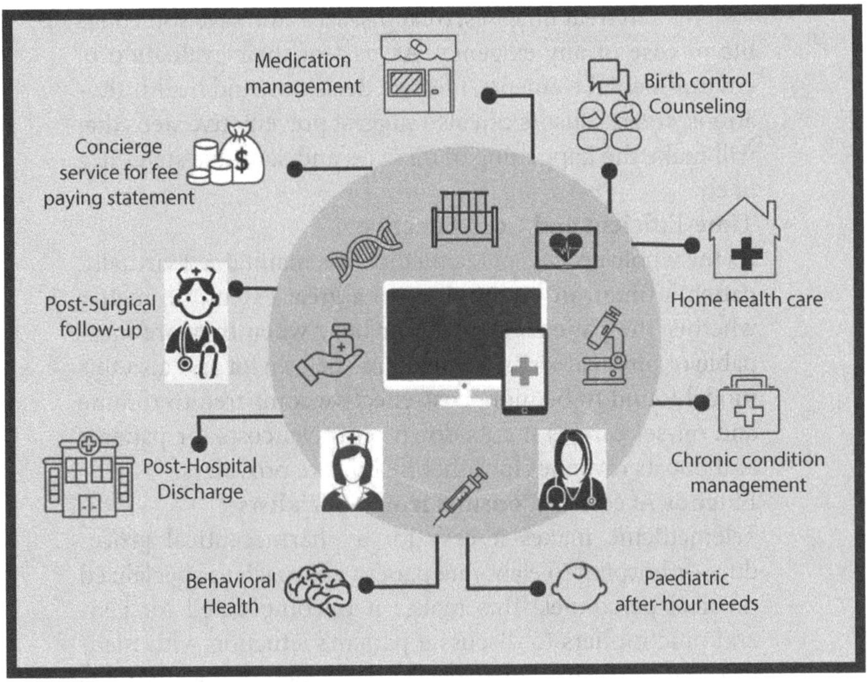

Figure 5.6 Various aspects of telemedicine.

the various factors has been developed and offers a number of advantages which Figure 5.6 vividly illustrates that includes more privacy, easier patient follow-up, flexible work hours, improved time and commute efficiency, no long waits are required for patients, aids in improved diagnosis, improved access to specialized medical opinions and many more, some of them are listed below:

- **More Practical and Reachable Patient Care**
  The practice's addition of virtual care provides patients with convenient, on-demand care. If a patient is occupied with the circumstances like being confined to a house, living in a faraway location, or being busy with important things they can obtain treatment virtually. With the aid of the telemedicine working paradigm, more patients and providers are connected than ever before thanks to video conferencing, mobile applications, and online management systems.
- **Remote Monitoring and Prevention**
  Similar to the model, various sensors are utilized to support live tracking via connected devices, and intelligent alarming

systems can treat illnesses, treat diseases, and save someone's life in case of any exigency. As part of their evaluation of environmental elements, lifestyle decisions, and health situations, smart sensors can also suggest preventative steps that will make the happening of diseases and sensitive states less likely.

- **Time-Efficient and Cost-Effective**
  As the whole process of telemedicine is maintained virtually, patient's time can be saved up to a great extent, no matter whether the patients are aged or busy when they are incapable of physical visits telemedicine helps a lot and also this model found to be more cost-effective compared to routine and on-set care as it cuts down on travel costs for patients and boosts revenues for other healthcare providers.
- **Extends Access to Consults from Specialists**
  Telemedicine makes it easy for a pharmaceutical procedure or hospital to elaborate patient approach to specialized medical personnel. This makes it uncomplicated for general practitioners to discuss a patient's situation with medical specialists and for patients—regardless of location—to consult a needed specialist for a rare sort of cancer. Another example is the capability of small hospitals to use telemedicine to contract out the interpretation of X-rays when they lack enough radiology professionals on staff.
- **Increasing Patient Engagement**
  The patient of today expects a new kind of care experience since they live in a world that is becoming more connected. By enabling more regular, easy communication between patients and their doctors, telemedicine engages patients. As a result, patients will ask and receive more questions, the doctor-patient connection will be better, and patients will feel more in control of their medical treatment.

5.5 Constraints of the Proposed Model

Everything has a perk and a drawback So as the field of Telehealth, as its reach broadens every day, challenges also do as well [20] as depicted in Figure 5.7 which illustrates the difficulties that both patients and doctors must overcome in order to practice telemedicine that includes a lack of regulations governing the operation of telemedicine, worries about the

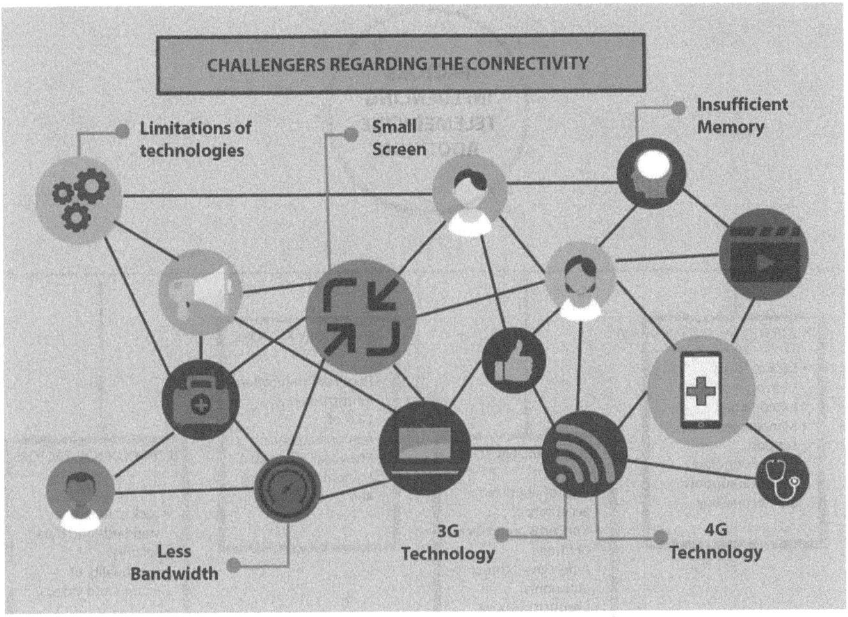

Figure 5.7 Challenges in telemedicine.

confidentiality of the sent medical information, technical issues with the equipment, a poor internet connection, and a lack of broad access to technology and some of them are listed below.

- **Not successful in every case**
  Certain visits cannot be done casually. For different processes like radiology studies and verification of blood, as well as for different diagnostic approaches physical assistance is required. Therefore, It may not be successful in every instance.
- **Database corruption**
  In spite of making a fully-fledged model, there is a chance of losing data due to system corruption, data storage failure, and accidental data deletion also data security is one of the major concerns for patients.
- **Heavily depends on external factors such as connectivity**
  Even in a functional model, there is a danger of losing connection or a delay in data transfer because of things like malfunctioning hardware or inadequate connectivity. It could be a problem in many situations when speed and accuracy are crucial.

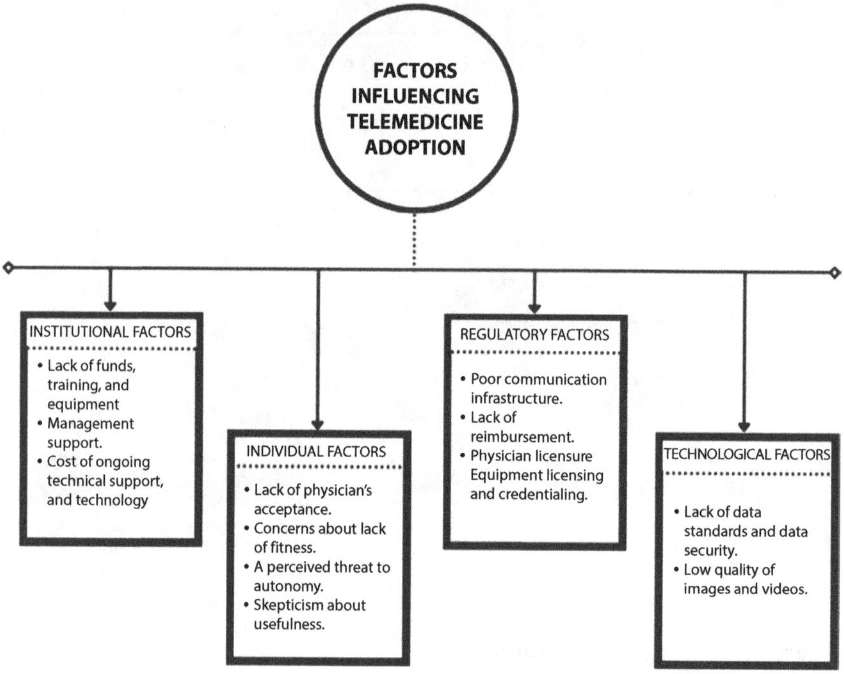

Figure 5.8 Factors influencing telemedicine adoption.

Several other factors are also responsible for the evolution of the telemedicine sector. The majority of them are shown in Figure 5.8 and are broken down into Institutional, Individual, Regulatory, and Technological elements.

Considering each factor individually,

- Institutional elements define internal and external environmental factors in accordance with which an organization's behavior may be revealed and studied, the practice of Telemedicine involves problems like financial issues with the technique, equipment availability, location accessibility, technical support costs, and equipment costs.
- Individual issues include the unwillingness of doctors to utilize the treatment, worries about losing face-to-face encounters, worries about fitness, patient accessibility to the process, and doubts about its value.
- Regulatory factors include the various laws, rules, and regulations that must be followed for any Procedure to be carried

out properly, the practice of telemedicine involves difficulties like fear of the law, malpractice responsibility, inadequate communication systems, inadequate reimbursement, equipment licensing, and more.
- The term "technological factors" refers to aspects of modern technology that make use of recent advanced technologies. The procedure is impacted by a number of variables, including communication infrastructure, image, and video quality, compatibility concerns with new technologies, data standards, and data security.

When creating a model or carrying out any activity in the field of telemedicine, all of these must be kept in mind.

5.6 Conclusion

There is no denying that telemedicine is currently expanding quickly. Technology is still developing, transforming every aspect of our existence. However, we cannot deny the necessity of potency and usefulness research. Nobody wishes to provide ineffective care, receive it, or pay for it. In this research, we have seen from Telemedicine's inception to its future possibilities, have witnessed the various aspects of telemedicine, its actual significance, and our comprehension of it. It is observed that the research began in the 1960s and 1970s, but as technology has advanced and the Internet of Things has become more prevalent, there has been an increase in demand for telemedicine. As physical contact between doctors and patients was outlawed under COVID-19, the use of telemedicine has increased. Additionally, it aids in the treatment of those who live close to remote islands and suffer from breast ailments. Given the significance of the suggested model, we developed a system for monitoring patient wellness that makes use of various sensors, including the ESP32 and MAX30100, to measure heart rate, oxygen saturation, and other wellness-related parameters. This system improves patient monitoring and uses a centralized database to deliver real-time data. The technology is also designed to make the process less expensive while still guaranteeing that it is carried out accurately and safely. Additionally, we discussed the numerous benefits and drawbacks of the proposed model as well as the various aspects of telemedicine that are crucial to its accuracy and correct operation. Such clarification could favor in creating factual instructions that could focus on particular categories of clinical issues. If the technology is seriously considered, It could have a variety of impacts on the way health services are provided. It is, however, extremely challenging to foresee the precise path that events will take given the pace of technological development and the unpredictability of the healthcare system.

5.7 Future Works

In this study, we looked at how telemedicine has evolved and will likely develop in the future. Throughout the world, Telemedicine is growing rapidly. Because of the advanced technology, Hospitals and healthcare facilities are developing faster in online mode also. It is predicted that by the year 2026, the market of telemedicine will increase by 14.9%. Due to the stimulation of more cost-effective utilization of hospital beds, patients tend to heal more quickly at home. It is used, especially in rural regions, to provide consulting services in a range of medical disciplines. Moreover, it focuses on how the Telemedicine Processing model functions, the instruments, and technologies involved, as well as its upshots. Telemedicine has a promising future, according to the below observations:

Telemedicine Will Become a Standard Service

In the current context, when patients are accustomed to the accessibility that telemedicine provides, there is no turning back for them. Over the past ten years, telemedicine has grown significantly, becoming a significant and strong force in different fields like social and economic issues areas along with technological areas since it provides answers to urgent issues. The declining population, the expanding demand for health care, the need to manage large volumes of data, etc., are some of these challenges.

Remote Patient Care Will be a Reality

Remote medical services provided by connections to professionals in other areas are one of telemedicine's most widely used uses. In most nations, specialized medical professionals are concentrated in a single area, forcing patients in remote areas to travel great distances to access the proper care they require. Telemedicine will assist us in primary care by enabling remote consultations with the referral hospital, which will ultimately decrease waiting periods for diagnosis and treatment as well as the number of referrals. Telemedicine was initially employed more and more in the delivery of care, but it is now enlarging into remote patient monitoring and check-out, with what appear to be separate apps serving the same function.

Tele-Surgery

Undoubtedly, one of the biggest advancements in telemedicine has been telesurgery, which is only expected to grow over time. It uses cutting-edge technologies like virtual reality and robots, which have increased the number of experimental telesurgeries over the past few years. A paradigm shift away from conventional in-person clinic appointments is not hindered by the high patient satisfaction with video sessions. To direct and implement the wider usage of telemedicine, a future study comparing various clinic visit quality metrics is required.

Robotics in the Industry of Telemedicine

In the era of Continuously evolving technology, The field of telemedicine will heavily rely on robotics. Medical personnel can confer on cases in other facilities using healthcare robotics, which also aids in remote patient monitoring. For instance, autonomous robots in telehealth can find patients in waiting areas or hospital rooms and communicate with them online. While the nurse responds to other patients and duties, an acute care patient can communicate with a nurse from their bedside. Some robots can follow physicians as they go about their rounds and send live feeds to faraway experts so they can offer on-screen consultations [21].

References

1. Orozco-Barbosa, L. and Karmouch, A., Multimedia inter-hospital communications systems for medical communications. *IEEE J. Sel. Areas Commun.*, 10, 7, 1145–1157, Sep. 1992.
2. Al-Kassab, M.H., Dongming, L., Yunhe, P., How telemedicine is being used today and tomorrow. *J. Syst. Eng. Electron.*, 10, 2, 55–64.
3. Melício Monteiro, E.J., Bastião Silva, L.A., Costa, C., CloudMed: Promoting telemedicine processes over the cloud. *7th Iberian Conference on Information Systems and Technologies (CISTI 2012)*, pp. 1–6, 2012.
4. Houtchens, B.A., Clemmer, T.P., Holloway, H.C. *et al.*, Telemedicine and international disaster response: Medical consultation to Armenia and Russia via a Telemedicine Spacebridge. *Prehosp. Disaster Med.*, 8, 57–66, 1993.

5. Cawthon, M.A., Goeringer, F., Telepak, R.J. *et al.*, Preliminary assessment of computed tomography and satellite teleradiology from Operation Desert Storm. *Invest. Radiol.*, 26, 854–857, 1991.
6. Adriaanse, B.M.E., Tromp, C.H.N., Simpson, J.M. *et al.*, Interobserver agreement in detailed prenatal diagnosis of congenital heart disease by telemedicine using four-dimensional ultrasound with spatiotemporal image correlation. *Ultrasound Obstet. Gynecol. Official Journal of the International Society of Ultrasound in Obstetrics & Gynecology*, 39, 2, 203–209, 2012.
7. World Health Organization, Opportunities and developments telemedicine in member states report on the second global survey on eHealth global observatory for eHealth series -volume 2.
8. Murphy, R.L.H., Fitzpatrick, T.B., Haynes, H.A. *et al.*, Accuracy of dermatologic diagnosis by television. *Arch. Dermatol.*, 105, 833–835, 1972.
9. Schneider, R.B. and Biglan, K.M., The promise of telemedicine for chronic neurological disorders: The example of Parkinson's disease. *Lancet Neurol.*, 16, 7, 541–551, 2017, doi: 10.1016/S1474-4422(17)30167-9.
10. Nadeem, A. *et al.*, Application-specific study, analysis and classification of body area wireless sensor network applications. *Comput. Netw.*, 83, 363–380, 2015.
11. Karan, O., Bayraktar, C., Gümüşkaya, H., Karlık, B., Diagnosing diabetes using neural networks on small mobile devices. *Expert Syst. Appl.*, 39, 1, 54–60, 2012.
12. Dongier, M., Tempier, R., Lalinec-Michaud, M., Meunier, D., Telepsychiatry: Psychiatric consultation through two-way television. A controlled study. *Can. J. Psychiatry*, 31, 32–34, 1986.
13. Pedersen, S., Hartviksen, G., Haga, D., Teleconsultation of patients with otorhinolaryngologic conditions. *Arch. Otolaryngol. Head Neck Surg.*, 120, 133–136, 1994.
14. Karamanoukian, H., Pande, R.U., Patel, Y., Freeman, A.M., Aoukar, P.S., D'Ancona, G., Telerobotics, telesurgery, telementoring. *Pediatr. Endosurg. Innov. Tech.*, 7, 4, 2003.
15. Satava, R.M., Virtual reality and telepresence for military medicine. *Ann. Acad. Med. Singap.*, 26, 118–120, 1997.
16. Lanata, A. *et al.*, Complexity index from a personalized wear-able monitoring system for assessing remission in mental health. *IEEE J. Biomed. Health Inform.*, 19, 1, 132–139, 2015.
17. Fisk, N.M., Bower, S., Sepulveda, W. *et al.*, Fetal telemedicine: Interactive transfer of realtime ultrasound and video via ISDN for remote consultation. *J. Telemed. Telecare*, 1, 38–44, 1995.
18. Yan, H.R., Huo, H.W., Xu, Y.Z., Gidlund, M., Wireless sensor network based E-health system - implementation and experimental results. *IEEE Trans. Consum. Electron.*, 56, 4, 2288–2295, Nov. 2010.

19. Craven, D., McGinley, B., Kilmartin, L., Glavin, M., Jones, E., Compressed sensing for bioelectric signals: A review. *IEEE J. Biomed. Health Inf.*, 19, 2, 529–540, March 2015.
20. Richter, G.M., Williams, S.L., Starren, J. *et al.*, Telemedicine for retinopathy of prematurity diagnosis: Evaluation and challenges. *Surv. Ophthalmol.*, 54, 6, 671–685, 2009.
21. Pupolin, S. and Greenstein, L., Performance analysis of digital radio links with nonlinear transmit amplifiers. *IEEE J. Sel. Areas Commun.*, 5, 3, 534–546, Apr 1987.

6

Crop Prediction by Implementing Machine Learning in an IoT-Based System

Vivian Rawade* **and Shubham Sahoo**

Kalinga Institute of Industrial Technology, Deemed to be University, Bhubaneswar, Odisha, India

Abstract

In recent years, the agriculture industry has undergone significant transformation due to the integration of data-driven and IoT-based technologies. The integration of computing techniques such as machine learning, data analytics, wireless sensor networks, and the internet of things has demonstrated the significant potential of these technologies in the agriculture industry by increasing both the quantity and quality of agricultural production. To enhance the potential of these technologies, a prediction model has been developed that uses machine learning to determine the most suitable type of crop for a given set of atmospheric and soil data. Furthermore, insights from farmers have been collected through a local survey to understand the impact of these technologies. Despite their potential benefits when incorporated into traditional farming practices, challenges such as high costs, a lack of knowledge among farmers, and compatibility issues with existing farming practices must be addressed to maximize the potential of these technologies. Thus, efforts need to be made to provide farmers with education, training, and incentives to adopt these technologies. In doing so, the agriculture industry can harness various technologies to meet the increasing demand for food while also improving efficiency, sustainability, and profitability. This integration of technologies has the potential to drive innovation in the agriculture industry, leading to a more sustainable and productive future.

Keywords: Agriculture, Internet of Things (IoT), machine learning (ML), precision agriculture (PA), data analytics (DA), crop prediction

**Corresponding author*: vivianrawade10@gmail.com

Hrudaya Kumar Tripathy, Sushruta Mishra, Minakhi Rout, S. Balamurugan and Samaresh Mishra (eds.) Optimized Computational Intelligence Driven Decision-Making: Theory, Application and Challenges, (107–130) © 2024 Scrivener Publishing LLC

6.1 Introduction

Agriculture has long been regarded as the primary source of supplies for meeting people's basic requirements. Due to India's rich soil fertility and vast network of irrigation supplies, it is also regarded as a primary occupation and one of the country's key economic pursuits. Although resources are well present, in terms of resource utilization according to their availability, do not offer similar outcomes. This is due to scarcity and deficiency of knowledge of the technology used, and a lack of awareness among the agrarians. It leads to reduced soil productivity and adversely affects the quality and quantity of the harvest. Ultimately leading to barren lands that are unsuitable for agricultural use. This has an adverse impact on both the environment and the livelihood of the farmers. Therefore, it is necessary to develop a model that can track environmental, soil, and plant parameters using spatially distributed sensors that remotely assist in predicting the average values of the monitored parameters in order to help agriculturalists select the crop variety best suited for a given time period on a specific farm. This will eliminate the need for any chemical use and significantly lower the likelihood of failure in a planned and predictable manner. Additionally, it would undoubtedly increase agricultural production, bridging the gap between rising food demand and population growth and also lowering the percentage of crops that are lost overall.

IoT and AI are two smart technologies that the agricultural sector is widely utilizing to increase the productivity of organic product cultivation on limited land and address farmers' long-standing difficulties. In order to minimize the depletion of soil nutrients as a result of poor crop selection or ineffective cultivation methods, IoT-based smart farming systems are designed to monitor soil moisture and soil nutrients using sensors. Site-specific agriculture, often known as precision agriculture (PA), is a technical strategy that creates high-quality agricultural production by utilizing contemporary data technology. This entire technology has been depicted in Figure 6.1, where a farmer is seen monitoring and controlling all the activities of his farm remotely from a mobile device using an interface. WSN is a practical method to increase crop yields. PA has produced good results and managed to recoup the expenditure required for its deployment. Crop identification and localization, monitoring performance, machinery, plant monitoring, and mapping are some of the PA-related operations. The widespread use of PA worldwide is a result of the availability of machine learning (ML) algorithms, fast internet, and effective computing hardware. In order to handle the growing difficulties brought on by climate fluctuations

Figure 6.1 Accessing crop data using IoT on mobile device.

like rain, humidity, temperature, etc., which are seriously harming data analytics, machine learning, and agricultural productivity approaches are crucial. Accurate data analysis must be improved in order to carry out production effectively. In order to achieve the greatest outcomes, IoT sensor devices and drones are combined to produce more food in both quantity and quality while reducing crop failure. The same devices monitor all metrics on a regular basis. PA can become a viable practice by combining machine learning and IoT data analytics. We first collect PA datasets that include information on soil types, soil nutrients, soil types and conditions, weather factors, environmental variables, farmer records, and data from agricultural equipment such as yield monitoring. This collection depends on the precision agriculture application specification. Therefore, IoT is the key to creating a network of devices that can monitor and collect data on all the key aspects that are necessary for maximum crop yield. ML algorithms are also explored for analyzing the collected data to determine the optimum crop variety to be sown, which is the most suitable for a given time period in a particular region, all while upholding the requirements and finances of the entire system according to the farmer.

Main Contributions of the Paper

- A predictive model using machine learning algorithms to determine the most suitable crops to grow in a given region based on atmospheric and soil data could help optimize crop yields and improve the efficiency of resources used.

- The incorporation of IoT-based sensor networks to monitor environmental variables such as temperature, moisture, NPK values, and pH levels of the soil can enable real-time monitoring and management of crops, leading to more precise and efficient agriculture practices.
- The integration of these technologies can lead to a more sustainable agriculture industry by reducing the use of resources like water, fertilizer, and pesticides and minimizing waste.
- The proposed model and technology could potentially be implemented in small-scale and large-scale farming operations alike, with the potential to benefit farmers and increase food production globally.

Overall, the paper's contributions could lead to a more efficient, sustainable, and profitable agriculture industry by using data-driven and IoT-based technologies.

6.2 Literature Review

There have been several studies that have investigated the use of machine learning techniques in the field of agriculture. Many have focused on predicting crop production while analyzing the impact of various factors on crop productivity. One study, conducted by Arun Kumar *et al.* [1] examined the use of machine learning algorithms to predict crop production. The authors used Artificial Neural Networks (ANN) to estimate the actual crop yields and anticipated expenses by classifying crop yields depending on the quality of the yield and using regression analysis. Nithin Singh and Saurabh Chaturvedi [2] conducted a study on using Machine Learning Algorithms for predicting the weather. The authors gathered previous weather reports from various weather monitoring stations to predict the weather in the upcoming season. Aakash Parmar and Mithila Sompura [3] concentrates on improving crop forecasts by applying Machine Learning to track rainfall. Sachee Nene and Priya [4] evaluated the application of Artificial Intelligence to forecast agricultural yields based on soil and atmospheric variables. Ramesh Medar and Anand M. Ambekar [5] used artificial intelligence to anticipate the crop production of sugarcane plants. To generate this forecast, they used three datasets namely rainfall, soil, and crop production. Andrew Crane Droesch [6] introduced an ML model for evaluating the impacts of global warming and climate changes on agriculture and subsequently forecasting crop output. The model is a complex neural network with semi-parametric features. Vinita Shah and Prachi Shah [7] peanut harvests based on soil, climatic, and inorganic data using machine learning. The RMSE known as root mean square error was used to evaluate the forecasts' accuracy. Renuka and Sujata Terdal [8] compared the accuracy of various Artificial Intelligence models for forecasting agriculture production based on data input from the soil and rainfall. The MSE, short for mean squared error, was used to make comparisons between the algorithms. P. Vinciya and Dr. A. Valarmathi [9] used deep learning to distinguish various soil types and evaluate data on numerous crops. They achieved this using a Decision Tree and Multivariate linear regression. Shivnath Ghosh and Santanu Koley [10] evaluated soil parameters and enhanced forecast accuracy. For this, they utilized the backpropagation algorithm to train artificial neural networks. Aerial hyperspectral and hyperspectral imagery along with vegetation indicators were used by Xujun Ye *et al.* [11] to estimate the production of the Satsuma Mandarin. A review research was carried out by De Ollas *et al.* [12] on the consequences of changes in the climate, on the productive yield of crops in the Eastern Mediterranean region. The crops studied were olives, tangerines, clementine and grapevines. Vogel *et al.* [13] examined how changes in the climate affect crop production worldwide. They conducted their research using a worldwide dataset of different crops put into the Random Forest Classifier which would analyze the data. Nagaz *et al.* [14] tested the impacts of restricting water to an orange tree by closely studying the growth, production, and quality of fruits of the tree. The findings demonstrated that insufficient watering had a severely detrimental effect on all the factors tested above mainly impacting the overall development of the tree. Akhter, R. and Sofi, S.A. [15] proposed a model using Machine learning and Data analytics in an IoT system that predicts whether Apple disease will affect the apple orchards in the Kashmir valley. A survey was also conducted using precision agriculture, by deploying their model and understanding its utility. Kumar R and Singhal V. [16] proposed a system that aims to automate irrigation by utilizing temperature and soil sensors, and a microcontroller board to control the water pump based on real-time data, while also providing farmers with an Android application that uses machine learning to predict suitable crop types based on soil parameters and weather conditions. Bakthavatchalam *et al.* [17] proposed an IoT-based system and selected a machine learning model using WEKA to predict suitable crops for different combinations of attributes, achieving a high accuracy rate of classification and aiming to enhance precision agriculture and high crop yield. Rezk *et al.* [18] proposed an IoT-based smart farming system utilizing a prediction method called WPART, which combines wrapper feature selection and PART classification techniques to accurately predict crop productivity and drought, outperforming existing algorithms with high accuracy rates. Ikram *et al.* [19] proposed Smart Crop Selection (SCS), utilizing machine learning algorithms and real-time sensory data analysis to select crops for maximal yield based on metrological and soil factors, achieving high accuracy in predicting multiple crops and incorporating rainfall prediction for improved crop selection. Senapaty *et al.* [20] proposed an IoT-enabled soil nutrient classification and crop recommendation (IoTSNA-CR) model that utilizes sensors, cloud computing, and machine learning algorithms to provide farmers with real-time data analysis, crop recommendations, and soil nutrient management, resulting in increased productivity and reduced expenditure.

6.3 Proposed Model for Crop Prediction

The entire Crop Prediction module comprises various components that have been diagrammatically defined in Figure 6.2. The datasets Scientific_names.csv, Crop_recommender.csv, and Soil.csv are utilized in the Crop Recommendation module. The website Kaggle was used to get all of these datasets. The other datasets used for crop prediction were Soil_Info.csv and Environment_Parameters.csv. All these datasets were a result of about

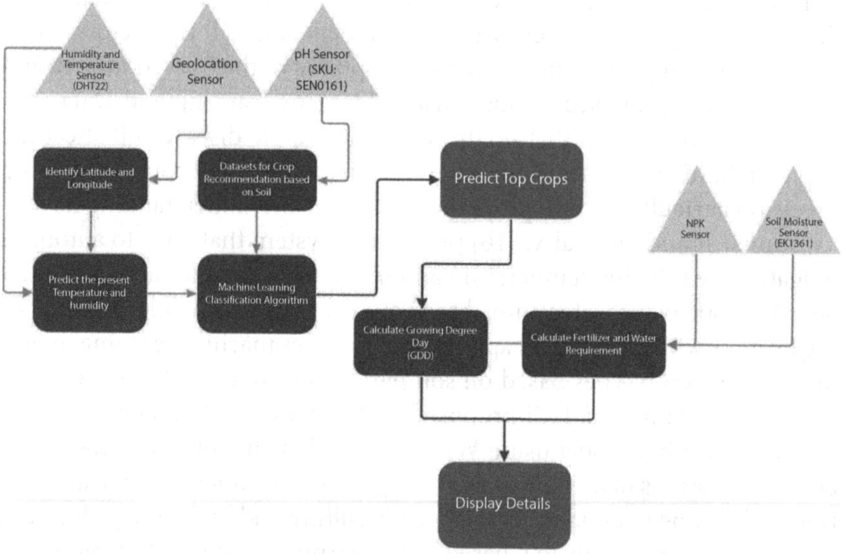

Figure 6.2 Proposed model for crop prediction.

60 days of monitoring and logging observations provided by distinct sensors, including the Geolocation Sensor, the Soil Moisture Sensor (EK1361), the Temperature and Humidity Sensor (DHT22), and the NPK Sensor. These tools separately gathered the environmental characteristics (geolocation, nitrogen, potassium, and phosphorus concentrations, temperature, soil moisture, and humidity). The inputs from various sensors were then processed, and the output was used as the data for the above-mentioned datasets.

In addition to forecasting harvests, the IoT system intends to assist farmers in making wise decisions. Along with the live data, the Geolocation sensor also collects and stores historical location data for the location's humidity and temperature from official sources in order to improve accuracy. For the project's crop prediction to be certain and accurate, it analyzes the field's temperature and humidity. Data is collected in real-time using a DHT-22 sensor, as well as historical data obtained from the Google Weather API and/or government website, the type of soil the farmer uses, and historical rainfall data. Both supervised and unsupervised machine learning algorithms are capable of analyzing it and are used for the training of datasets. To determine which machine learning approach produces the most accurate output, which will then be provided to the user, the accuracy attained by several ML techniques is compared. The method suggests fertilizer for the crop in addition to the most suitable crop.

The Crop_recommender.csv dataset was chosen for the training model because it includes characteristics that are crucial for crop prediction, including soil pH, potassium requirement ratio, average rainfall, nitrogen requirement ratio, temperature, humidity, and phosphorus requirement ratio. Post-prediction analyses are performed to identify the kinds of scientific names and the soil of the anticipated plants using datasets like Scientific Names.csv and Soil.csv, respectively.

Step 1: Importing Libraries and Dataset

The use of preprocessing tools and ML algorithms requires the loading of certain libraries that facilitate model creation and prediction. The required libraries imported were Train Test Split, Matplotlib, Pickle, Seaborn, Pandas, and Label Encoder. LGBM Classifier, Logistic Regression, Bagging Classifier, Decision Tree Classifier, SVM, Random Forest Classifier, XGBoost Classifier, AdaBoost Classifier, Gradient Boosting Classifier, KNN, and Naive Bayes are the models that were imported. The algorithms were first trained and then tested against the crop_recommendation.csv dataset. A sample of the crop recommendation dataset is shown

```
In [4]: df
Out[4]:
```

	N	P	K	temperature	humidity	ph	rainfall	label
0	90	42	43	20.879744	82.002744	6.502985	202.935536	rice
1	85	58	41	21.770462	80.319644	7.038096	226.655537	rice
2	60	55	44	23.004459	82.320763	7.840207	263.964248	rice
3	74	35	40	26.491096	80.158363	6.980401	242.864034	rice
4	78	42	42	20.130175	81.604873	7.628473	262.717340	rice
...	...	...	...	...	...	...	...	...
2195	107	34	32	26.774637	66.413269	6.780064	177.774507	coffee
2196	99	15	27	27.417112	56.636362	6.086922	127.924610	coffee
2197	118	33	30	24.131797	67.225123	6.362608	173.322839	coffee
2198	117	32	34	26.272418	52.127394	6.758793	127.175293	coffee
2199	104	18	30	23.603016	60.396475	6.779833	140.937041	coffee

2200 rows × 8 columns

Figure 6.3 Dataset to recommend crops.

in Figure 6.3: Temperature, humidity, rainfall, nitrogen, potassium, and phosphorus values are specific to each crop.

Step 2: Descriptive Analysis

Descriptive analytics helps you come up with fresh ideas and gives you a notion of how your dataset should appear. Missing values for each attribute are examined after the dataset has been imported. The attributes for the crop recommendation dataset don't have any missing values.

Step 3: Evaluate the data through Data Visualization

Once the essential details of the dataset have been acquired, data visualization is used to evaluate it visually. In essence, a correlation matrix is a relational grid that tabulates the correlation coefficients between features. Figure 6.4 shows the distribution of crop occurrences in the dataset. According to the figure, the distribution of each crop in the data set is equal. The dataset's crop instance distribution is seen in Figure 6.4. The graph depicts an equal distribution of all the crops in the data set. The label column, resembling the dependent variable, is plotted against each characteristic in the dataset. The analysis indicates that cotton requires more nitrogen for growth than any other crop when considering the nitrogen requirements for each crop. Similar to how the potassium requirement for each crop – in which orange requires the least and grape and apple require the highest. The phosphorus requirement is also the highest in grapes and apples. In Figure 6.6, the details are further illustrated.

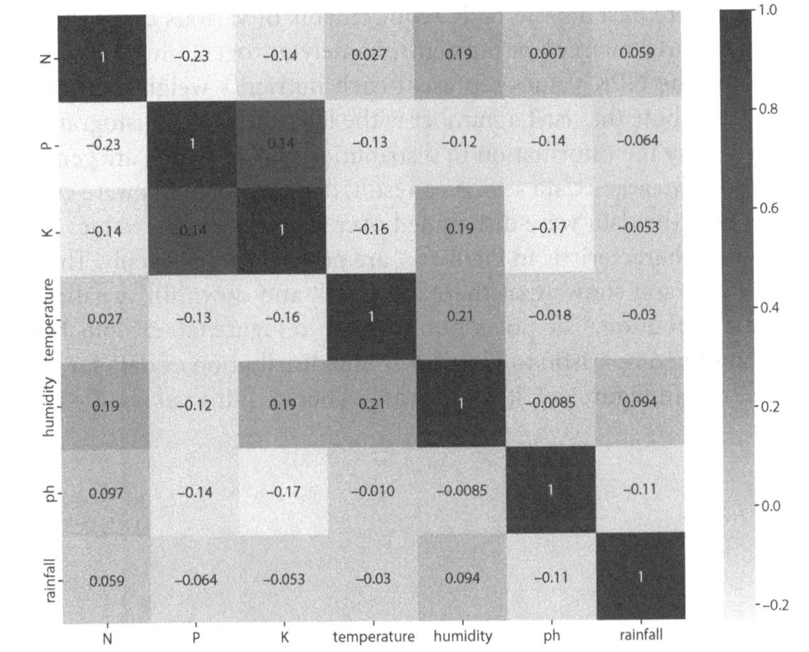

Figure 6.4 Correlation matrix between all parameters related to crops.

In terms of temperature requirements, papaya requires a higher temperature than other crops, rice requires more rainfall than other crops, and coconut needs more humidity than other crops. On analyzing the soil pH requirement of a crop, it is found that most crops require a higher pH value. All the above data is visualized in Figure 6.5.

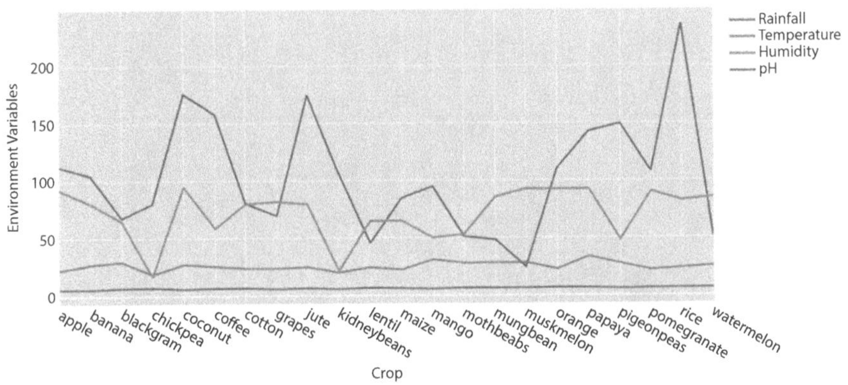

Figure 6.5 Crop-wise environment requirements.

Figure 6.6 represents the NPK requirements of various crops. It is the ratio of the three main plant nutrients, namely nitrogen, phosphorus, and potassium. The NPK values represent each nutrient's weight proportion. The only attribute that isn't a number is the label attribute. Histograms are used to display the information in distribution charts, which are generally utilized for univariate data sets. As a result, distribution plots were created to show how the data were distributed over the dataset. The values of the phosphorus characteristic in Figure 6.7 are not distributed evenly. The blue line on the graph shows that there are uphill and downhill variations in the data distribution. The potassium column in Figure 6.8 exhibits higher NIL values in comparison to Figure 6.7. The distribution of data for nitrogen is shown in Figure 6.9. It mostly ranges between the values 20–45 and 75–120.

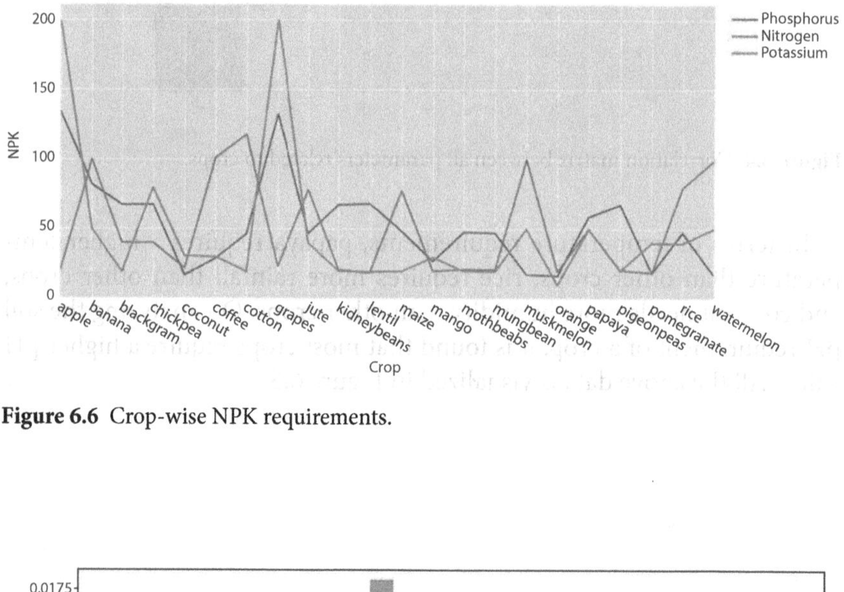

Figure 6.6 Crop-wise NPK requirements.

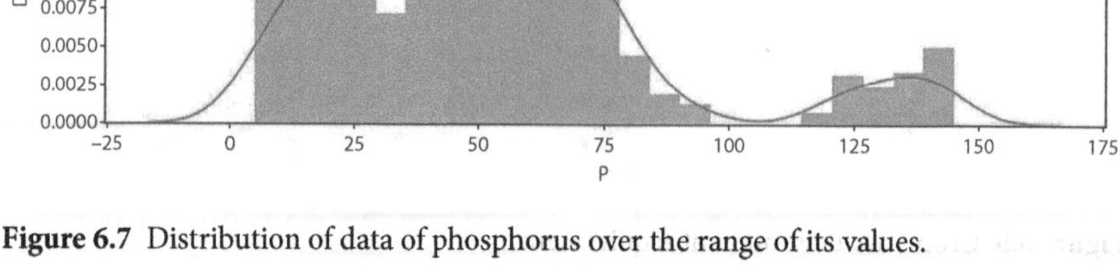

Figure 6.7 Distribution of data of phosphorus over the range of its values.

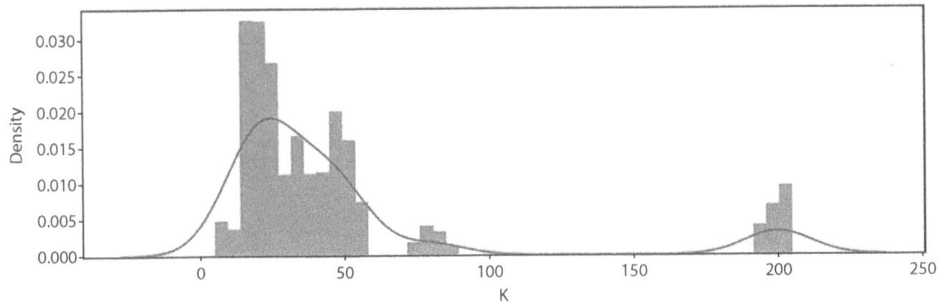

Figure 6.8 Distribution of data of potassium over the range of its values.

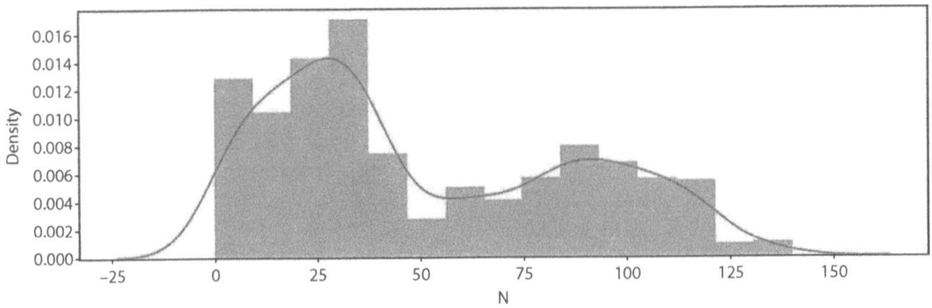

Figure 6.9 Distribution of data of nitrogen over the range of its values.

Step 4: Detection and Treatment of Outliers

The data distribution for the Temperature Column is shown in Figure 6.10. Most values fall between 20 and 35. The data's normal distribution is also demonstrated. The data in Figure 6.11 are not widely spread according to rainfall. In comparison to Figure 6.12, the values for the soil pH column are more evenly distributed throughout the range of 5.5–7.5. The data in each column is not normally distributed, as shown by all distribution maps, and their numerical values differ from region to region.

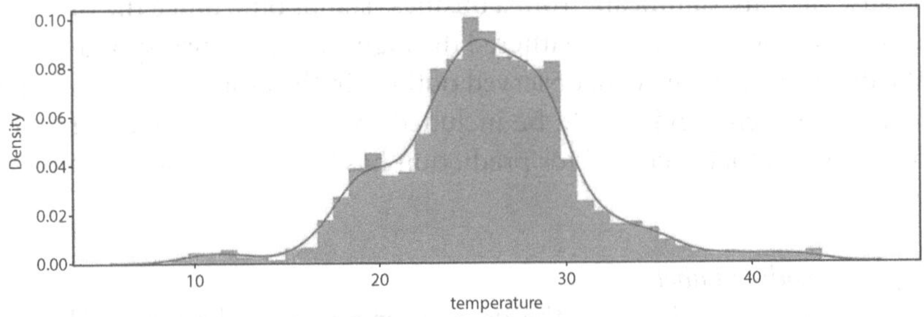

Figure 6.10 Distribution of data of temperature over the range of its values.

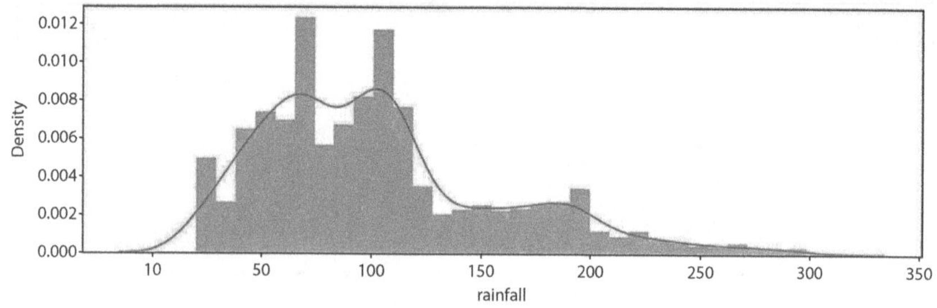

Figure 6.11 Distribution of data of rainfall over the range of its values.

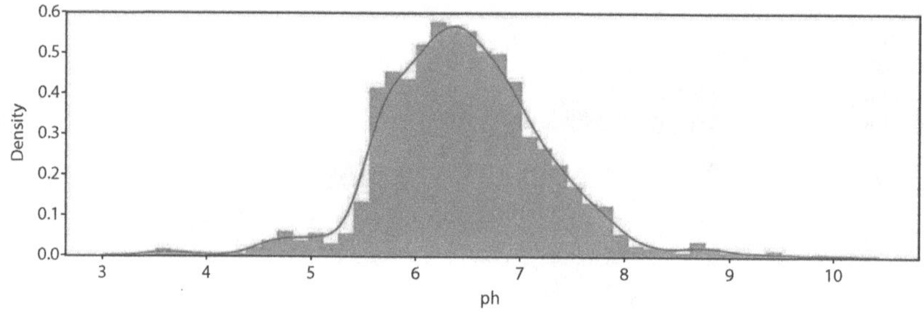

Figure 6.12 Distribution of data of pH over the range of its values.

Data points known as outliers diverge significantly from other values in a given set. These anomalies can taint data sets and produce inaccurate forecasts. Therefore, outliers should be detected and tested before the data is passed to the model. The suggested model employed a boxplot and the IQR technique to identify outliers. Box charts make it easier to see an attribute's outliers.

Quintile ranges, which define the fraction of data that goes outside the 0.75–0.25 quintile range, are used to define outliers in the Interquartile Range Technique (IQR). Outliers may occasionally be kept or excluded because they are significant from a business standpoint. Since the tuples or readings in the dataset were gathered through trials and correspond to specific details of crop growth, observed outliers in the crop recommendation dataset are useful and should be included. As a result, every observation in the dataset that is crucial for prediction has been incorporated into this work.

Step 5: Encoding Labels

The dependent attributes of the dataset, or a variable known as "labels," are label-encoded. This attribute essentially contains the name of the crop and has categorical and non-numeric values. Many classification models discourage the use of non-numeric values, so you should encode them to numeric values first. We then input the data into the model in order to predict the future.

Step 6: Partitioning the data into Training and Testing data sets
Splitting is done within the dataset into training and testing datasets, which is the primary stage in model creation after outliers have been located and their attributes are shown. The testing and training sets are split in half evenly. Training data refers to the initial data needed to develop a machine learning algorithm. The training dataset is provided to ML algorithms to teach them to make precise predictions as well as carry out specified actions. A data set known as a testing set is used to assess how accurately the final model matches the training sample.

Step 7: Building a Model
This aims to utilize Machine Learning Classification methods to identify the crops. To pick the ideal model for making future predictions, ten classification algorithms are employed in this case. The following are the steps in producing a model:

- The library is accessed to import the model, which is then defined.
- Training data and test data are made to fit into the model.
- The model is tested over the testing dataset after training the model.
- Evaluation metrics and the confusion matrix are derived.

Results for the Random Forest Classifier are shown in Figure 6.13. Following the training set's accuracy score, the accuracy score of the test set, and the overall accuracy score for this model of crop suggestion dataset, the test set predictions are shown as a list which is pointed to by y predictions. The accuracy, recall, and F1 scores of the used model are displayed in the classification report. To produce the matching predicted values, the accuracy of the training set, test set and the report of the classification is used. An identical process is carried out for each of the 11 algorithms stated above. The random forest classifier's confusion matrix is displayed in Figure 6.13. The confusion matrix diagonal reveals how frequently the predictions were accurate. A formula, along with the ones included in the confusion matrix, is used to calculate accuracy, as portrayed in Figure 6.14.

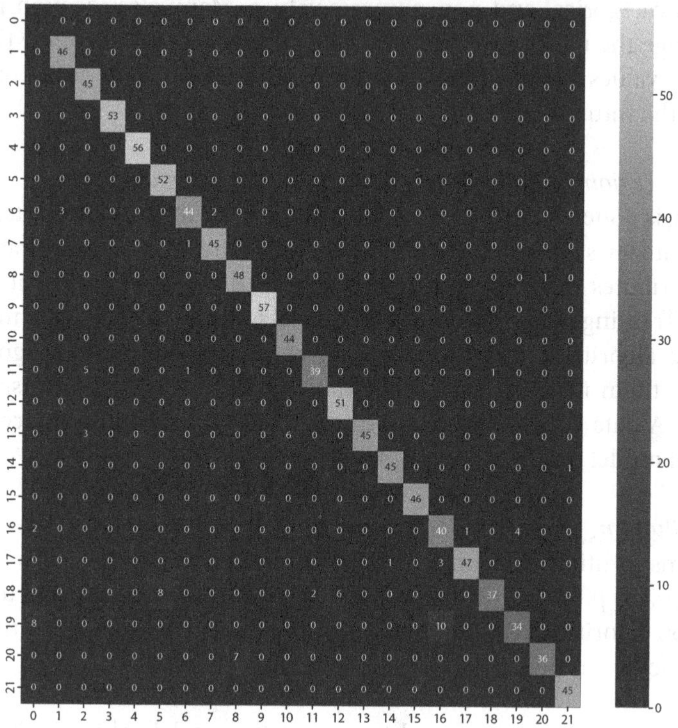

Figure 6.13 Random forest classifier's confusion matrix.

	Algorithm	model	Train_score	Test_score	accuracy
0	KNN	knn	89.636364	84.272727	84.272727
1	NaiveBayes	nb	96.363636	94.727273	94.727273
2	LogisticRegression	lr	66.454545	63.909091	63.909091
3	SVM	svm	69.454545	65.181818	65.181818
4	DecisionTreeClassifier	dt	69.454545	65.181818	92.181818
5	BaggingClassifier	bg	99.454545	92.545455	92.545455
6	RandomForestClassifier	rf	100.000000	94.727273	92.545455
7	AdaBoostClassifier	ad	14.363636	12.909091	12.909091
8	GradientBoostingClassifier	gb	95.727273	90.454545	90.454545
9	XGBClassifier	xg	96.363636	91.727273	91.727273
10	lbgmClassifier	lbgm	100.000000	93.454545	93.454545

Figure 6.14 Accuracy of different Algorithms based on results of training and testing data set.

Step 8: Hyperparameter Tuning

Hyperparameters are a key determinant of the model architecture. Hyperparameter tuning is the procedure for finding the optimal model architecture. Hyperparameters tune the algorithm that was filtered. Before training an algorithm, some parameters must be defined that are specific to that algorithm. Once these parameters are defined, they are added to the model and are defined using either randomized CVS or even a grid search CV algorithm can be used. Using these parameters, the hyper-tuned model is trained using training data before being tested using the test set. The Hypermatch algorithm's accuracy is computed and utilized as a metric to select the best algorithm for further predictions.

As illustrated in Figure 6.15, some algorithms' efficiency was reduced after hyperparameter tuning, however, the accuracy of the Random Forest Classifier was enhanced to 95.4545% after being hyper-tuned using RandomSearchCV, as depicted in Figure 6.15. The accuracy comparison of several algorithms with that of hyperparameter tuning is shown in Figure 6.15. This demonstrates that most algorithms tend to deliver higher accuracy when they are hyper tuned. As a result, the Random Forest Classifier, which is hyper-tuned and predominantly utilizes the randomized search CV, is chosen for deeper evaluation.

Step 9: Prediction of the most suitable crop

The model that performs the best crop prediction is then saved inside a pickle file. Python object architectures are serialized using pickle files, a procedure that turns an object of the memory into something known as a byte stream, which is stored in the form of a binary file on a device. Then, by importing it into a Python code, this binary file may be deserialized into a Python object. As a result, this model is then kept in a pickle file, which may help in determining the most suitable crop.

	Algorithm	model	Train_score	Test_score	accuracy
1	NaiveBayes	nb	96.363636	94.727273	94.727273
4	DecisionTreeClassifier	dt	69.454545	65.181818	92.181818
5	BaggingClassifier	bg	99.454545	92.545455	92.545455
6	RandomForestClassifier	rf	100.000000	94.727273	92.545455
8	GradientBoostingClassifier	gb	95.727273	90.454545	90.454545
9	XGBClassifier	xg	96.363636	91.727273	91.727273
10	lbgmClassifier	lbgm	100.000000	93.454545	93.454545

Figure 6.15 Accuracy of the top 7 models with hyperparameter tuning.

To predict yield, the model needs to input values such as temperature, moisture level, percentage of average rainfall, and pH of the soil. The geographic latitude and longitude of the present location are determined via Web Scraping to be able to acquire the climatic conditions, like temperature and humidity. To accomplish this, the program is requested to launch a website that retrieves details about your current location using your device's IP address. The extracted values of longitude and latitude are then supplied to the OpenWeatherAPI, which returns the location's details such as area, site, and temperature. The user then inputs the soil pH and the average rainfall using a web interface. All these parameters are integrated before being sent into the model for predictions. The crops with the greatest model prediction probabilities are chosen, their values are decoded to reveal their names, and the probabilities are then ranked in descending order. The model then exhibits the best five plants that are most suitable to cultivate in that region. Along with providing recommendations for crops, this work also recommends the GDDs (Growing Degree Days) as well as the NPKs (Nitrogen, Potassium, and Phosphorus) that are necessary for crop growth.

Case 1: $\text{Min}_{\text{temp}} <= \text{Base}_{\text{temp}}$

$$\mathbf{GDD} = \frac{(\text{Base}_{\text{temp}} + \text{Min}_{\text{temp}}) - \text{Base}_{\text{temp}}}{2} \tag{6.1}$$

Case 2: $\text{Min}_{\text{temp}} > \text{Base}_{\text{temp}}$

$$\mathbf{GDD} = \frac{(\text{Min}_{\text{temp}} + \text{Max}_{\text{temp}}) - \text{Base}_{\text{temp}}}{2} \tag{6.2}$$

Case 3: $\text{Max}_{\text{temp}} > \text{Base}_{\text{temp}}$

$$\mathbf{GDD} = \frac{(\text{Min}_{\text{temp}} + \text{Base}_{\text{temp}}) - \text{Base}_{\text{temp}}}{2} \tag{6.3}$$

$$\textbf{Nitrogen concentration for 200 lb. Fertilizer} = \frac{200 * \overline{\text{Nitrogen ratio}}}{100} \tag{6.4}$$

$$\textbf{Phosphorus concentration for 200 lb. Fertilizer} = \frac{200 * \overline{\text{Phosphorus ratio}}}{100} \tag{6.5}$$

$$\textbf{Potassium concentration for 200 lb. Fertilizer} = \frac{200 * \overline{\text{Phosphorus ratio}}}{100} \tag{6.6}$$

GDDs are indicators that are weather-based and can monitor a crop's progress. This calculates the crop production and measures temperature levels, which help in predicting the growth rate of plants and pests, along with the approximate time when a crop will be ready for harvest. The method used to determine the value of GDD for every crop that's predicted is shown in Equations (6.1), (6.2), and (6.3). The equivalent base temperature values are acquired from the dataset for the crop. Every crop has a maximum and minimum baseline temperature. These readings are then collected as a dataset and taken from verified agricultural websites. Based on minimum, base, and maximum temperatures, the GDD value is produced for the crop according to the case it satisfies, as given, in Equations (6.1), (6.2), and (6.3).

The presence of nitrogen, phosphorus, and potassium is essential for crop development. In order to calculate the concentration of nutrients for 200 pounds of fertilizer for 1 hectare, the average values of phosphorus, nitrogen, and potassium of a given crop were extracted, subdivided by 100, and then multiplied by 200. These calculations are shown in Equations (6.4), (6.5), and (6.6). Hence, all six equations mentioned above are crucial to finding the most appropriate crops. Thus, the results are presented in Table 6.1.

6.4 Results and Analysis

For crop recommendation, eleven algorithms were considered. As seen in Figure 6.14, an accuracy score of 90% or above was chosen. The optimal model with the maximum accuracy was derived from these chosen methods by using Hyperparameter Tuning, as shown in Figure 6.15. The best model, as determined by Figure 6.15, is the hyper-tuned Random Forest classifier using Randomized CV, which has an accuracy of 95.45% and is saved as a pickle file for future examination.

Table 6.1 Top crops suitable for the derived location and soil type.

Crop details					
Crops	pigeonpeas	apple	banana	blackgram	chickpea
Scientific name	Cajanus cajan	Malus Pumila	Musa	Vigns mungo	Cicer arietinum
Soil type	['sandy' 'loamy' 'clayey']	['loamy']	['loamy']	['sandy' 'loamy']	['sandy' 'loamy' 'black' 'clayey']
Crop type	['Rabi Crop', 'Kharif Crop']	['Kharif Crop']	['Rabi Crop', 'Kharif Crop']	['Rabi Crop', 'Kharif Crop']	['Rabi Crop', 'Kharif Crop']
Growing degree days	9.648524160000001	15.516694609999998	12.459534894999997	5.021994705000001	9.010003044999998
Nitrogen requirement for 200lb fertilizer	41.46	41.6	200.45999999999998	80.04	80.18
Potassium requirement for 200lb fertilizer	40.58	399.78	100.1	38.48	159.84
Phosphorous requirement for 200lb fertilizer	135.46	268.44	164.02	134.94	135.58

The IoT system offers information on average rainfall as well as soil pH. Various other sensors calculate the other variables. Initially, a Geolocation Sensor is used to capture the latitude and longitude data from the current location, which are then sent as inputs to a particular Weather API. This then gives the past information on the current location's temperature and humidity. This data is also compared with the data present in the dataset Environment_Parameters.csv, which helps us in comparing and finding the mean value of temperature and humidity to achieve a better degree of accuracy. The chemical composition and pH values of the soil are monitored by the various sensors and stored in the Soil_Info.csv dataset. This dataset is used to provide the system with the additional required information. The algorithm predicts the best crops and offers information depending on the provided parameters.

The top recommendations of crops are included in Table 6.1 below, along with their scientific names. The model then identifies the crop's scientific name once it has estimated the name of the best-suited crop from the database. The soil.csv dataset is responsible for determining the type of soil for a certain crop. Crop types are divided into groups like Rabi, Kharif, and Summer Crops. These categories are based on factors including temperature, humidity, and rainfall, the crop has been divided into several categories.

The model first determines the crop, then calculates the crop's needs for nitrogen, potassium, and phosphorus and its growing degree days (GDD). The formulas for these calculations are provided in Equations (6.1), (6.2), (6.3), (6.4), (6.5), and (6.6). Apple has the highest GDD, whereas black gram has the lowest. Apple and red gram both need the same quantity of nitrogen; however, bananas require more nitrogen to be incorporated into the fertilizer. Apples have the most phosphorus and potassium requirements. The projections, coupled with the specifics, are extremely useful to a user when analyzing and selecting a crop. Furthermore, it does not supply any new values. The suggested method predicts top harvests and provides data that the user may utilize to carry out their task more efficiently.

6.5 Challenges Faced

Site Description: Latitudes and longitudes must be used to base the module on a specific area of the Indian Subcontinent. Because of the wide range in height, each location has a variable climate, which leads to uncertainty in grain production and pricing. As a result, a detailed conversation with the farmer regarding his property is required.

Budget: The objective of the researcher should always be to boost production while minimizing the cost needed. Since there are so many different gateways, sensors, and base infrastructure components, there should be a cost study performed before building a smart agricultural system. Maintenance is an important factor to consider in cost analysis.

Lack of Exposure to the Technology era: The majority of farm work is performed in rural areas by uneducated farmers. They fail to perceive and incorporate new technologies in changing times. As a result, IoT development in rural regions is lagging.

Interoperability: It refers to a piece of equipment, product, service, or system's capacity to function in tandem with others. An IoT device must be able to be put anywhere and remain linked globally in order to function properly. Other benefits include the capacity to support roaming and give the system dependability for correct outcomes.

Supervisory challenges: The organizing, managing, and authorizing foundations must be sorted out between the field owners and the data corporations. Regulations governing data privacy, data security, technological challenges, available resources such as bandwidth, and so on vary by nation. Differentiating these supervisory aspects has an impact on agricultural use cases. The variety of supervisory procedures thus poses a challenge for most farmers [21].

Technicality-based challenges: It may be quite harmful at times, even if just one component of the system fails. When food safety is jeopardized, the system becomes ineffective. To satisfy the demands of precision agriculture, the most durable and dependable technology must be developed. Every technologically based task is becoming automated through the use of IoT and other mechanized methods. In PA, the IoT sensory devices are vulnerable to environment related interference, such as animal assaults and theft [22].

When employed in agriculture, the IoT system is frequently attacked by device capture assaults. Existing systems and protocols should handle a high number of IoT nodes. Device physical safety must be assured in order for massive IoT data precision systems for agriculture to be correctly built. Maintaining a precision agricultural system in wide areas is a difficult task. Powerful gateways, efficient sensors, cloud services, and other IoT-related equipment might result in increased energy usage. The benefits of precision agriculture have previously been highlighted. Various problems arise while creating a smart PA system that uses IoT and big data.

Challenges based on data analytics: It is challenging to meet the ever-increasing need for solutions to integrate data from multiple sources efficiently, as well as for knowledge mining and data mining. Data mining is the process of developing efficient and suitable solutions for Big IoT data sets that can subsequently be used for other data types. Given the characteristics of Big IoT data, data mining operations become challenging tasks. Information extraction and data exploration are two problems that have arisen as a result of cloud computing and big data platforms.

6.6 Advantages of the Proposed Model

- The capacity of the model to use sensors to monitor environmental, soil, and plant data can aid in predicting the crop variety that will be most effective within a specific time frame.
- In this manner, the yield will be maximum while retaining the nutrients of the soil and preserving the environment.
- The model will specify the requirements of the soil by intimating the percentage of nutrients and composition of the soil, which will prevent the land from getting unbalanced or ultimately leading to barren land.
- The model learns according to its user, hence can be customized according to the requirements of the farmers and the factors affecting their land. This reduces the chances of failure in a predictive and controlled manner. Overall, reducing the loss percentage of crops.

6.7 Disadvantages of the Proposed Model

- The farmers using the modules will be from rural backgrounds. Indicating that they may not be aware of the technology used and would find it difficult to understand the basic functions of the model.
- Environmental factors like storms, floods, the sun and wild life may damage the sensors and the hardware used, to a

nonfunctional state. Hence, installing it will cause inconvenience and increase the overall cost.
- After-sales service and maintenance of the entire system can pose a huge issue. The project has to cater to the size of the field, which may raise issues of scalability and customization.

6.8 Conclusion

Agriculture is a vital sector of the Indian economy, but it is not without its challenges. Farmers in the traditional agricultural industry often struggle with inadequate crop growth due to a range of environmental and climatic issues. In order to address these challenges, innovative solutions using IoT (Internet of Things) and ML (machine learning) techniques are being developed to assist farmers in making informed decisions about the crops that are most suitable for cultivation in a particular region. By using real-time sensor readings and applying ML algorithms, farmers can also receive recommendations for fertilizers based on soil conditions and weather patterns. Among the various ML algorithms tested, XGBoost has proven to be the most effective, providing the highest accuracy in its recommendations. These IoT and ML-based solutions can help farmers understand the various factors that impact their costs, yields, and decision-making, leading to a scalable and reliable solution that could benefit millions of people in India. By providing farmers with the tools and knowledge they need to optimize their operations, we can help ensure the success and sustainability of the agricultural industry in India.

References

1. Kumar, A., Kumar, N., Vats, V., Efficient crop yield prediction using machine learning algorithms. *Int. Res. J. Eng. Technol. (IRJET)*, 05, 06, June 2018, e-ISSN: 2395-0056, pISSN:2395-0072.
2. Singh, N. and Chaturvedi, S., Weather forecasting using machine learning. *2019 International Conference on Signal Processing and Communication (ICSC)*, vol. 05, Dec 2019.
3. Parmar, A. and Sompura, M., Rainfall prediction using machine learning. *2017 International Conference on (ICIIECS)*, Coimbatore, vol. 3, March 2017.
4. Nene, S. and Priya, R., Prediction of crop yield using machine learning. *Int. Res. J. Eng. Technol. (IRJET)*, 05, 02, Feb 2018.

5. Medar, R. and Ambekar, A.M., Sugarcane crop prediction using supervised machine learning. *Int. J. Intell. Syst. Appl.*, 3, August 2019.
6. Droesch, A.C., Machine learning methods for crop yield prediction and climate change impact assessment in agriculture. 05, Oct 2018, Published by IOP Publishing Ltd.
7. Shah, V. and Shah, P., Groundnut prediction using machine learning techniques. *IJSRCSEIT*, March 2020, UGC Journal No : 64718.
8. Renuka, and Terdal, S., Evaluation of machine learning algorithms for crop prediction. *Int. J. Eng. Adv. Technol. (IJEAT)*, 8, August 2019.
9. Vinciya, P. and Valarmathi, A., Agriculture analysis for next generation high tech farming in data mining. *IJARCSSE*, 6, 5, 2016.
10. Ghosh, S. and Koley, S., Machine learning for soil fertility and plant nutrient management using back propagation neural networks. *IJRITCC*, 2, 2, 292–297, 2014.
11. Ye, X., Sakai, K., Manago, M., Asada, S.I., Sasao, A., Prediction of citrus yield from airborne hyperspectral imagery. *Precis. Agric.*, 8, 111–125, 2007.
12. De Ollas, C., Morillón, R., Fotopoulos, V., Puértolas, J., Ollitrault, P., Gómez-Cadenas, A., Arbona, V., Facing climate change: Biotechnology of iconic Mediterranean woody crops. *Front. Plant Sci.*, 10, 427, 2019.
13. Vogel, E., Donat, M.G., Alexander, L.V., Meinshausen, M., Ray, D.K., Karoly, D., Meinshausen, N., Frieler, K., The effects of climate extremes on global agricultural yields. *Environ. Res. Lett.*, 14, 054010, 2019.
14. Nagaz, K., El Mokh, F., Ben Hassen, N., Masmoudi, M., Ben Mechlia, N., Baba Sy, M., Belkheiri, O., Ghiglieri, G., Impact of deficit irrigation on yield and fruit quality of orange trees (Citrus sinensis, L. Osbeck, cv. Meski Maltaise) in southern Tunisia. *Irrig. Drain.*, 69, 186–193, 2020.
15. Akhter, R. and Sofi, S.A., Precision agriculture using IoT data analytics and machine learning. *J. King Saud Univ.-Comput. Inf. Sci.*, 34, 8, 5602–5618, 2022.
16. Kumar, R. and Singhal, V., IoT enabled crop prediction and irrigation automation system using machine learning. *Recent Adv. Comput. Sci. Commun. (Formerly: Recent Pat. Comput. Sci.)*, 15, 1, 88–97, Jan 1 2022.
17. Bakthavatchalam, K., Karthik, B., Thiruvengadam, V., Muthal, S., Jose, D., Kotecha, K., Varadarajan, V., IoT framework for measurement and precision agriculture: Predicting the crop using machine learning algorithms. *Technologies*, 10, 1, 13, Jan 20 2022.
18. Rezk, N.G., Hemdan, E.E., Attia, A.F., El-Sayed, A., El-Rashidy, M.A., An efficient IoT-based smart farming system using machine learning algorithms. *Multimed. Tools Appl.*, 80, 773–97, Jan 2021.
19. Ikram, A., Aslam, W., Aziz, R.H., Noor, F., Mallah, G.A., Ikram, S., Ahmad, M.S., Abdullah, A.M., Ullah, I., Crop yield maximization using an IoT-based smart decision. *J. Sens.*, 2022, May 17 2022.

20. Senapaty, M.K., Ray, A., Padhy, N., IoT-enabled soil nutrient analysis and crop recommendation model for precision agriculture. *Computers*, 12, 3, 61, Mar 12 2023.
21. Mohapatra, R., Mishra, S., Mohapatra, T., Coverage problem in wireless sensor networks. *Comp. Cytogenet.*, 2, 1, 67–72, 2012.
22. Mishra, S., Mishra, S., Kayal, A., Chudi, S.R., Simulation in wireless sensor networks. *Int. J. Electron. Commun. Comput. Technol. (IJECCT)*, 2, 4, 176, 2012.

7

Relevance of Smart Management of Road Traffic System Using Advanced Intelligence

Koustab Chowdhury* **and Rishabh Kapoor**

Kalinga Institute of Industrial Technology, Deemed to be University, Bhubaneswar, Odisha, India

Abstract

Traffic control mechanisms are always an intensive discussion domain, but due to cutting-edge innovations like sensory computing and machine analytics, intelligent traffic monitoring remains a high priority. Most traffic prediction techniques focus on urban and highway traffic. There is less research on collector roads and closed campuses, which have lower traffic volumes but can still experience congestion. This research paper proposes models for Traffic Lighting System, Smart Parking System and Vehicle Theft Detection System using IoT. Smart traffic lights can use a realistic traffic tracking model to dynamically adjust traffic signals based on traffic density. In smart parking an IoT module is deployed on-site to monitor the availability of individual parking spaces. This module collects data on the availability of each individual parking space. By improving the efficiency of the parking facility, smart parking systems can enhance the overall parking experience for drivers. In vehicle theft detection systems, as soon as the vehicle is stolen or moved without the owner's knowledge, the system provides security by sending the exact location of the vehicle. The system consists of a sensor like a camera (in case of traffic lighting system), PIR motion sensor (in case of theft detection), RFID based cards (in case of smart parking) and a high speed processor Raspberry-pi and a cloud storage database for data maintenance. The goal is to deliver real-time traffic reports on traffic congestion, parking and theft.

Keywords: Internet of things, artificial intelligence, raspberry-pi, RFID-based cards, PIR motion sensor, traffic lighting system, smart parking system, vehicle theft detection system

**Corresponding author*: 2028020@kiit.ac.in

Hrudaya Kumar Tripathy, Sushruta Mishra, Minakhi Rout, S. Balamurugan and Samaresh Mishra (eds.) Optimized Computational Intelligence Driven Decision-Making: Theory, Application and Challenges, (131–150) © 2024 Scrivener Publishing LLC

7.1 Introduction

The robustness and intelligence of the smart city idea depend on the technology employed to improve the quality of life. Governance of smart city is a key element of programs that will promote the techniques for improved decision-making. The public value produced by smart services is one of the fundamental components of the governance framework for smart cities. The government must work on a number of smart city solutions, including smart parking, smart transit, smart building management, and smart traffic management, in order to produce public value for the service being provided. The concept of smart cities has evolved with the development of the internet of things (IoT). The urban infrastructure of the city is outfitted with smart gadgets which continually create multidimensional data in multiple areas in a smart city scenario. After that, these data are analyzed to give the infrastructure insight. Finally, intelligence is utilized to improve social and economic occurrences within society.

In developed and cosmopolitan areas around the world, traffic congestion is still a serious issue for municipal governments, commuters, and urban planners, especially during rush hour. Numerous negative effects and repercussions of traffic congestion include slower economic growth, higher fuel usage, and greater air pollution. According to a United Nations study, 56% of people resided in urban areas in 2015, and that percentage has been constantly rising year by 1.84%. The forecasts state that by the year 2050, nearly 64% of the developing society and 86% of the industrialized society would be urbanized, and there will be gradually more automobiles on the road. As the number of automobiles on the roads rises, traffic congestion will only serve to provide more obstacles and hurdles for economic advancement while also having a negative impact on quality of life. Therefore, people will invariably prefer to relocate to areas that are safer, more effective, and less congested with traffic. For urban planners, this poses a huge issue in terms of creating a traffic management system that is effective, can serve the requirements of a large metropolitan area, and is easily adaptable to population increase. In the conventional traffic management system, police officers manually operate inefficient traffic lights with predefined timers (Fixed Cycle (Static) TLS). Because Fixed Cycle TLS increases traffic density and congestion, especially on routes with high traffic volumes, it is unsuccessful. This is especially clear in countries like Malaysia where the use of Fixed Cycle TLS is widespread. Without taking real-time traffic data into account, it is possible for a "green light" to be given to an empty lane while several vehicles are backed up at a "red signal" on the other lanes since each lane receives green lights at the same interval. By adjusting traffic patterns based on the amount of vehicles on the road at any given time, a dynamic TLS, on the other hand, can significantly minimize this congestion. The proposed model analyzes the current traffic scenario of a particular instruction and then shows the traffic lights accordingly, so there won't be any traffic congestion on the roads. There is also a smart parking system which assigns a parking space for the user (if available). The smart theft detection system which helps to detect theft and sends the vehicles live location to the user. Traffic management is very essential for today's world. Not only it saves our time but also saves lives by avoiding accidents. Although we have made much advancement today in managing traffic, there are still limitations.

Over the past few years, the human population in metropolitan areas has increased notably over the world, outpacing the similar figure in rural areas. The world population trends from 1950 to 2050 are shown in the graph below, with the abscissa denoting the years and the ordinate denoting the number of people in millions. Out of the 7.4 billion people who make up the world's population as of 2016, 4.1 billion lived in metropolitan areas. In the future, out of the 9 billion people on the earth, 6 billion would be living in cities. This is explained in Figure 7.1 through a graphical representation.

As the population grows, so does the need for transportation, which has resulted in an increase in the number of automobiles on roads. It is crucial because the traditional traffic management system, which employs traffic lights and manual police supervision, is ineffective due to population expansion. This causes major traffic issues and jams, which raises vehicle emissions and wastes our valuable time. There are other factors

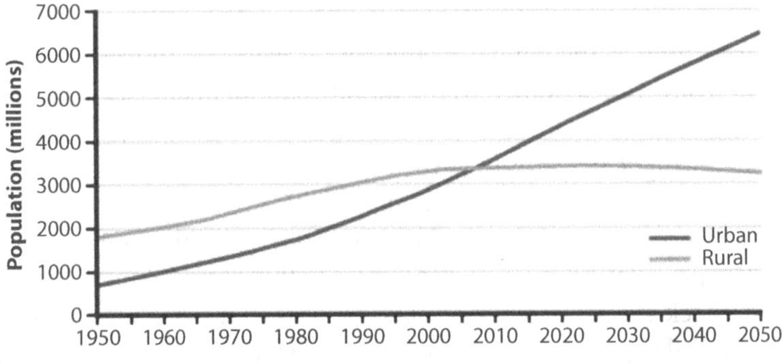

Figure 7.1 Rural and Urban population growth from 1950 to 2050.

contributing to traffic congestion, such as a lack of planning to deal with the increased number of cars on the roads. For instance, waiting for the light to turn green even when there are no vehicles on the road is rather common. This could make the traffic jam worse and necessitate human intervention from police officers. Hence, it is necessary to improve public transportation systems. Additionally, it's important to promote other modes of transit, enhance street safety, boost public security, and improve pedestrian mobility. This alternative would be unwelcome, necessitating the need for an effective traffic management system to address the shortcomings of the current system.

So, the main aim of the modeling presented here is to enhance the management of traffic with the help of technologies. The primary objective of this paper is to introduce a system which will control the traffic lights based on the amount of traffic present. There is also a smart parking system that helps in easy parking without human interference and the theft detection system that helps in detection and tracking of stolen vehicles. This study aims to develop a meaningful metric for assessing the importance of implementing traffic routing and management. IoT is a new paradigm for study, and it appears that we are just beginning to learn about its collection of knowledge. Nowadays the traffic management system is mostly manually operated or partially operated but here we aim to propose models which are fully automated. Here, we will be focusing on mainly 3 ways of traffic management: Smart traffic lighting system, Smart parking system and Smart vehicle theft detection system. In the proposed model of the lighting system we have designed a model which will tend to avoid traffic congestion at all costs. We have employed sensors to identify the vehicles on the road, and the collected data will be sent to a processor who will process it before sending it to a database where it will be used to train a model to more accurately predict changes in traffic signal timing and density. The controller operates the traffic lights by switching between red, yellow, and green, according to the database. The data is also sent to the cloud server from where it gets distributed to the mobile handheld devices. In the parking system we have designed a system in which the user has to interact with the system using RFID card then the arduino checks if there are any empty slots available in the parking lot and if available it assigns a slot for the user and sends a positive response to the user, if not available then a message is sent to the user stating space is not available through GSM module. Users can also check the availability of parking slots in a particular parking lot through mobile applications. In the vehicle detection system the vehicle is equipped with PIR, gas and pressure sensors which can detect if it is stolen. Then the camera captures an image and the GPS module saves the live location of the device. The user's registered email is then used to send them the device's image and current position.

Main contribution:

This paper proposed three models:

- ❖ Intelligent traffic lighting system which enables intelligent traffic routing through a machine learning model and lets the user know about the traffic density of various sections in the city.
- ❖ Smart parking system which enables fully automated management of a parking lot without interaction. It has a user friendly interface and through a mobile it can let the user know about the availability of parking space.
- ❖ Smart vehicle theft detection system secures the vehicle by detecting theft and sending the user the current location of the vehicle. The system keeps on sending location details of the vehicle to the user after every 120 m of displacement.

7.2 Related Works

In this section we will discuss recent developments of traffic management systems and flaws in them.

7.2.1 Traffic Lighting System

In an effort to address some challenging traffic issues, numerous studies on traffic light systems have been conducted. However, the majority of these studies occasionally fall short in their ability to effectively handle challenging, time-varying traffic conditions, and controllers are unable to satisfy the real-time nature of traffic signals. Intelligent wireless sensors are used in smart city infrastructures to detect traffic flow, predict crowding, and flexibly control traffic routes. This is known as autonomous traffic sensing. This effectively raises awareness, allowing for more systematic use of resources and infrastructure [1]. The initial phase in the traffic management process is identifying and measuring congestion [2]. Density is a measure of traffic congestion that is frequently utilized, and it is typically produced from photographs or videos that vision systems have first acquired [3]. Traffic warning messages are transmitted via smart phones, radio, television, light signals, dynamic variable message signs, or screen displays based on these measurements. Researchers were particularly interested in mobile-based web applications [4, 5]. The majority of recent advancements in providing immediate traffic updates utilized the dynamic control of traffic signals using congestion predictions [6]. For the dynamic handling of traffic signals based on traffic congestion, an IoT-based real-time traffic monitoring system is proposed. The proposed system utilizes a dual set of ultrasonic sensors and consists of two components: one for priority management and another for vehicle monitoring. Ultrasonic sensors are employed to detect vehicle positions, and congestion levels of specific roads are relayed to a display unit and stored on a server for future reference.

7.2.2 Smart Parking System

[7] In IoT-based smart parking systems, the sensors store and retrieve data from remote locations through the cloud, leading to the emergence of the Cloud of Things (CoT). This allows for remote viewing and management of the nodes from any location. With the aid of the mobile application, users from distant locations may reserve parking spaces using the system we offer, which gives information about the availability of parking spaces. [8] A cloud-based parking system's efficiency is increased by using an algorithm and network architecture technologies. The methodology used to identify the cheapest parking spot is described in this algorithm. Taking into account the quantity of parking spaces available and the parking area's distance from the user. The user may immediately contact the cloud-based server to retrieve the parking spot information. To retrieve this information, the user may also download an app on their portable devices. This algorithm can reduce the amount of time a person has to wait to get a parking spot, it does not address security issues [9]. To find the parking place, a smart phone application and a wireless sensor node are employed. This system uses wireless technology, which results in excellent precision and proficiency. In this system, the components are utilized for vehicle-to-vehicle communication. The customer parks their car in one of the many available bays, then a mechanical lift lifts it out. The user is issued a ticket key and an ID that can only be used by them to collect the car. Given an RFID card, the user does not need to carry a paper ticket. This technique is cost-effective. To preserve the user's privacy, security characteristics must be upgraded [10]. The author of the report on smart parking systems has classified car sensors and detector systems into two categories: intrusive sensors and non-intrusive sensors. Tunneling underneath the road allows for the installation of intrusive sensors in holes in the road's surface.

Installation and maintenance of non-intrusive sensors are simple, and they have no impact on the road's surface. Smart parking systems cut emissions from cars while also assisting us in finding solutions to the root causes of traffic congestion [11]. A paper offers a time-saving, effective method of addressing the issue of parking availability in the actual world. In this, the data is delivered locally while being filtered by several devices. The signal is transmitted through the cloud for both the procedure and evaluation involving machine learning algorithms. A mobile phone application is utilized in this study that leverages the Google API to give users access to real-time traffic information. resulting in less traffic congestion. The ability to reserve parking spaces is not offered by this publication [12]. IoT technology for smart parking aids in the design and development of a true smart parking system that informs users of available spots and directs them to the closest ones. To increase security, this research uses computer vision to recognize vehicle license plates. Before getting inside the automobile, the user can utilize mobile payment to pay for the parking spot. ensuring the parking reservation in the process. The user is informed of the parking location, the quantity of open spots, and all other pertinent information. The license plate text extraction methods and algorithms used in this study are efficient. An algorithm determines the user's minimal cost based on the ultrasonic sensor's recognition of the car entering the parking space. [13] A smart parking system based on reservations enables the reserve of an open spot and uses smart parking system based on reservation (SPSR). This entails managing the main parking database, which gathers and stores data on the driver's identification and the exact location of the parking spot. As the parking reservation time approaches its expiration, the user will receive a notification through the web service administered by the admin. The biggest disadvantage is that another user could take up a spot that has been booked; to prevent this, QR scanners are employed to recognize the user [14]. By indicating the destination and the kind of automobile with the aid of mobile devices, it enables us to suggest a method in which the user may book his parking place. By leveraging GPS, the user's location is constantly refreshed in the cloud, while the booking data is stored in the cloud, allowing for the computation of the shortest route from the user to the parking spot. The user's RFID is scanned when they arrive at the parking lot, and they are then granted entry to the space. The billing is done by the cloud server. The primary drawback is that a parking space cannot be used by the user until it has been enrolled in the smart parking system [15]. This article outlines the process of constructing wireless sensor networks (WSN) for a car parking system, utilizing a server powered by XBee ZigBee technology. The vehicle parked in the designated space can be identified by the vehicle parking system. Making this project both affordable and user-friendly is its goal. The user is assisted by a car parking system in maintaining data that is 90% accurate [16]. For both users and parking area administrators, a smart auto parking system offers a complete parking solution. It offers the ability to identify reserved users and offer a feature for a reserved parking space. Based on the size of the automobile, the user might utilize this to locate the nearest parking place. The user has the option to make hourly, daily, weekly, or monthly reservations for parking spaces. Based on size, a formula is developed to identify the closest parking space. Reservations and pay-as-you-go service are made through the user's smartphone application.

7.2.3 Vehicle Theft Detection System

The drawback of mechanical devices is that they may easily be dismantled with other mechanical tools. We examined some of the most cutting-edge alternatives to conventional theft detection systems, including fingerprint detection, which on two-wheelers seems impossible and prohibitively expensive to deploy but records the driver's image and matches it to the owner's recorded image to indicate a breach. In addition to taking up more room, modern technology provides a significant obstacle for two-wheelers. Criminals have created strategies to get beyond the system's present defenses. The development of a system with size, cost, and power restrictions is necessary. The report's [17] section outlines the intention behind controlling and keeping an eye on the fleet. The security unit, which is the main device in the system, includes a current sensor, GSM, relay, GPS, and microcontroller. The current sensor sends an analogue signal to the controller when the car is in motion, and the owner gets a text message as confirmation. In [18], the framework acknowledges the message and transmits it to the whole vehicle network through CAN Bus. Within a short period of time, the owner will get the precise location. The message is received by the GSM modem connected to the microcontroller, and as a result, a flow sensor-based component that disables vehicle ignition is activated, which causes the car to halt. In [19], the researcher described a technique for alerting the user when someone unauthorized tries to leave the vehicle, turn off the ignition, or deactivate the device. [20] made the recommendation for a GPS and GSM-enabled car security system. Through this substructure, users may connect to automobiles, utilize Google Earth to find out where they are right now and how they are doing, and follow the whereabouts of alerted automobiles [21]. Uses a microcontroller and GSM to operate the motors of the cars. For the car to start, the announced secret phrase needs to be figured out. The ignition of the car will start precisely at the moment the secret word coordinates. Whenever the secret key fails to match within three attempts, the system will activate the siren and transmit a notification to the owner through the GSM system.

7.3 Proposed Model of Traffic Management System

Nowadays there is no efficient automated traffic management system available. The timer based traffic lighting system is very inefficient as there may be a situation when the green light is shown to a lane which is not having any traffic on the other hand red light may be shown to a lane which has a lot of traffic. The vehicle parking system which is used these days is totally operated manually and this makes the parking system very expensive and there may be problems in the system due to human error. There is also no theft detection system installed in the vehicles.

7.3.1 Traffic Lighting System

It has a cloud-based central server and some of its components are installed by the side of the road to make the system communication model. The setup at the roadside consists of sensors, a CPU, a database, and a controller. At each junction, a unit is placed up. The main server has interfaces, cloud services, and data storage. Wi-Fi enables the components to communicate with one another. A sensor layer, processing layer, network layer, and application layer are the four main layers of an IoT-based system framework. Data is acquired from objects by the sensing layer, processed and analyzed by the processing layer, then the processed information is sent to the cloud server and from the cloud to mobile devices by the network layer, and then displayed by the application layer as the user interface (mobile device). In Figure 7.2, this layered design is shown.

The camera is the system's initial and most important component. While the camera data is transferred to the database for training the module for subsequent prediction, the cameras interact with the actual world, including the presence or absence of cars. In accordance with the presence of cars nearby, the cameras send status. Additionally, it features magnetic detectors that determine the portion of the road that each car occupies in order to display the traffic density. The processor (a Raspberry Pi), which analyzes the data and sends it to the controller, which receives data from the camera at specified intervals. Through Wi-Fi communication, the controller

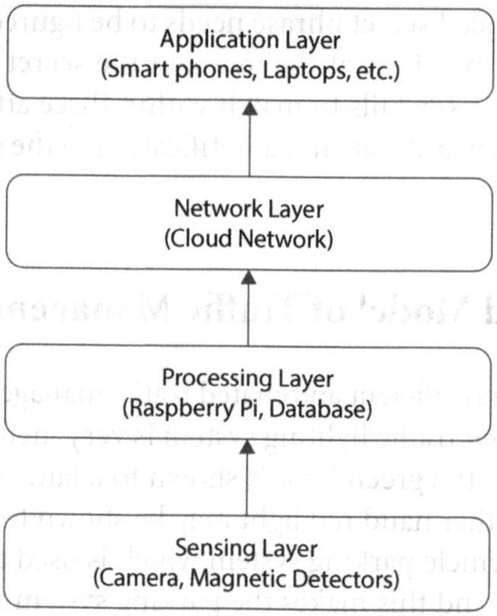

Figure 7.2 Layered architecture of traffic lighting system.

receives the calculated data from the Raspberry Pi. The controller uses the gathered information to carry out the intelligent traffic routing. The main goal of this system is to track moving vehicles and provide them with a clear path to their destinations. Traffic signals should automatically adjust to offer this clear path for these vehicles. The cameras are present and the traffic lights in this planned system are LEDs. Using physical wire, both blocks are linked to a raspberry pi. After receiving the acquired camera data, the Node MCU switches between green, yellow, and red to operate the traffic lights. Based on the distances detected by the camera and the intervals between those readings, the raspberry pi calculates the number of cars on the street of the intersection it is responsible for. The raspberry pi then updates the database every minute with the number of automobiles. The database is used to train the module in order to improve its ability to forecast changes in traffic signal timing and density. Wi-Fi is used in this connection. More specifically, the cloud server employs an equation to calculate the time gap between LEDs required for a smooth traffic flow based on the data received (number of automobiles). The current LEDs' real time is then contrasted with this estimated time (this data is saved in a database). A decision is made by the processor. The decision is to raise the green time, or reduce the green time, if the present real green time is smaller than the calculated time.

Figure 7.3 represents the diagrammatic representation of a single unit of a traffic lighting system. In each unit there is are sensors (Camera,

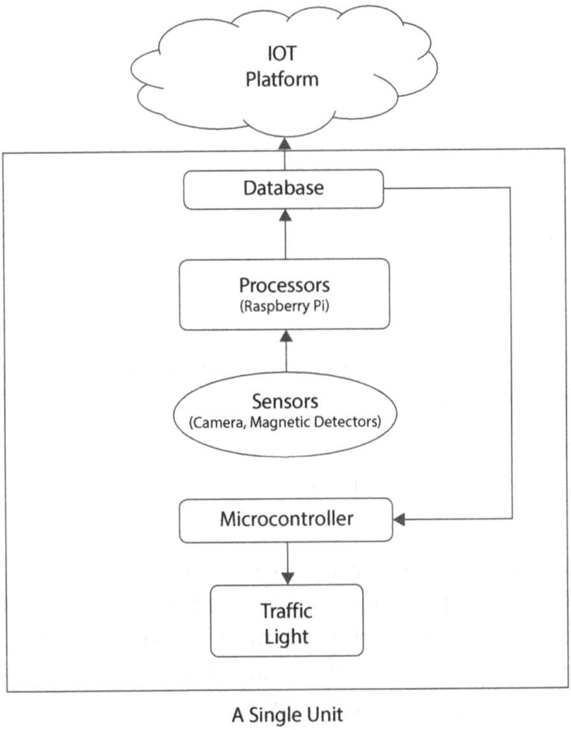

Figure 7.3 Representation of a single unit of traffic lighting system.

Magnetic detectors) which takes pictures of the traffic and sends them to the raspberry-pi for the purpose of processing the images and the obtained is sent to the database present in each unit. The database also analyzes the data and sends instructions to the microcontroller which activates the traffic lights. Each unit's database sends the analyzed data to the cloud. These units are placed in an intersection which it monitors. The immediate data from the database is uploaded to the cloud server where the data is interpreted and converted to map format and distributed to mobile devices like smartphones, laptops to show the traffic congestion of different intersections in the city.

In Figure 7.4 the roadside setup of the traffic lighting system is represented. Each of the units are set at each intersection of the roads and each unit is connected to an IOT cloud and from the cloud the mobile devices of the user. The status of the traffic situation is sent from the cloud platform to the mobile devices. The connection between each unit with the cloud and from the cloud to the mobile devices is completely wireless and achieved with the help of the Internet.

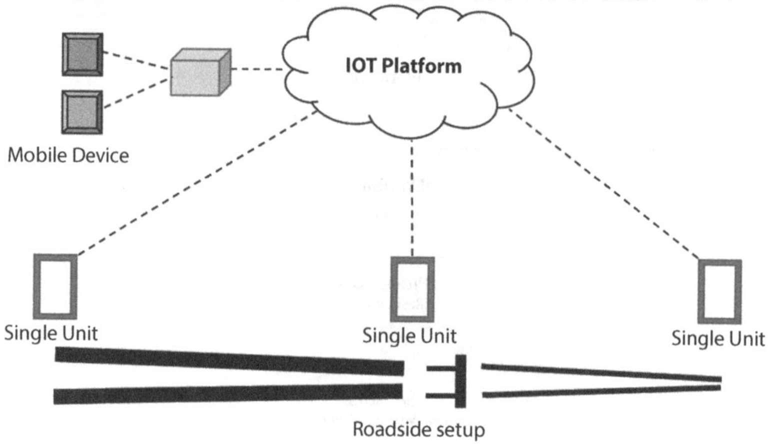

Figure 7.4 Roadside setup of traffic lighting system.

The advantages of the model are as follows. The proposed model helps to regulate the traffic properly. The timing based traffic lighting system which is used in the Indian roads nowadays are solely based on giving the same amount of green light to each and every lane present in the intersection which means there may be a situation when there is red light shown to a lane which is full of vehicles whereas green light is shown to a lane where there are no vehicles. This leads to traffic congestion. But the proposed model does not depend on timing but it gives green light to the lane where it is needed the most. The proposed model can also show the traffic density of a particular area through Google maps which helps the users to find a proper path to their destination in order to avoid traffic congestion. This can be viewed on multiple handheld devices like smartphones or laptops. Despite having many advantages the model has a few disadvantages. The proposed model employs Wi-Fi to connect devices, however this study does not take into account how much energy these devices require or how to recharge them. These kinds of devices may consume a lot of energy which may cause problems as the maintenance cost would rise. Alternative options like solar power or recharging from street lights can be investigated, although they might not be very effective. Only the single-lane road will allow the suggested concept to function as envisioned. However, in order to identify the erroneous detection, it can be helpful to test the model in a multi-lane situation.

7.3.2 Smart Parking System

The proposed model of the smart parking system consists of three portions:

- The first portion consists of the parking lot, which incorporates Arduino gadgets and an IR sensor. These gadgets enable the user to interact with the parking area. The user cannot access the parking area unless they have an RFID card.
- The cloud-related online services that serve as a mediator between the customer and the parking area are included in the second section. Depending on whether a parking space is available, the cloud is updated. It assigns a free parking space to the user (if available). The administrator is in charge of the cloud services, which is also accessible by the user to verify the availability.
- The user's side is the third portion. The user receives a notification via the GSM module based on the availability of space in the parking lot.

In Figure 7.5 the workflow of the smart parking system is shown. Each person using the parking space has an RFID card with their personal information on it. Reader module scans the user's RFID tag when

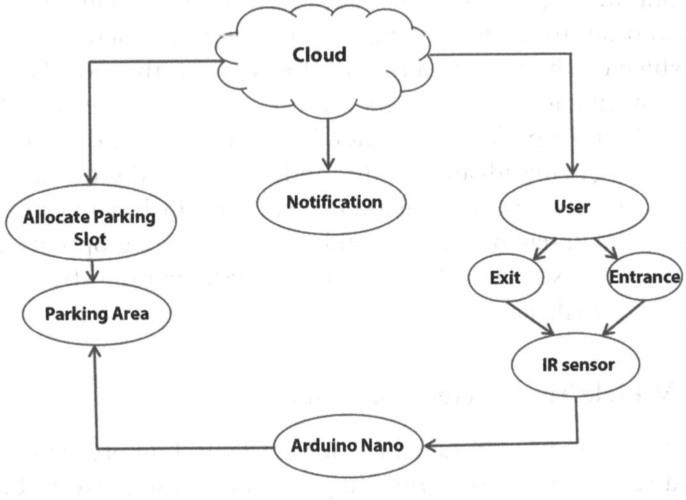

Figure 7.5 Flowchart of smart management system.

the car goes into the parking lot. The information is transmitted to the Arduino so that it may concurrently verify the parking lot's availability. The parking barrier gate won't open if there are no spaces available. Along with the aid of the GSM module, a message is delivered to the user in response to the accessibility and inaccessibility of the parking place. Depending on whether a vehicle is there, the IR sensor delivers signals. In order to support the system, the WIFI module stores all of the data in the cloud. The devices are linked to the cloud server by it. An RFID card is scanned by the user that has been given to them here. The user will see the word "Welcome" and the barrier gate will open if there is room for them to park their car. The RFID must be scanned once more by the user before leaving the parking area, and the word "Thanks" is shown. The cloud database will keep track of the user's actions in the parking space. With the support of the mobile applications, the user will be able to decide if a certain area is accessible using the cloud status. The cloud status changes from 0 to 1 when the car is parked and from 1 to 0 when it is driven away due to an IR sensor that identifies the occupancy of a car. The user can then leave their vehicle parked where the cloud status is 0. Every two minutes, the cloud status is refreshed. The advantages of the smart parking of the are given below. The proposed model reduces a lot of manpower which is needed in the current parking system which are operated manually. The parking system which is used currently requires a manager and a number of staff members in order to manage the parking lot. But the proposed system requires almost no human interference; it helps in order to reduce the maintenance cost and there won't be any errors which can be accidentally made by humans. The model proposed is very convenient because it lets the users know beforehand if there are slots left in the parking lot or not. This saves a lot of the user's time. Despite having many advantages the model has a few disadvantages. The proposed model may consume a lot of energy which may cause problems as the maintenance cost would rise. Alternative options can be explored further, such as solar charging or charging from street lighting but may not be efficient.

7.3.3 Vehicle Theft Detection System

The proposed model is installed in the vehicle itself in order to monitor theft and report to the user. The setup includes sensors like PIR sensor, Gas sensor and pressure sensor, camera to capture images, GPS module to keep the user informed about the location of the vehicle, a buzzer for alerting and a processor (Raspberry Pi). When the theft detection system is turned off, the system may initially operate in user mode. Theft detection can be activated after the car has been parked. The first and primary elements of this system are the sensors (PIR, Gas sensors, pressure sensors) they constantly monitor for any abnormal changes. If any of the sensors gives a HIGH output then the other modules are activated. The camera module takes an image of the person driving the vehicle and the GPS module tracks down the present latitude and longitude of the vehicle along with the date and time of the system. If the vehicle is displaced by 120m from the initial location then the revised coordinates of the vehicle are saved by the GPS module. The camera image as well as the link to the current location of the vehicle is sent to the user through the registered Email. Through the link the user can track the live location of the vehicle in Google maps. The user can switch back to user mode once they get to the vehicle.

Figure 7.6 shows the architecture of the vehicle theft detection system. As mentioned earlier when the system is enabled then if abnormal activity is detected by the sensors (Gas Sensor, PIR Motion Sensor, Piezo Sensor) then the camera module takes an image of the intruder (the person sitting on the driver's seat) and the GPS module tracks the current location of the car and sends the location and the picture of the thief to the user through Email with the help of Wi-Fi Dongle. If the vehicle is displaced more than 120 m then the user is sent the location of the vehicle again. The system is powered by a battery which is charged when the vehicle is running. There is also a buzzer present in the system which makes a sound to alert the passersby about the theft.

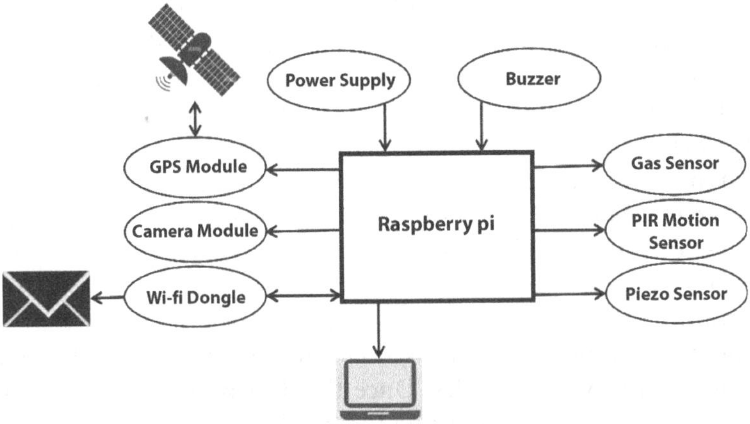

Figure 7.6 Vehicle theft detection system using IoT.

The advantages of the vehicle theft detection system are as follows. The proposed model helps in detection and tracking of vehicles that have been stolen. Nowadays there is no smart theft detection system available in vehicles which makes it difficult if there is theft and people have to totally depend on the police for recovery of the vehicle. But as proposed in the model, it can detect if the vehicle has been stolen and is capable of sending the live location of the vehicle to the user through the registered Email.

Despite having many advantages the model has a few disadvantages. The proposed model works well in the places with good internet connectivity in the cities and in most of the towns. But the theft detection may not work in the places with any connectivity and in the year 2022, 9.31% of villages in India are not equipped with network connectivity. The model is not optimized so it may consume much of the vehicle's fuel and can be inefficient.

7.4 Role of AI in Traffic Management

Intelligent Traffic Management Systems, which have been revolutionized by the rapid development of artificial intelligence (AI), are used to manage traffic on roads. It is an applied field of computer vision that focuses on the classification of vehicles, the detection of traffic violations, and the study of traffic flow. By keeping an eye out for things like the space between two moving cars and people at crossroads, you may frequently assist to lessen the amount of congestion. It combines IoT and AI to assist traffic flow more smoothly in order to increase mobility, lessen pollution, and increase life expectancy. AI has revolutionized the prediction and control of the movement of people, objects, vehicles, and cargo across different points on the transportation network. Its remarkable accuracy enables the prevention of accidents through optimized flows at intersections and improved safety when roads are closed for construction or other purposes. Moreover, citizens now experience unparalleled service thanks to AI, which also facilitates the efficient operation of mass transit options such as ride-sharing services. This is achievable due to AI's remarkable ability to analyze and evaluate vast amounts of data. Road traffic management utilizes AI to evaluate real-time data from various modes of transportation such as vehicles, buses, and trains. The AI analyzes the data to identify trends that may indicate security vulnerabilities. Once the data is collected, recommendations are provided on mitigating these risks and reducing the number of accidents. Numerous traffic operations can be greatly enhanced. There are many advantages of using AI to keep traffic moving in response to the present situation:

- The environment benefits greatly from the smooth flow of traffic and the absence of congestion. Furthermore, achieving this requires a focus on ongoing software development rather than relying on hardware, which adds another important environmental aspect.
- It makes it possible to standardize a variety of corporate operations, including deliveries, which is very advantageous for the economy.
- Minimizing human error, which is the most common cause of accidents, can be achieved by effectively controlling the flow of traffic. The number of accidents might be significantly decreased by removing the human factor.
- The concept of "Truck Platooning" involves electronically connecting multiple trucks to form a convoy on the highway. This presents intriguing opportunities in the transportation industry, where only the first truck in the convoy is operated by a human driver. All other truck's controls are taken over by AI.

All of these elements help to optimize the transportation system as a whole. Using AI for traffic control presents a number of challenges.

- The gathering of data and comprehension of the problem under consideration.
- Data analysis involves extracting features for predictive modeling.
- The model is deployed, monitored, and updated.
- Feedback is analyzed, and lessons are learned from mistakes.
- Handling ambiguity and noise in the data.
- Integrating several types of data (e.g., video, image, GPS)
- Scalability – can the system support more users as cities grow in size??
- Privacy concerns – how will personal data be utilized or distributed?
- Cost-effectiveness – Is it possible to employ AI without making substantial investments in brand-new infrastructure?

7.5 Conclusion and Future Works

The primary objective of this synthesis was to identify and synthesize practices and devices for establishing an efficient traffic control system. This research suggests a model for an Internet of Things-based system to collect, evaluate, and store real-time traffic data. Through roadside messaging devices, this research enabled monitoring of traffic in real-time for updates. In order to assist the public in decision-making and reduce travel time, the traffic authorities utilize appropriate messaging units to broadcast alerts regarding VIP visits, medical situations, accidents, and other incidents. Real-time data is processed by Wi-Fi equipped microcontrollers, which then transmit it to an IoT platform for further processing. Residents receive priority alerts from traffic management, effectively mitigating traffic congestion resulting from unexpected circumstances. A smart traffic management system is deployed to address congestion and implement rerouting at road junctions. This research offers a practical solution to the challenge of increasing traffic flow, particularly in rapidly expanding cities where conventional traffic management methods exhibit significant limitations. By dynamically adjusting signal timing based on traffic volume at specific intersections, the system adeptly controls traffic flow. Moreover, the decentralized approach ensures the system's efficiency and effectiveness, even in the event of a local or centralized server failure. Additionally, the information that can be used to improve road design is made available to higher authorities by technology, assisting in the effective use of resources. The suggested approach might be further enhanced by taking various factors into account for future developments. On the basis of real-time data, the first dimension suggests an ideal path for the drivers. Future work will also incorporate capability for dynamic traffic signal control. In this instance, it is necessary to create traffic signal communications. The development of the prototype into a comprehensive end-to-end system with server communication as the central component is imperative. Additionally, the real-time implementation of the system involves integrating the elements of IoT security within the communication layer. Future research will also explore the inter-display and IoT security integration connectivity, and traffic signal technology. New possibilities for smart cities have emerged due to advancements in IoT and cloud technology. Intelligent parking facilities have traditionally served as the cornerstone for the development of smart cities. These systems offer real-time data on available parking spaces, enhancing efficiency and saving users' time when searching for suitable parking. They also contribute to tackling the escalating problem of traffic congestion. In the future, users will have the ability to remotely reserve parking spaces well in advance. Furthermore, potential future enhancements may include the incorporation of GPS, reservation capabilities, and license plate readers. The Raspberry Pi is used to build a device that is small, efficient, and reasonably priced for detecting car theft. The technique can eventually be used to create a stable product. By raising the component costs, we may increase accuracy. In order for them to get to that area quickly, we may provide the phone numbers of the police stations and ambulance services. As for the future, an application can be developed that can construct a voice-based alert for the user. The complete system may be connected with an Android app, and for the convenience of users, the relevant details might be recorded in a database.

References

1. Lomax, T.J., Quantifying congestion (issue 398), Transportation Research Board, 1, 1–112, 1997.
2. Rao, A.M. and Rao, K.R., Measuring urban traffic congestion e a review. *Int. J. Traffic Transport. Eng.*, 2, 4, 286–305, 2012.
3. Dubey, A., Akshdeep, Rane, S., Implementation of an intelligent traffic control system and real time traffic statistics broadcasting, in: *Proceedings of the International Conference on Electronics, Communication and Aerospace Technology*, 2017, ICECA, 2, pp. 33–37, January 2017.
4. Atta, A., Abbas, S., Khan, M.A., Ahmed, G., Farooq, U., An adaptive approach: Smart traffic congestion control system. *J. King Saud Univ. - Comput. Inf. Sci.*, 32, 1012–1019, 2018.
5. Islam, M.R., Shahid, N.I., Ul Karim, D.T., Al Mamun, A., Rhaman, M.K., An efficient algorithm for detecting traffic congestion and a framework for smart traffic control system, in: *International Conference on Advanced Communication Technology, ICACT*, 18, pp. 802–807, 2016.
6. Nagmode, V.S., An intelligent framework for vehicle traffic monitoring system using IoT, 7–10, 2017.
7. Abhirup Khanna, R.A., IoT based smart parking system. *International Conference on Internet of Things and Applications (IOTA)*, IEEE, Pune, 1, p. 5, 2016.
8. Deng, D., A cloud-based smart-parking system based on Internet-of-Things Technologies. *IEEE*, 11, 2015.
9. Orrie, O. and S., B., A wireless smart parking system. *IECON*, IEEE, Yokohama, 41, p. 5, 2015.
10. Khaoula Hassoune, W.D., Smart parking systems: A survey, *IEEE*, p. 6, 2016.
11. Wael Alsafery, B.A., Smart car parking system solution for the Internet of Things in smart cities, *IEEE*, 1, p. 5, 2018.

12. Rachapol Lookmuang, K.N., Smart parking using IoT technology, *IEEE*, 5, p. 6, 2018.
13. Mohit Patil, R.S., Smart parking system based on reservation. *Int. J. Sci. Eng. Res. (IJSER)*, 6, 2014.
14. Vishwanath Y, A.D., Survey paper on Smart Parking System based on Internet of Things. *Int. J. Recent Trends Eng. Res. (IJRTER)*, 5, 2016.
15. Dr.V. Kepuska, H.A., Smart car parking system. *Int. J. Sci. Technol.*, 7, 2016.
16. Cynthia, J. and B., C., IOT based smart parking management system. *Int. J. Recent Technol. Eng. (IJRTE)*, 6, 2018.
17. Ramadan, M.N., Al-Khedher, M.A., Al-Kheder, S.A., Intelligent anti-theft and tracking system for automobiles. *IJMLC*, 2, 1, February 2012.
18. Siyal, K. and Gugapriya, G., Anti-theft vehicle locking system using CAN. *IJST*, 9, December 2016.
19. Zhou, P., Zheng, Y., Li, M., How long to wait? Predicting bus arrival time with mobile phone based participatory sensing. *IEEE Trans. Mob. Comput.*, 13, 6, 1228–1241, June 2014.
20. Jiang, D., Taliwal, V., Meier, A., Holfelder, W., Design of 5.9 GHz DSRC-based vehicular safety communication. *IEEE Wirel. Commun. Mag.*, 8, 107399–107408, October 2006.
21. Nagaraja, B.G., Rayappa, R., Mahesh, M., Patil, C.M., Manjunath, T.C., Design & development of a GSM based vehicle theft control system, *IEEE*, pp. 148–152, 2008, 978-0-7695- 3516 6/08, DOI 10.1109/ICACC.2009.154.

8

Visualization of Textual Corpora Using Social Network Analysis

Indu Rodda* and Durga Bhavani S.

University of Hyderabad, Hyderabad, Telangana, India

Abstract

As more people lean toward movies and television shows, the practice of reading novels is gradually dwindling. This study uses social network analysis to make novel visualization more enjoyable. A character network is generated using the book's main characters as nodes and the interactions between them as edges. Cumulative sub-graphs of the chapters in the novel are used to generate network snapshots. The main goal of the visualization is to depict how the main characters' interactions change as the plot progresses by applying the community discovery for dynamic networks methodology. Additionally, using a similarity score, a mapping algorithm is developed to map the communities between two successive snapshots. This algorithm recognizes nearly invariant communities, community mergers, and the emergence of new communities. Additionally, the communities are given creative color assignments, so the viewer can easily detect community changes. The implementation is carried out for one of J.K. Rowling's well-known novels, and the outcomes are displayed for chapters highlighting essential plot points. The implementation's results demonstrate that the color scheme and similarity score both successfully depict the dynamics of community transitions.

***Keywords*:** Dynamic networks, community discovery, similarity measure, visualization

**Corresponding author*: roddaindu@gmail.com

Hrudaya Kumar Tripathy, Sushruta Mishra, Minakhi Rout, S. Balamurugan and Samaresh Mishra (eds.) Optimized Computational Intelligence Driven Decision-Making: Theory, Application and Challenges, (151–172)

8.1 Introduction

Today, many of us would rather watch a vivid screen than read a book. Even though viewing movies is enjoyable, reading a novel is more enjoyable when the reader understands the flow of the tale. It may aid the reluctant reader by providing a book summary or a visual representation of the text. As a result, graphic novels are becoming increasingly popular. This work proposes visualization techniques based on social network analysis to increase interest in a novel.

There are numerous ways to visualize the text. It is possible to do so using graphs, charts, or word clouds. Text visualization provides a fast overview of the text, essential insights, hidden trends, and patterns. The graphical depiction of the text is the primary emphasis of this paper.

8.1.1 Importance of Character Networks

Characters are important in any work. Characters can be found in fiction, biographies, the news, and professional meetings. The analysis of character networks reveals a lot about them. Regardless of genre, the plot of the novel revolves around characters. In movies, there are a few main characters around which the plot revolves, but secondary characters who have an impact on the plot are often overlooked. A plot's character network, on the other hand, can capture such things.

Static networks, as shown in Figure 8.1, do not evolve over time. These networks' nodes and edges remain constant throughout their existence. Dynamic networks are those that change over time. Furthermore, there are two types of dynamic networks: offline dynamic networks and online dynamic networks [1]. All network states, including future states, are available in offline dynamic networks, whereas only the past and current states

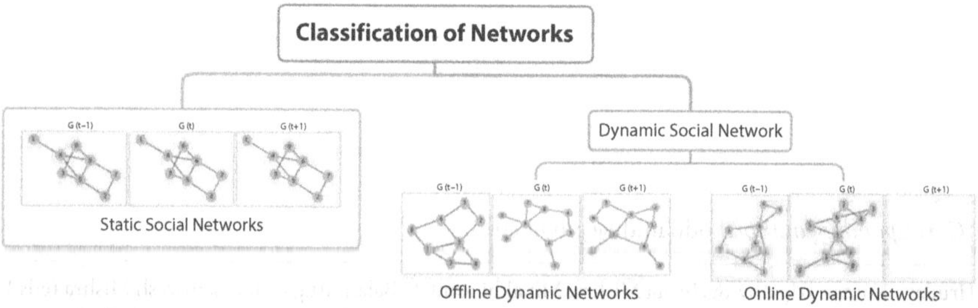

Figure 8.1 Classification of networks.

are available in online dynamic networks. Offline dynamic networks make more sense for character networks than static networks.

8.1.2 Visualization of Dynamic Networks

There are two ways of representing Dynamic Networks:

1. Temporal Network
2. Snapshot Network

Temporal Network Temporal Network is defined as $G = (V,E,T)$ where V is list of (v,t_s,t_e) and E is list of (u,v,t_s,t_e) and T is set of all t_s and t_e where t_s is entry time of that node or edge to the network and t_e is exit time of that node or edge from the network.

Snapshot Network Snapshot Network is defined as G with $G = g_1,g_2,\dots g_t$ where each snapshot $g_i = (V_i,E_i)$. In this paper, snapshot networks are used to visualize the text.

One of the most important aspects of visualizing dynamic networks is observing network growth. This expansion can be visualized using centrality measures, community detection, k-core, and other metrics. In this paper, a dynamic character network is visualized in terms of community change over the course of a novel or book.

8.1.3 Contributions

- The significance of the characters in a novel keeps changing as the story progresses. We propose a visualization algorithm that captures the changes in the interactions among the characters in the novel. Based on the events in the story, characters tend to change groups. Community Discovery (CD) is applied to each chapter's network to see how characters form communities among themselves.
- Once we have independent communities of snapshots, a mapping algorithm is proposed to identify the flow of interaction among the groups across the chapters by mapping the changes between consecutive time stamps through operations of merge, split, growth, etc.
- Further, we propose a scheme for coloring the groups of characters across the different time stamps such that the movement of a character across groups is clear visually. The novelty of the approach is that it maintains the consistency of the colors of communities across time stamps so that the reader can easily observe the changes.

8.2 Related Literature

As discussed earlier, in fiction novels, stories generally revolve around the main characters. So there are different tools in Social Network Analysis(SNA) [2, 3] to help us visualize the character networks. In 2020, Duval *et al.* [4] explored the novels "Harry Potter and the Philosopher's Stone" and "Lord of the Rings" using centrality measures and community discovery on a static character network.

8.2.1 Visualization of Social Networks

Social networks can be represented in many ways; broadly, there are three kinds of visualization, as shown in Figure 8.2.

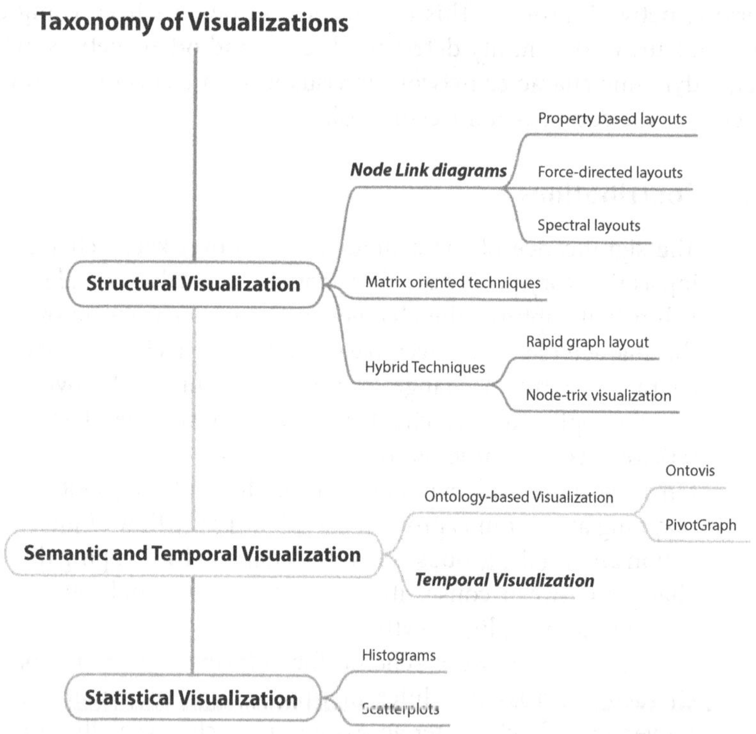

Figure 8.2 Visualizing a network.

This paper is focused on the **Structural Visualization** of the character network. Hence, a Node link diagram with Property Based Layout [5] is used to represent the network.

8.2.2 Community Discovery (CD)

Social networks have a modular structure, and to understand the network, it is essential to analyze the communities of the network. On the static entity graph, Duval *et al.* [4] used the CD algorithm proposed by Girvan and Newman [6]. By applying CD to the entire graph, the groups formed by the novel can be seen, but the changes that happen throughout the novel cannot be visualized. So for these changes to be visualized, the CD should be performed on a dynamic entity network.

8.2.3 Community Discovery in Dynamic Networks

The visualization of dynamic networks has always been an area of interest among researchers. Applying CD algorithms to the same network again may not give the same community representation. For this reason, when a change in community representation occurs in a dynamic network, it is difficult to know if that change is because of changes in the network or the algorithm itself. Many solutions have been proposed to solve this problem. Mainly, there are three approaches, namely

- Independent Community Detection and Matching (ICDM).
- Informed Iterative Community Detection (IICD).
- Global Community Detection on all Snapshots (GCDS).

Independent Community Detection and Matching In this approach [7], a CD algorithm (any algorithm of the users' choice) is applied to individual snapshots of the graph and then, using a matching measure, communities are mapped. This kind of approach has its advantages and disadvantages. One of the main advantages of this approach is that each snapshot is taken as a static network, and existing CD algorithms are applied. Then mapping is done between consecutive snapshots.

This kind of approach's disadvantage is that running CD on the same network may not provide the same results. So therefore, it is not possible to know if the change in community structure between snapshots is because of changes in the network or the CD algorithm.

Informed Iterative Community Detection The communities at g_t are detected using the community information of g_{t-1}. In this approach, [8] there will not be a need for mapping, but the traditional CD algorithms should be modified before use.

Global Community Detection on all Snapshots In this approach, [9], community detection is applied to the whole network, and communities of snapshots are taken from this. Here, it is difficult to see the merge or split operations of the communities.

So, for this paper, we use the **ICDM** technique for visualization and analysis. Louvain Community Detection (LCD) [10] is applied to individual snapshots, and a matching algorithm and color assignment algorithm are proposed to map communities of consecutive timestamps.

8.3 Proposed Method

This work presented here is done as part of the M.Tech project work[1]. The characters of the novel are used to analyze the content of the book. For each chapter, a character graph network is created and analyzed. The novel's insights are extracted through community discovery. The high-degree nodes serve as the book's major characters. Based on the communities formed and the movement of the nodes between the communities in the network, the novel is explained.

8.3.1 Basic Idea of Algorithm

The concept of matching communities of snapshots taken at t and $t + 1$ is the foundation for the suggested algorithm. A specific text or novel is regarded in this as a dynamic network. At the end of each chapter, network snapshots are taken, and each snapshot is subjected to independent community detection [11]. Then communities of g_t and g_{t+1} are matched until the end of the text. The assignment of color to the communities as they undergo a transition as the story progresses is another essential aspect of the algorithm.

8.3.2 Life Cycle of Dynamic Communities

Every community goes through the stages mentioned below. It is called the life cycle of the community [12]. The stages that a community may go through are shown in Figure 8.3.

---

[1] Rodda Indu, Visualization of Textual Corpora using Social Network Analysis, Submitted to University of Hyderabad, July 2022

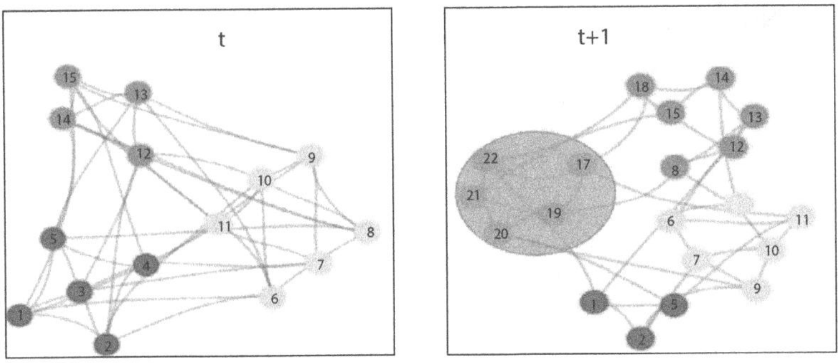

Figure 8.3 Snapshots of a graph to explain the community life cycle.

- **Birth:** New communities take birth: this happens when new nodes form a community or a very small percentage of existing nodes from different communities merge. The **blue** community was born in the Figure 8.3 with all the new nodes joining the network.
- **Growth:** For the existing community, new nodes get added without disrupting the properties of the existing community. The **green** community is growing in the Figure 8.3, with one existing node and a few new nodes contributing to the network's expansion.
- **Merging:** A community is formed by more than one existing community.
- **Splitting:** The community is split into more than one community, and they satisfy the threshold value, which makes it different from contraction.
- **Contraction:** Some of the nodes from the existing community move out or get deleted from the network. In Figure 8.3 nodes from **red** community got deleted, which led to the contraction of the community.
- **Death:** A community is said to be dead if most of the nodes get deleted from the network or the nodes disperse among other communities.

Keeping in view the community life cycle, an algorithm is proposed to visualize a novel in an effective and colorful way.

8.3.3 Notation

Notation	Explanation
T	Number of snapshots or chapters
N	Number of communities by the end of the novel
C_t	Set of communities at t
N_t	Number of communities at t
G	Graph of the novel
g_t	Snapshot of the Greatest Connected Component of the subgraph of novel until time t
n	Total number of nodes in the network
n_t	Count of nodes in g_t
node-list	Node list with attributes node id and node label
edge-list$_t$	g_t edge list with source, target, and weight attributes
K	Maximum number of communities formed among all the snapshots
COLHSV	Color Dictionary in HSV format
SC(A,B)	Score of correspondence between communities A and B where $A \in C_t$ and $B \in C_{t+1}$

8.3.4 Algorithm

The algorithm 1 consists of 5 stages, namely pre-processing text, graph generation, community discovery on snapshots, community mapping and visualization. Below is a brief description of each step, followed by the implementation.

Algorithm 1 Visualization of a text(Novel)

Input: Text file
Output: Visualization of text
1: node-list = node-gen(novelname)
2: T =chapterize(novelname) {Returns number of chapters the text is divided into}
3: **for** t $\in$ T **do**
4: edge-list$_t$ = edge-list($_{t-1}$ $\cup$ pre-process(t.txt)
5: N_t = community-disc(g_t)
6: **end for**
7: K = max(N_t)
8: Generate COLHSV having K colors.
9: **for** $g_t \in G$ **do**
10: Match communities of consecutive timestamps i.e., between g_t and g_{t+1}
11: **end for**
12: Draw graph with nodes grouping by community Id, Color for nodes by community id.

Pre-Processing and Graph Generation The given text is pre-processed to be divided into chapters (snapshots), and then network generation is done with the novel's characters as nodes. An edge between two nodes exists when there is an interaction between the characters in the story. The weight of the edges represents the number of times characters interact.

Community Discovery CD is applied to each snapshot of the graph. There are many community detection algorithms available [13], namely Louvain, Label Propagation, Leiden Community Detection, Walktrap Community Detection, etc. The fast Louvain community detection technique [10] is applied to individual graph snapshots in this work. Each snapshot is taken as a static network, and community detection is applied.

Score - Similarity Measure Using Formula 1, the score value for the ith community of g_t and the jth community of g_{t+1} is calculated after applying community detection to snapshots of G.

$$SC(A,B) = \frac{|A \cap B|}{|A|} \tag{8.1}$$

Algorithm 2 Score Matrix of $g, g_{t, t+1}$

Input: List of Community id's of g_t and g_{t+1}

Output: $SC_{N_t \times N_{t+1}}$

1: **for** each pair of communities in g_t and g_{t+1} **do**

2: Calculate pair wise Score values using formula 1

3: Generate $SC_{N_t \times N_{t+1}}$

4: **end for**

If algorithm 2 generates a matrix as shown below, it can be seen that there are three communities at t and four communities at $t + 1$.

0.08 0.41 0.41 0.08

0.00 0.60 0.20 0.20

0.81 0.06 0.10 0.00

If a given network has 12, 10, and 16 nodes in three communities of t and 14, 16, 10, and 6 nodes in four communities of $t + 1$, and the community structures change as shown in Figure 8.4, the above score matrix results.

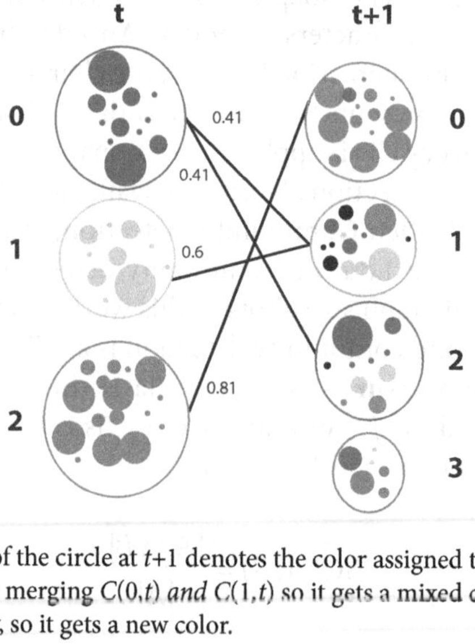

Figure 8.4 The color of the circle at t+1 denotes the color assigned to the community. $C(1,t + 1)$ is formed by merging $C(0,t)$ *and* $C(1,t)$ so it gets a mixed color, whereas $C(3,t + 1)$ is a new community, so it gets a new color.

Algorithm 3 Mapping communities of g_t with communities of g_{t+1}

Input: $SC_{N_1 \times N_2}$

Output: Updated colors for communities at t+1

temp $\leftarrow N_2$
for j = 0 to temp **do**
list = {}
    for i = 0 to N_1 **do**
    if $SC(i,j) > \beta$ **then**
    flag = 1
    $list \leftarrow list \cup \{i\}$
    end if
    end for
    if $|list| > 1$ **then**
    sum color(*list,j*)
    else if $|list| = 1$ **then**
    COLHSV[j] $\leftarrow$ COLHSV[i]
    end if
    if flag = 0 **then**
    COLHSV[j] $\leftarrow$ COLHSV[N2+1]
end if
end for

Mapping Communities The color of a node in a subgraph g_t is determined by the community to which it belongs. Each community will have a unique color associated with it. To determine the color of the communities at time stamp t+1, the algorithm 3 is used. When a community is formed by merging the majority nodes of more than one community, then a new color is given by the function in algorithm 4. If a community is formed by majority nodes of one community, then the same community color is given to the community at $t + 1$. Another possibility is that a community is formed by minority nodes of t communities or by new nodes, in which case the "j" community is given a new color.

There are three cases in the process of mapping communities. Throughout the process, "i" is used to index t communities and "j" for $t + 1$ communities.

- **Case 1:** The majority of nodes in $C(i,t_1)$ are migrating to $C(j,t_2)$.
- **Case 2:** Nodes of $C(i,t_1)$ scattered among communities of t_2.
- **Case 3:** $C(j,t_2)$ is formed by combining more than one t_1 community that meets the *beta* threshold.

Because $C(0, t{+}1)$ is formed by the majority of $C(2,t)$, in Figure 8.4 the color of $C(0,t + 1)$ is the same as the color of $C(2,t)$. Whereas $C(1,t + 1)$ is formed by merging of $C(0,t)$ and $C(1,t)$ color is determined by the algorithm 4. $C(3,t{+}1)$ is composed of new nodes and t nodes that are less similar to *beta*. Hence, a new color is assigned to this community. Any color given to a community is always unique. If a new community is formed, only primary colors (not mixed colors) are assigned to it; otherwise, a color is assigned based on t. The new color of the community formed by merging more than one $t{+}1$ community is calculated by adding the products of the scores and hues of the involved t communities, as specified in the algorithm 4.

Algorithm 4 New color sum color(list,j)

Input: list, j

Output: New color for community

1: **while** list$\neq \varnothing$ **do**

2: $\quad sum \leftarrow sum + COLHSV(com\_id) * SC(com\_id, j)$

3: $\quad list \leftarrow list \setminus \{com\_id\}$

4: **end while**

5: COLHSV(j) $\leftarrow$ sum

Visualizing Dynamic Network The primary goal of this paper is to depict how characters in the novel form communities and how those communities change throughout the novel. The color of nodes in a community is generated using the proposed algorithm 4.

After community mapping is completed, colors are assigned to all communities in a given snapshot. The network's visualization is critical for algorithm representation and implementation. To visualize the network, a node link diagram is used, and [5] is used with the community id as a property to group nodes in a community. More information is provided in the following section.

8.4 Implementation and Results

Social Network Analysis(SNA) is a new area of Computational Intelligence where the data is modeled as a graph. In this paper, a network of character interactions in a novel is modeled as a graph to extract interesting insights. Using SNA tools, the overall novel is visualized. This work aims to summarize a novel colorfully and visually explain the behavior of characters throughout the novel.

J.K. Rowling's Harry Potter and the Prisoner of Azkaban [14] is used to test the proposed algorithm. This is the third book in the Harry Potter series, with 317 pages and 22 chapters in the first edition. The implementation is explained in two sets of snapshots. One set contains Chapters 1, 2, and 3; the other contains Chapters 17, 18, and 19.

8.4.1 Pre-Processing the Data

The dataset is generated from the novel, and firstly, character names are extracted from the text. An open software library called SpaCy [15] is used to extract the character names. Each character entity will have a first name, and some entities will have a last name also. Once we have a list of characters, it is further modified for ease of use.

Before generating the edge list, the chapterize function [16] is applied to the novel, which divides the novel "Harry Potter and the Prisoner of Azkaban" into chapters. This novel has 22 chapters. The above-cited chapterize function correctly divides the novel into 22 chapters (as verified by going through the novel). After this, the edge lists are generated.

Matching of entities Extracted entities are matched based on their first or last name. Given the entity Harry Potter, if the extracted text is Harry, it should be matched with Harry Potter, and Mr. Potter should also be matched with Harry Potter. One problem that might occur is that sometimes Mr. Potter can be used to represent Harry's father, James Potter. For now, this problem has not been taken care of, as this paper concentrates more on visualizing the social network.

Edge list Once we have chapters of the text, the network after each chapter is considered a snapshot of the graph. For graph generation, entities are extracted from the text. Each unique character is a node. Nodes are connected if the number of words between a pair of nodes is 10 (10 is my choice for this novel). For extracting the relationships and edges spaCy [15] is employed. The sum of all the edge lists of $(edge - list)_1$, $(edge - list)_2$,...until $(edge - list)_{t-1}$ is $(edge - list)_t$.

8.4.2 Generating Graph

As mentioned earlier, the Character Network of the novel is taken as a dynamic social network with 22 snapshots. By the end of the novel, there are 55 active nodes, i.e., the network has 55 characters in the Greatest Connected Component (GCC) of Graph G. A g_t subgraph is the aggregate of all sub-graphs

$$(g_1, g_2, \ldots g_t - _1).$$

After obtaining edge lists for all of the Graph G snapshots ($g_1, g_2, \ldots g_{22}$), Louvain Community Detection (LCD) is applied to each snapshot independently.

8.4.3 Community Detection

On each GCC of subgraph g_t, the Louvain [10] community detection function is applied, and this function returns the dictionary of key-value pair node ids and the community id to which the node belongs to. *node list* = $\{(node\ id : C_{id}), (node\ id : C_{id}) \ldots.\}$.

The GCC of the subgraph g_t is subjected to CD. So node list contains information about GCC nodes; by default, if a node is not in the GCC, its community id is -1.

8.4.4 Score-Similarity Measure

Once we have a list of communities, we must compute the pairwise similarity between i^{th} community of t_1 and j^{th} community of t_2 using the Formula 1. It should be noted that the score matrix has a size of (number of communities at t) × (number of communities at t + 1).

8.4.5 Visualization of Network

After calculating the score matrix, the proposed algorithm's final step is to compute the colors for communities to ensure consistency. After observing the values of the score matrix, the threshold β is set to 0.5 for this novel.

HSV [17] color encoding is used to assign colors to the nodes of a community. As explained earlier, when a community is formed by merging more than one community, the mixing of colors is easy using HSV encoding. The coloring is decided by the algorithm 3 based on the classification

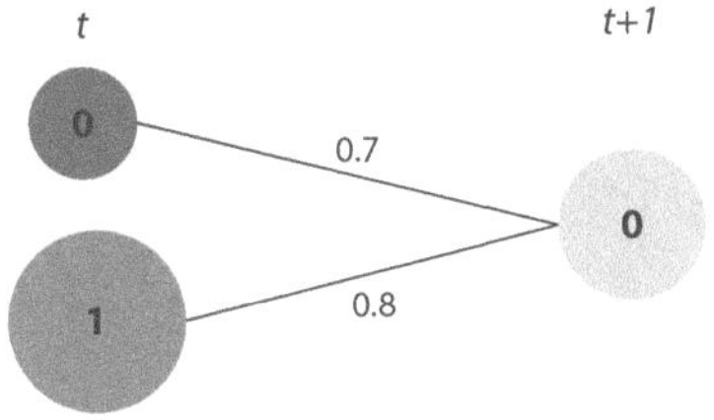

(a) Example showing merging of communities

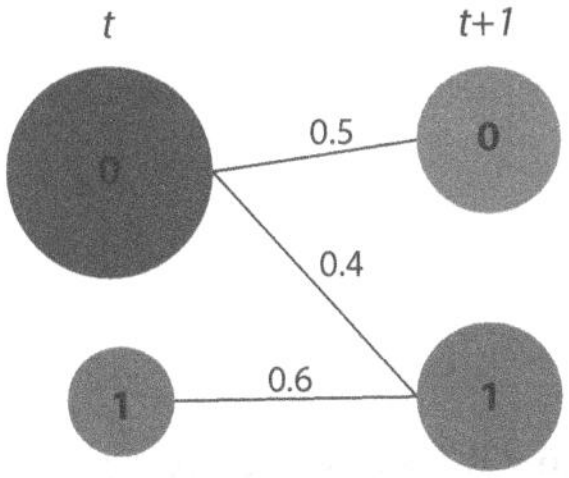

(b) Example showing splitting of communities

Figure 8.5 Splitting and shrinking of communities.

of cases, resulting in new colors. When a community is formed by new nodes that are added to GCC, it is given primary colors, making sure that no color is repeated. When a community is formed by the merging of more than one community, it is given a color that depicts the merging of colors.

For visualizing, drawing graph "**netgraph**" package is used. In netgraph, "**community**" layout is used to show nodes of a community together.

Primary colors are assigned to communities at *t*, as shown in Figure 8.5a, and similarly in Figure 8.5b, which has primary colors at *t*. Colors formed at *t*+1 is calculated from the existing colors at *t*. The color implementation is explained in further sections.

8.4.6 Visualization of Snapshots

An edge list is generated after the text has been pre-processed and chapterized. The generated edge list shows that Chapter 1 has 23 edges, Chapter 2 has 36 edges, and Chapter 3 has 50 edges. Whereas Chapter 17 has 258 edges, Chapter 18 has 262 edges, and Chapter 19 has 273 edges. It can be observed from Figures 8.6 and 8.7 that the number of new edges that are getting added to the network is gradually decreasing because, as the story progresses, the number of interactions between the same characters increases, resulting in an increase in edge weight and a decrease in the frequency of the formation of new links.

Once the edge list is generated, the graph is generated. For each snapshot, the node list remains the same if CD is applied to the graph; the disconnected components form isolated communities that do not impact visualization. As a result, the experiment is limited to the subgraph g_t's Greatest Connected Component (GCC). GCC of g_1,g_2,g_3 is shown in Figure 8.6 and GCC of g_{17},g_{18},g_{19} is shown in Figure 8.7.

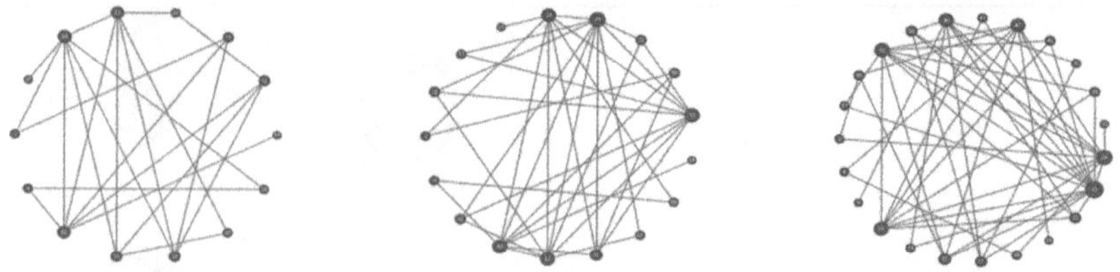

Figure 8.6 GCCs of sub-graphs g_1, g_2, and g_3.

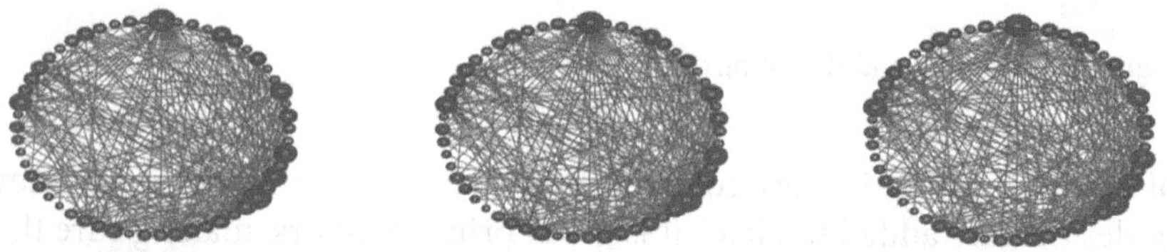

Figure 8.7 GCC of sub-graphs g_{17}, g_{18}, and g_{19}.

CD is applied to the GCC of sub-graphs of the novel. The Python **community** package is used in the implementation. This module uses the LCD method of Blondel *et al.* in [10]. This **community** module has "**best partition**" function that returns a dictionary list of (node id community id). It is observed that among all the snapshots, there are no more than five communities. Chapter 1 has four communities, which are reduced to 2 communities in Chapter 2, and then formed into 3 communities in Chapter 3. Whereas Chapters 17, 18, and 19 have 4, 4, and 5 communities, respectively. To give colors to the nodes of a community, HSV [17] color encoding is used. As explained earlier, when a community is formed by merging more than one community, the mixing of colors is easy with HSV encoding. According to the algorithm 3 depending on the classification of cases, the coloring is decided, resulting in the new colors given as shown in Figure 8.8.

As previously stated, there are four communities in g_1 and two communities in g_2. Based on the score matrix calculated by algorithm 2, the two communities of g_2 should be matched with four communities of g_1. The following pair of graphs, g_2 and g_3, should be matched later. Two g_2 communities should be paired with three g_3 communities, and so on.

The communities are grouped and colored according to their community ID. Figure 8.8 shows that the 1st community of Chapter 2 is mostly made up of red and yellow. Also, the 0th community of Chapter 2 is formed by the merging of blue and green.

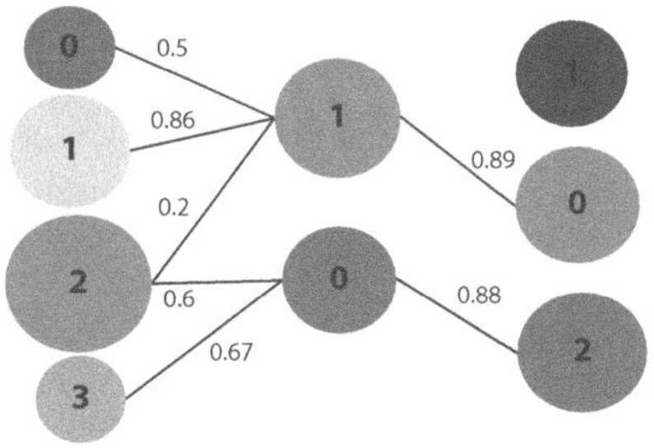

Figure 8.8 Mapping communities based Score on Score Values of (g_1, g_2) and g_2,g_3).

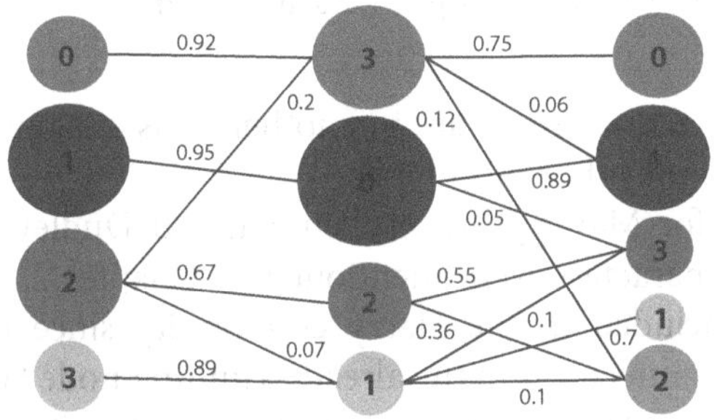

Figure 8.9 Mapping communities based onValues of (g_{17}, $g18$) and ($g18$,$g19$).

Whereas from Chapter 1 to 2, there is a shift in community colors, and a new community is formed by new nodes that are entering the GCC of the network. As shown in Figure 8.8, the community ids should be mapped in order to visualize the graph. Four g_1 communities are to be mapped to two g_2 communities.

By looking at Figure 8.9, it can be seen that the four communities of Chapter 17 are to be mapped with four communities of Chapter 18, and then these four communities are mapped with 5 communities of Chapter 19. It may be seen in Figure 8.8 and Figure 8.9 that the community's colors change completely from light to dark throughout the narrative.

8.4.7 Analysis of Results

From the Figure 8.10, the following observations are made:

1. Aunt Marge visits Dursly's house in Chapter 2, joins the network, and forms a community with Harry and Vernon. The active characters of the chapter are in one community, while

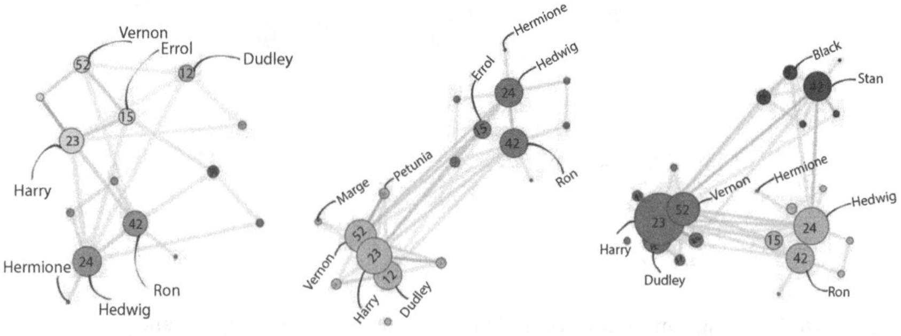

(**a**) Chapter 1 (**b**) Chapter 2 (**c**) Chapter 3

Figure 8.10 Character network of Chapter 1, 2 and 3 of HPPA.

the less active people are in another. This analogy may not apply to the other sub-graphs.

2. Except for Marge, Vernon, Petunia, and Dudley, no additional characters were mentioned in Chapter 2. And they were included in the Harry community since they have strong connections to the already-running node Harry.
3. Arthur and Molly are in one separate community in Chapter 1, but in Chapter 2, they get merged into Harry's community.

The following observations can be made from Figure 8.11.

1. As previously stated, significant disclosures occur in Chapter 17 of the narrative. Before Chapter 17, Professor Trelawney describes her vision of Voldemort returning. So she's been mentioned several times in the following chapters, and by the end of chapter 19, Prof. Trelawney has joined the community that includes Seamus of Dumbledore's Army.
2. According to Figure 8.11, Hermione and Harry belonged to the same community in Chapter 17, but Hermione moved to the community with Ron in Chapter 18. One explanation for this type of shift is that Harry's interactions with Hermione have decreased while his interactions with Tom have increased. As a result, Tom enters the community alongside Harry.
3. Sirius Black was in a community with Malfoy after Chapter 17, but by the end of Chapter 19, he has joined a community

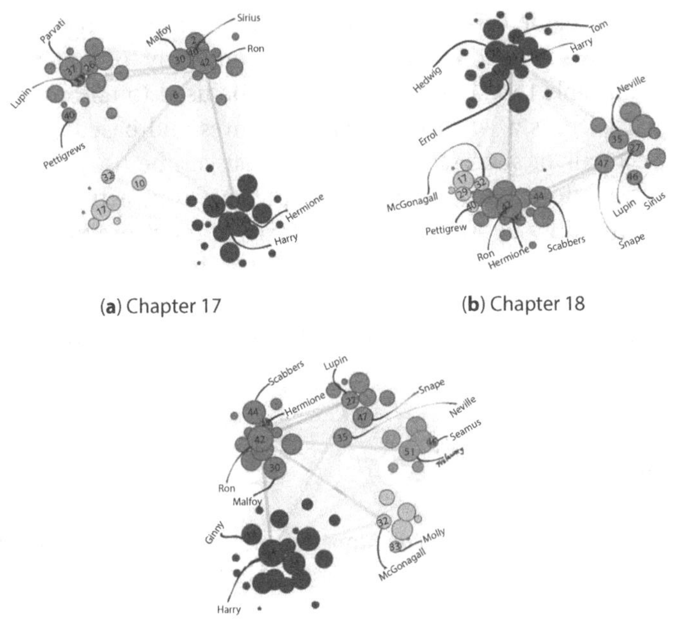

(a) Chapter 17 (b) Chapter 18

(c) Chapter 19

Figure 8.11 Character network of Chapter 17, 18, and 19 of HPPA.

with Lupin and Neville. This is because, up until Chapter 17, Harry believes Sirius is the murderer. Figure 8.11 depicts the migration from community **3** of g_18 to **4** of g_19 after confronting Sirius.

8.5 Conclusion and Future Work

A dynamic character network is created, and each network snapshot is a compilation of all previous snapshots. By the end of the novel, we have a dynamic network on which community detection is applied to each snapshot and colors are assigned to communities to preserve consistency. This coloring is accomplished by the use of a score matrix, which is a metric used to track the movement of nodes from one community to the next in consecutive timestamps. Our method of blending colors in the ratio of the similarity measure accurately showed communities merging. When a community is separated, the saturation of the color is reduced in order to obtain the color of the community. This overall pattern effectively depicts the color change.

This color scheme can be further modified by modifying not only the hue but also the saturation and value correspondingly for better visualization. A changeable threshold value can also be used to tailor community mapping. If there is a way to calculate the threshold based on the Score Matrix for a pair of successive snapshots, that can be used to gain more precise information about the merging, splitting, contraction, and expansion of communities by watching the patterns in the Score Matrix.

Making a text novel into a graphic novel that displays the primary story among the novel's central characters would be a particularly difficult task. We feel that our method is a first, small step toward meeting this enormous challenge.

References

1. Crnovrsanin, T. *et al.*, Staged animation strategies for online dynamic networks. *IEEE Trans. Vis. Comput. Graph.*, 27, 1–1, Oct. 2020, doi: 10.1109/TVCG.2020.3030385.
2. Newman, M.E.J., Modularity and community structure in networks. *Proc. Natl. Acad. Sci.*, 103, 23, 8577–8582, 2006, doi: 10.1073/pnas.0601602103. url: https://www.pnas.org/ doi/abs/10.1073/pnas.0601602103.
3. Wasserman, S. and Faust, K., *Social network analysis: Methods and applications*, Structural Analysis in the Social Sciences. Cambridge University Press, 1, 1–825, 1994, doi: 10.1017/CBO9780511815478.
4. Duval, A. and De Léséleuc, G., Novel comprehension via a dynamic social network of characters, 18, 1–12, Aug. 2020, doi: 10.13140/RG. 2.2.13445.17128.
5. Aris, A. and Shneiderman, B., Designing semantic substrates for visual network exploration. *Inf. Visualization*, 6, 4, 281–300, 2007, doi: 10.1057/palgrave.ivs.9500162. url: https://doi. org/10.1057/palgrave.ivs.9500162.
6. Girvan, M. and Newman, M.E.J., Community structure in social and biological networks. *Proc. Natl. Acad. Sci.*, 99, 12, 7821–7826, 2002, doi: 10.1073/pnas.122653799. url: https://www.pnas.org/doi/abs/10.1073/pnas.122653799.
7. Greene, D., Doyle, D., Cunningham, P., Tracking the evolution of communities in dynamic social networks, in: *2010 International Conference on Advances in Social Networks Analysis and Mining*, 1, pp. 176–183, 2010, doi: 10.1109/ASONAM.2010.17.
8. Rosvall, M. and Bergstrom, C.T., Mapping change in large networks. *PLoS One*, 5, 1, e8694, 2010.
9. Yang, T. *et al.*, Detecting communities and their evolutions in dynamic social networks—A Bayesian approach. *Mach. Learn.*, 82, 2, 157–189, 2011.
10. Blondel, V.D. *et al.*, Fast unfolding of communities in large networks. *J. Stat. Mech.: Theory Exp.*, 10, 10008, 2008, Oct. 2008, issn: 1742-5468.

11. Cazabet, R. and Amblard, F., Dynamic community detection, in: *Encyclopedia of Social Network Analysis and Mining*, R. Alhajj and J. Rokne (Eds.), pp. 404–414, Springer New York, New York, NY, 2014, isbn: 978-1-4614-6170-8. doi: 10.1007/978-1-4614-6170-8_383. url: https://doi.org/10.1007/978-1-4614-6170-8_383.
12. Cazabet, R. and Amblard, F., Dynamic community detection, in: *Encyclopedia of Social Network Analysis and Mining*, R. Alhajj and J. Rokne (Eds.), pp. 404–414, Springer New York, New York, NY, 2014, isbn: 978-1-4614-6170-8. doi: 10.1007/978-1-4614-6170-8_383. url: https://doi.org/10.1007/978-1-4614-6170-8_383.
13. Algesheimer, R., Tessone, C., Yang, Z., Comparative analysis of community detection algorithms on artificial networks. *Sci. Rep.*, 6, 30750, 2016, doi: 10.1038/srep30750. url: https://doi.org/10.1038/srep30750.
14. Rowling, J.K., *Harry Potter and the Prisoner of Azkaban*, 1, 1–317, 1999.
15. *Industrial-strength natural language processing*, url: https://spacy. io/.
16. Reeve, J., *Chapterize*, 2018, url: https://github.com/JonathanReeve/ chapterize.
17. *Graph workflow - crafting truthful data graphs*, url: https: //graphworkflow.com/retinal/value/.

9

Autonomous Intelligent Vehicles: Impact, Current Market, Future Trends, Challenges, and Limitations

Kamalanathan Shanmugam, Muhammad Ehsan Rana* and Felix Ting Yu Hong

Asia Pacific University of Technology & Innovation (APU) Bukit Jalil, Kuala Lumpur, Malaysia

Abstract
There are many levels of automation in cars, whereas most cars are equipped with at least level 2. Semi-autonomous vehicles like Tesla have level 3 automation. There must be much more testing, research, development, and regulation before society can sell and adopt fully autonomous vehicles. The Autonomous Vehicle (AV) industry impacts our world and communities in several sectors, which include traffic, urban infrastructure, logistics, health, job market, environment, and society. The AVs come with an Advanced Driver Assistance System (ADAS). ADAS has helped reduce car accidents significantly on the road. ADAS is essential and has many benefits, such as improving safety and comfort, enhancing productivity, boosting the economy, and many more. For the large-scale implementation of fully autonomous vehicles, industries and companies must address the current limitations and issues such as data privacy, cybersecurity, lack of policies and regulations, and, most importantly, how to deal with ethical concerns. We still have a long way to go in the large-scale implementation of fully autonomous vehicles, but the future holds many exciting innovations and developments in the AV industry.

Keywords: Autonomous cars, intelligent vehicles, driver assistance systems, artificial intelligence, machine learning

**Corresponding author*: muhd_ehsanrana@apu.edu.my

Hrudaya Kumar Tripathy, Sushruta Mishra, Minakhi Rout, S. Balamurugan and Samaresh Mishra (eds.) Optimized Computational Intelligence Driven Decision-Making: Theory, Application and Challenges, (173–194) © 2024 Scrivener Publishing LLC

9.1 Introduction

Advanced Driver Assistance System, or ADAS for short, is an example of ubiquitous computing we can find in our vehicles. ADAS improves our comfort and safety while driving, including features like parking assistance, cruise control, reverse cameras, blind sport monitoring, lane departure warning and centering, etc. ADAS is the foundation of introducing automation and self-driving vehicles into the automotive industry [1].

ADAS are classified into several levels of automation, with levels 0-2 being support features and self-driving, whereas autonomous features start from level 3 to level 5 as described in Table 9.1 [2].

ADAS began decades ago in the 1940s with the invention of simple cruise control and anti-lock brake systems. However, the introduction of ubiquitous computing into driving assistance systems started in the 1990s when adaptive cruise control and collision warning using sensors were introduced. In addition, Lane departure features were introduced in the 2000s, where systems use cameras that capture lane markings in real time, which are processed in the embedded computer software. The system detects if the vehicle is close to encountering the lane markings and then gives feedback to the driver through steering vibrations or displaying on the screen interfaces [3].

As computer vision and computing power improved over the years, these features improved to include lane correction and centering support. In addition, many established automobile companies like Audi, Mercedes, and Tesla have introduced level 2 and above systems that combine multiple features into a centralized system in the vehicle. Tesla has also introduced autopilot and auto parking features to enhance the driver experience and provide a smooth and comfortable ride, reducing stress and fatigue from drivers [3].

Most car accidents are caused by human error, and only a small portion is due to external environmental factors. ADAS features like blind spot monitoring, lane-keeping assistance, emergency braking, and assistive media help assist and mitigate the mistakes and bad judgments drivers can make [1].

The aim of this research is to explore the potential impact of Advanced Driver Assistance Systems (ADAS) and autonomous vehicles on the automotive industry, economy, and society. The study emphasizes the importance of implementing these safety systems across all levels of driving automation in cars. To achieve this goal, the authors will conduct an analysis of various companies and organizations operating in the industry, with

Table 9.1 Five levels of autonomy for autonomous vehicles.

Level	Criterion	Description
0	Fully Manual	The human driver is in complete control of the vehicle at all times.
1	Driver Assistance	The vehicle can assist the driver with steering, acceleration, and braking, but the human driver must remain engaged and ready to take control at any time.
2	Partial Autonomy	The vehicle can perform some driving tasks autonomously, such as steering and acceleration/braking, but the human driver must be ready to take control when necessary.
3	Conditional Autonomy	The vehicle can perform all driving tasks autonomously. Still, the human driver must be available to take control when certain conditions are met (e.g., in certain weather conditions or if the system fails).
4	Highly Autonomous	The vehicle can perform all driving tasks autonomously, with human intervention, in certain defined circumstances (e.g., on a designated stretch of highway).
5	Fully Autonomous	The vehicle can be autonomous at all times, in all conditions, and does not require a human driver.

a particular focus on their products and innovations. This analysis will provide insight into how different entities are approaching the development of ADAS and autonomous vehicles and the impact they could have on the industry as a whole. Moreover, the authors will also discuss the current limitations facing developers and companies in this field, as well as the future prospects and features that could emerge. Ultimately, this research will provide a comprehensive overview of the potential impact of ADAS and autonomous vehicles, and highlight the importance of ensuring that safety systems are integrated at all levels of driving automation.

9.2 The Global Impact of the AV Industry

The autonomous vehicle industry has the potential to impact globally in several ways. Some of the potential impacts of the industry are outlined below:

- **Transportation:** Autonomous vehicles have the potential to change the way we move people and goods significantly. They could increase the efficiency and safety of transportation, reduce the need for personal vehicle ownership, and improve accessibility for people unable to drive due to age or disability.
- **Employment:** The widespread adoption of autonomous vehicles could significantly impact work in the transportation sector. Some jobs, such as truck and taxi drivers, may be automated, while new jobs may be created in areas such as vehicle maintenance and data analysis.
- **Energy:** Autonomous vehicles could reduce the demand for fossil fuels by increasing transportation efficiency and using electric or alternative-fuel vehicles.
- **Infrastructure:** Autonomous vehicles could require changes to how our roads and infrastructure are designed and used. For example, there may be a need for more charging stations for electric vehicles, and traffic patterns may change as autonomous vehicles become more prevalent.
- **Insurance:** The increased safety of autonomous vehicles could lead to declining accidents and insurance claims, impacting the insurance industry.
- **Manufacturing:** The production of autonomous vehicles and related technology will likely create new manufacturing jobs and impact the supply chain for traditional automotive manufacturers.

Many automotive companies have started testing autonomous vehicles and taxis on actual road conditions, which could change how we think about transportation and the current traffic and road infrastructure. Furthermore, as the government and private sector scale their investments into AV technology, we could see fully commercialized autonomous vehicles adoption in cities and personal ownership of AVs. This revolutionary technology will transform our lives in many positive ways, improve our

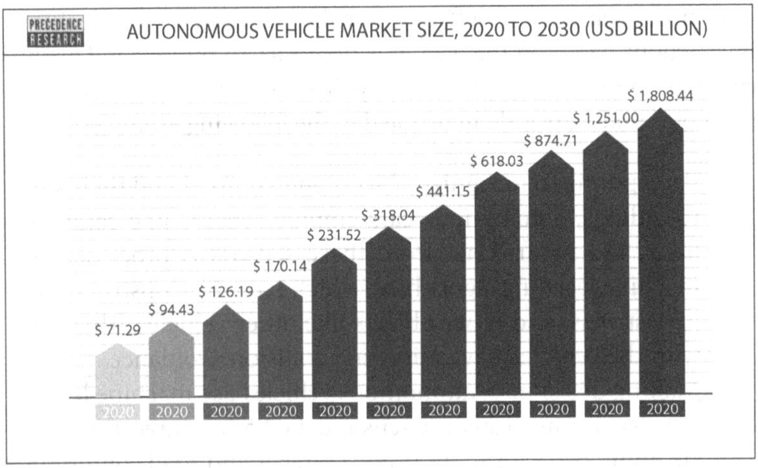

Figure 9.1 The global autonomous vehicle market [5].

lifestyles and increase efficiency and productivity. The impact of AVs can be seen in several sectors, which include the economy, road and urban infrastructure, logistics, health, general traffic, and the job market. The following sections will briefly go through each impact area [4].

As shown in Figure 9.1, the global autonomous vehicle market was estimated to be worth USD 94.43 billion in 2021 and is projected to hit around USD 1808.44 billion by 2030 due to the increasing demand for autonomous vehicles and the growing adoption of advanced technologies such as artificial intelligence, machine learning, and computer vision. The increasing investments in the development of autonomous vehicles, the growing demand for convenience and safety, and the rising government initiatives to promote the adoption of autonomous vehicles are some of the major factors driving the growth of the global.

9.3 Role of Machine Learning in Autonomous Vehicles

In autonomous vehicles, machine learning allows the vehicle to learn from data and make predictions about its surroundings. Machine learning systems, for instance, can recognize objects, people, and other vehicles on the road and forecast their behavior. While creating and using autonomous vehicles, machine learning is essential. In order to observe and comprehend their environment, make action plans, and regulate their emotions, autonomous vehicles rely on various sensors and software. For the vehicles to make decisions and conduct actions based on the data, it is necessary to process and analyze the enormous amounts of data produced by these sensors. Using machine learning in autonomous cars is demonstrated in the following examples:

a) Perception: To recognize objects and comprehend their surroundings, autonomous cars employ sensors like cameras, lidar, and radar. Data from these sensors is processed by machine learning algorithms, which recognize obstructions, pedestrians, and other objects like other vehicles. This aids with the vehicle's navigation and collision avoidance
b) Making decisions: Autonomous vehicles use machine learning algorithms to analyze sensor data. For example, the vehicle might use data about traffic patterns and road conditions to determine the optimal route to take or use data about the behavior of other vehicles to predict their movements and avoid collisions.
c) Control: Autonomous vehicles use machine learning algorithms to control their movements, such as accelerating, braking, and steering. These algorithms use data from sensors and other sources to adjust the vehicle's real-time movements, ensuring safe and efficient operation.

Encouraging shared autonomous vehicles could reduce the number of cars on the road, helping to alleviate congestion. Figure 9.2 demonstrates the features of a system created by Berkeley engineers that applies machine learning to traffic management, especially when human, semi-manned, and autonomous cars share the road [6].

Figure 9.2 The system created by Berkeley engineers applies machine learning to traffic management [6].

In short, machine learning is a crucial component of autonomous vehicle technology, enabling vehicles to operate safely and effectively in complex environments [7].

9.4 Significance of the AV Industry in Various Sectors

The impact of AVs can be seen in several sectors, which include traffic management, road and urban infrastructure, logistics, healthcare, and the job market. The following discussion will briefly go through each impact area.

9.4.1 Traffic Management

There are a few ways that autonomous vehicles could help alleviate traffic jams; in autonomous vehicles, there will be little interaction between the car and the driver. This will significantly reduce the percentage of accidents and traffic jams caused by human error and impaired judgment. As autonomous vehicles roll out into society on large scales, we will see well-trained and effective deep-learning systems which fully control the traffic dynamics in highways, intersections, and more. Heavy traffic congestion is a vast and concerning issue in many countries and cities.

It can easily be caused by stopping or slowing down cars which, one by one, creates a domino effect that can stretch kilometers. They could communicate with one another and traffic infrastructure to coordinate their movements and reduce the likelihood of bottlenecks forming. AV could use real-time traffic data to choose the most efficient route, helping reduce road congestion. It could be designed to be more efficient in size and shape, allowing more vehicles to fit on the roads and increasing capacity. AV could be programmed to drive more smoothly and predictably, which could help improve the overall traffic flow.

Reducing accidents with autonomous vehicles is one of the reasons why we should start investing in and adopting autonomous cars. With AVs, traffic can be improved by having more consistent, optimized, and regular speed. For example, on highways, an AI-powered car can be programmed to set buffer distance between the front and back whilst driving [8]. As a result, we can expect less braking, resulting in smoother and faster car speeds. These are just a few ways autonomous vehicles could help alleviate traffic jams. These technologies will likely take some time to develop and deploy fully.

9.4.2 Roads and Urban Infrastructure

We can expect profound urban and road infrastructure changes as we slowly adopt large-scale AV. People do not have to own a license or a personal vehicle to travel around and get to places. A single autonomous vehicle or taxi can navigate around a town without a driver and bring many passengers a day around the city for work or leisure. This will vastly increase the demand for public transport while reducing the need for car ownership. Owners of autonomous vehicles can also open up their cars for sharing, where they can drop them off while they go to work by setting it to taxi mode. Taxi mode will allow their car to pick up passengers just like a taxi to fully utilize the autonomous vehicle capabilities. Their cars can be utilized while their owners are busy, so they can always be on the go. Platforms and systems in place, such as a car-sharing service with autonomous vehicles, can reduce the need for parking spaces. Open parking spaces can be converted or used for other infrastructure or projects. Roads and urban infrastructure can be expected to change to support the fully autonomous transport world. As autonomous vehicles can communicate with each other through a high-speed network like 5G, infrastructures such as network towers, road sensors, and electrical charging stations are expected to be built to support this revolutionary technology [9].

Safety is a top priority for all autonomous vehicles, regardless of where they are being used. Autonomous vehicles have the potential to significantly reduce the number of accidents and fatalities on the roads, as they can continuously monitor their surroundings and make decisions based on that data. However, these vehicles must be thoroughly tested and proven safe before being deployed on public roads. In Finland, as in other countries, there are likely to be specific regulations and guidelines to ensure

Figure 9.3 Trust and safety are among the top priorities for autonomous vehicles [10].

autonomous vehicles' safety [9]. Figure 9.3 highlights the skeleton prototype of safety in autonomous vehicles.

9.4.3 Logistics

AV will also have a significant impact on the logistical sector. Autonomous trucks and vans in the delivery system can help improve efficiency, save costs, increase delivery speed, and reduce delivery accidents or delays. We have seen autonomous drone delivery in Amazon for short-distance deliveries. Implementing large-scale autonomous delivery in logistics will allow a huge opportunity to scale the operation for many businesses. Autonomous delivery, shipment, or cargo transport can start with semi-autonomous features to reduce driver fatigue, especially for long distances, and assist the driver in traffic chauffeuring. We are familiar with plane autopilot, where the pilot takes control of taking off and landing. However, the aircraft will start auto-cruising towards its destination once it has reached a certain altitude. This helps remove the strain and fatigue from pilots. The same principles and systems can apply to ground vehicles with autopilot, which can help drivers in the same way as pilots. Long-distance deliveries such as driving for hours significantly strain drivers' health energy. Overworked and tired truck drivers can cause massive accidents, loss of life, and damaged cargo. With autonomous truck transport, vehicles can switch between manual and autonomous driving at any time, allowing the drivers to rest. This also improves the speed of delivery as they do not have to stop as often for drivers to rest. Accidents can be hugely reduced, and the scalability for deliveries can be increased, compensating for the shortage of active truck drivers. Suppose companies and developers can implement delivery transport that does not require human drivers. In that case, we can eliminate human error in accidents, and companies do not have to hire drivers, reducing costs [11]. Autonomous trucks will also be able to communicate with each other, just like passenger vehicles, to optimize traffic flow and reduce fuel consumption.

9.4.4 Healthcare

The autonomous vehicle revolution has the potential to improve health and well-being in a number of ways significantly. For example,

- **Improved safety:** AVs have the potential to significantly reduce the number of accidents on the roads, which could lead to fewer injuries and fatalities.

- **Increased accessibility:** Autonomous vehicles could make it easier for people who cannot drive, such as the elderly or those with disabilities, to get around and maintain their independence.
- **Reduced air pollution:** Autonomous electric vehicles could help reduce air pollution, a significant contributor to various health problems, including respiratory and cardiovascular diseases.
- **Improved public health:** Autonomous vehicles could potentially be used to transport goods, which could help to reduce the number of delivery trucks on the roads and improve air quality in cities.

Considering the factors stated above, it is essential to carefully consider the potential health and well-being impact of autonomous vehicles and ensure that they are developed and deployed in a way that maximizes their benefits and minimizes any potential negative impacts, as shown in Figure 9.4. Human health will also significantly improve with the introduction and mass adoption of autonomous cars and infrastructure. As full AVs will not require a driver, people without licenses, disabled, elderly, or potentially, children can also access the autonomous taxis to commute around their neighborhood. Safer travel with lesser accidents will not only save and improve lives, but it will also save a lot of costs in healthcare for injuries due to road accidents. In healthcare, we can see the adoption of autonomous medication and emergency supplies delivery to patients and those in need.

Figure 9.4 The autonomous vehicle revolution for our health and well-being [12].

This can reduce the escalation of disease or injuries and prevent them from reaching fatal levels, which would be a higher probability in today's systems with slower systems and communication. With robust systems of emergency contact using mobile phones or smart devices such as watches, patients can contact hospitals for emergencies, and autonomous vehicles can send them the medications or supplies needed. In addition, we can see the implementation of mobile pharmacies or clinics, allowing even faster and more revolutionary healthcare systems to improve our health and lifestyle [12].

9.4.5 Job Market

The general sentiment towards automation in any field, including the automotive industry, is harmful regarding job opportunities and replacement. Although this is the case, the system and infrastructure built to support large-scale AV adoption will create many job opportunities which we do not have today. Although drivers will be replaced by the mechanical and digital brains of the cars and AI algorithms to drive the car, we could see jobs like AV specialists, technicians, and maintenance jobs arise to support the industry and society. Existing and routine jobs like salesman, customer service, managers, and others would also be given considerable opportunities in this sector [4]. We could also see new qualifications and certifications for students and trainees to obtain, which would land them specialized and high-paying jobs in this industry and boost the education sector.

9.4.6 Environment and Society

As efficient and optimized traffic flow and driving introduced by AV are in place, cars would use less fuel, and there would be less wastage of resources. AVs are generally more eco-friendly as they are less affected by the driving style of the human driver. The driver's fuel usage is heavily affected by repeated braking and changing speeds between high and low, especially in traffic congestions, which would be reduced with AV traffic running and effective communication between vehicles, as mentioned in a prior section. With better route calculation and efficient fuel consumption, emissions and energy can be reduced. Large-scale AV transition will also reduce the number of vehicles on the road. AVs can be shared amongst communities' families, meeting the necessities of a small family that would otherwise require several vehicles to do the same [12]. Less land would be demanded in parking spaces, and local communities or governments could use the open spaces which would otherwise be used for parking to build green infrastructure or parks to make towns and cities greener.

Large-scale implementation of AV infrastructure into our society would significantly increase the accessibility to efficient and effective transportation, even for those who are unqualified or unable to drive. This means that there would be an increase in vehicle usage and the mobility of people. Many people rely on personal vehicles to commute to work or school. AVs will be able to open opportunities for people while improving productivity and practical time usage during travel. People can work on more critical tasks such as work, meetings, or other leisure activities during the journey and leave the driving to the vehicle. People would be more productive, boosting the economy even further [13]. We could also see on-board retailing or dining, currently present in long-distance travel like aeroplanes or trains, which would transform how we view personal vehicles or taxis today [4]. Auto parking is also a highly talked about feature of AVs where they can drop their owners to their destination, such as a mall or workplace and park themselves without the owners worrying.

Although machines are powerful in performing automated calculations like identifying objects and making certain decisions, it is still difficult to model the ethical and moral choices that vehicles must make to abide by our world's laws and driving ethics. We still have a long way to go to gain trust in the technology towards AI or Machine Language (ML) algorithms and security from customers and society. There will need robust training and tweaking of the algorithms to make decisions that society would accept. Cybersecurity of autonomous vehicles is also a massive part of the social impacts as significant accidents and irreversible tragedies can occur if large numbers of AVs are hacked and manipulated. Programmers and companies would have to deal with a wide variety of situations set by regulations before AVs can penetrate society on a large scale.

9.5 Current Market and Future Trends in AV Industry

We have mentioned in the previous section that AVs can help reduce accidents by alerting and preventing collisions and car crashes. Safety is the highest priority in introducing AVs on a large scale, and we would like to improve the quality of life of people. Large-scale AV adoption and infrastructure will help save the overall cost of society hugely in terms of fuel, healthcare, and money can be invested in other areas, such as making greener cities [14]. Society and humanity, in general, need to constantly innovate and move forward together to improve quality of life, stay creative and productive and contribute more towards society. Better access to transportation and better traffic flow and efficiency will free up much time from people, remove unnecessary traffic jams, and eliminate reckless driving. AVs will be more sustainable in reducing emissions, less fuel consumption and can make our cities clean through electric engines rather than traditional fuel combustion engines [15]. We should innovate to preserve our communities and environment; the market for AV is still in its early stages, but it is expected to grow significantly in the coming years. Here are some current trends and developments in the autonomous vehicle industry that can play a huge part:

- **Increasing investments:** Many companies within and outside the automotive industry are investing heavily in autonomous vehicle technology. These investments are expected to accelerate the development and deployment of autonomous vehicles.
- **Partnerships and collaborations:** Many partnerships and collaborations are forming within the autonomous vehicle industry as companies look to combine their expertise and resources to advance the technology.
- **Adoption of electric and connected vehicles: Autonomous** vehicles are often electric and connected. The growing adoption of these vehicles is expected to drive the growth of the autonomous vehicle market.
- **Testing and pilot projects:** Many companies are conducting testing and pilot projects to gather data and refine their autonomous vehicle technologies.
- **Regulation:** Governments worldwide are starting to put regulations for autonomous vehicles in place, which will help ensure their safe deployment and operation.
- **Growing demand for advanced driver assistance systems:** The increasing adoption of advanced driver assistance systems (such as lane departure warnings and automatic emergency braking) is expected to pave the way for deploying more fully autonomous vehicles.

9.5.1 Tesla and Waymo: Two Key Players in the Autonomous Vehicle Industry

Tesla is an automotive company that manufactures electric vehicles with advanced driver assistance systems and self-driving capabilities. Tesla cars are equipped with eight external cameras with high-tech computer vision processing capabilities with Tesla Vision and Autopilot instead of relying on traditional radar systems. Tesla cars come with several levels of autopilot: Basic Autopilot, Enhanced Autopilot, and full self-driving capabilities. The basic autopilot features traffic-aware cruise control, which detects and calculates its speed relative to all the other vehicles on the road. Like traditional cruise control, tesla cars can maintain a constant speed while adjusting automatically based on the overall traffic speed without the driver's intervention. Another feature included is the Autosteer feature [16]. Autosteer assists in steering within a marked lane and helps the car stay in the lane while on cruise control.

The enhanced autopilot level includes both features while adding several more advanced features. One of which is the navigation on autopilot. This feature assists the driver by suggesting lane changes, navigating interchanges, automatically engaging turn signals, and taking correct exits. In addition, the enhanced autopilot system also sports several other auto and intelligent features like auto lane changing, auto parking, summoning, and smart summon, allowing the car to navigate through complex environments to find its owner within a parking area [16].

Full self-driving capabilities include traffic and stop sign detection and control that slows the car upon approaching. Autosteering within city streets is currently being worked on by the company and is upcoming to be released. In the current state, Tesla cars are considered semi-autonomous as the systems have limited features that only apply to specific situations. Active driver supervision is still required, and there is still much training, testing, and regulatory approval before Tesla cars can start moving into large-scale autonomous vehicle systems development and adoption. Standard active safety features in Tesla cars include forward and side collision warnings, automatic emergency braking, obstacle awareness and warning, blind spot monitoring, lane departure, and avoidance features [16].

Waymo by Google is another company involved in the innovation and industry of the autonomous car. Unlike Tesla, Waymo does not manufacture vehicles but specializes in equipping vehicles with self-driving capabilities through radar, lidar, cameras, and advanced computing systems. Waymo, at its core, focuses on ride-hailing services, bringing people to where they want and when they want. Waymo helps remove people's frustrations and concerns about driving and commits to improving the safety and experience of passengers and people outside vehicles [17].

One of the products of Waymo is Waymo One, a ride-hailing service that allows people to hail a car through their mobile app. As a result, passengers can get in a car without worrying about driving and can get most of their time on the commute. People can focus on meaningful things while the autonomous driver handles the driving. Passengers can also add multiple stops and do their daily errands without worrying about the vehicle. Passengers can feel safe and secure while commuting with Waymo cars, and an emergency button is always available to call for agent support for questions or assistance. The mobile application also identifies the Waymo car, and each Waymo car is well-maintained to provide the best experience for customers [18].

Waymo has also projected itself into the logistics industry by providing its Waymo Via product. Waymo Via provides autonomous trucking solutions for moving goods while targeting truck driving challenges, such as safe turning, braking, lane changing, and more. Safety is the company's core value, and Waymo focuses on delivering solutions that make our roads safe and reduces injuries and fatalities. Waymo trucks aim to provide greater efficiency in uninterrupted driving, flexible scheduling, and routing while maximizing speed and asset utilization. The Waymo system always stays alert without getting distracted or tired, so goods arrive safely. Companies with Waymo trucking solutions can lower operations costs by freeing up time and energy, allowing them to focus on meaningful customer experiences [19].

Waymo vehicles enable self-driving capabilities by mapping their territory with the help of custom maps and real-time sensor data and can identify exact road locations at all times. In addition, Waymo vehicles keep an eye on everything on the road, predict situations before it happens, and plan for the safest outcome. Waymo vehicles also have active safety backup features like collision and avoidance systems, redundant braking and steering systems, and robust cyber security [17].

9.5.2 AI Datasets and ML-Based Development

Some companies focus on providing traffic situation datasets to train AI and ML systems. A good quality dataset is significant for computers to learn and make decisions on navigating roads. A bad-quality dataset will highly result in bad decisions, which will cause accidents [20].

The woven plant level 5 dataset is one such dataset that includes sensor and input maps, detected traffic agents and objects, predicted agent motion, and the path taken by the autonomous vehicle. The prediction dataset contains logs of the movement of cars, cyclists, pedestrians, and other traffic agents collected by their autonomous fleet. These data come from radar, lidar, and cameras. The perception dataset uses the raw camera and lidar inputs to generate human-labeled 3D bounding boxes of traffic agents to visualize the environment around the vehicle giving a top-down view and helping build autonomous vehicles without human intervention. The datasets include high-quality semantic maps of manually annotated lane segments, pedestrian crosswalks, parking zones, speed bumps, and more. The dataset also includes various scenarios like intersections and multi-lane traffic to help strengthen the dataset's quality [21].

Another dataset with the name BDD100K provides a large-scale, diverse driving dataset with comprehensive annotations of video and images. The dataset includes diverse scene types, such as city scenes, residential areas, and highways. The videos also have weather conditions, such as rainy, cloudy, clear, and foggy, at different times of the day. The diverse dataset allows researchers and developers of self-driving capabilities to work on tasks such as object detection, instance segmentation, multi-object tracking, segmentation tracking, semantic segmentation, lane markings, drivable area, and more. A diverse dataset used in AI and ML training would significantly improve the performance in different situations and would be less likely to be surprised by new conditions. The lane marking dataset includes several categories like road curb, crosswalk, double white, double yellow, and other attributes like full or dashed lanes and the direction. The drivable area dataset is split into two categories. The directly drivable area refers to the lane the driver is currently on. It also represents the region where the car has priority over other vehicles. Alternatively, a drivable area is a lane that the driver is not currently on but can switch over to. Objects for detection include pedestrians, cars, trucks, motorcycles, bicycles, and more [22].

9.5.3 Use of Sensors and Other Hardware

Sensor manufacturing and development is also a massive component of autonomous driving. Sensors are the eyes of the cars, and the data gathered are used in AI and ML development. One of the companies involved in this technology is Luminar.

Luminar develops advanced perception technology such as lasers, scanners, receivers, and processing electronics. Their sensor technology contributes to detecting and classifying objects out to 250 meters from cars to lanes, even in low light conditions. Their sensors also feature velocity detection in three dimensions allowing the detection of the vehicle changing lanes and the relative speed of all objects, including pedestrians around. Luminar products enable cars with advanced ADAS, confident decision-making, braking, and accelerating even at high speeds [23].

Another company, Innoviz Technologies, also manufactures and develops lidar technology like Luminar. Innoviz technologies lidar products enable accurate and real-time 3D sensing of surroundings to consumer vehicles, Robo taxis, logistics, industrial drones, heavy machinery, and more [24]. In addition, the lidar products are resilient to weather conditions and reliable but affordable, enabling self-driving cars' capabilities [25].

9.6 Challenges and Limitations

The current state of innovation and technology in the Autonomous vehicle industry and limitations in various sectors are still being addressed before society can launch a large-scale adoption of AV infrastructure. One such drawback or limitation industries currently face technological efficiency. Although we have seen huge progress in the past decade, a fully autonomous vehicle still needs to be available in the market. Currently, many AV manufacturers offer semi-autonomous vehicles that are only reliable in certain road conditions with minimal clutter and within good weather. Extensive libraries of datasets that include all possible weather, situation, objects on the road, and decisions to make depending on the combination of objects the car perceives have to be obtained before society can truly accept autonomous vehicles. Communications systems between cars and road infrastructure should also be robust, constantly online, with stable performance should be in place [26, 27].

Several challenges and limitations facing the autonomous vehicle industry include data privacy, cybersecurity, policies, regulations, ethical issues, technical challenges, public acceptance and infrastructure.

9.6.1 Data Privacy

Data privacy is also one of the enormous issues in our current world, where the world has never been as connected. Sensitive data collected from cameras or lidars of such autonomous vehicles may include their workplace, home, and more may not be consented to or comfortable for customers.

Other data such as location data, passenger or owner information, vehicle-to-vehicle, or other infrastructure data also need clear regulations before large-scale implementation of AVs. Policies and agreements must be in place to regulate how AV manufacturing and AI development companies store and use these data [28].

9.6.2 Cybersecurity

AVs are vulnerable to cyber-attacks, which could compromise their safety and reliability. Ensuring the cybersecurity of autonomous vehicles will be an essential challenge for the industry. Data privacy comes from cybersecurity, which is one of the significant concerns of autonomous vehicles being hacked. With the rise of the internet, the world has gone through several slight to large-scale cyber-attacks that have cost billions of dollars of damage, information leakage, and people's mental health. Threats and attacks to driverless cars or massive AV infrastructure and network hackings can cause severe or irreversible damage that we cannot even comprehend today. Hackers may take control of cars and use them for criminal activities or terrorism. Robust cybersecurity technology must be implemented in networks and AVs before they can roll out to society [29].

9.6.3 Policies and Regulations

There currently needs to be more regulations specifically for autonomous vehicles, and creating a consistent regulatory framework is a complex task. Governments must balance the need for safety with encouraging innovation in the industry. Government policies, insurance, and legislation are also one of the current issues that the AV industry has to face. Since AVs are still part of emergent technology, little legislation outlines the rules and laws compared to established technologies. There may be a grey area regarding who is responsible for AV-related accidents, as the passengers or owners are not actively involved in driving the car. Manufacturers of the AVs may also not be responsible if their vehicles have passed all the tests and rules before shipping to the market. Policymakers must ensure every industry legislation cranny is addressed and outlined [30].

9.6.4 Ethical Issues

Ethical issues are among the most complex parts of AV development faced by companies and researchers. The culture, ethics, and moral values instilled in every human being have been with us since we stepped foot in the world, and we have had many centuries to develop and normalize the right and wrong of each situation. In traffic conditions, humans can act instinctively and react to all situations based on the rules and moral values we have. These societal rules and ethics must be pre-programmed into the AV systems to mimic our decisions in all traffic situations. However, very complex situations can often result in uncertainty or cloudiness in decision-making [26].

There may be situations where the car has to avoid a collision with a pedestrian crossing the road, but steering to the side, may cause a crash to an adjacent car or even injure more pedestrians in the walkway or even crash onto a wall, injuring or killing the passenger. Any vehicle choice can be controversial to different groups of people. Many factors may play into decision-making, including the levels of damage, types of people or property involved in an accident resulting from the decision, and many more. People need help to write these laws and program them into AV systems in the industry's current state. The industry still has a long way to go before addressing these issues, and large-scale infrastructure can roll out to society. The media and general sentiment towards AVs are positive and considered safe and secure; people can still feel comfortable sitting inside a driverless car for a long time. Therefore, we may see worldwide adoption of AVs, as the technology and infrastructure are already in place [26].

9.6.5 Other Common Challenges

The following lists some of the other common challenges in the AV industry:

- **Technical challenges:** Developing reliable autonomous vehicle technology is complex, and many technical challenges must be overcome. These include sensor accuracy and reliability issues, mapping and localization, and decision-making under various conditions.
- **Public acceptance:** Autonomous vehicles will only be successful if the public is willing to use them. Building trust in the technology and addressing safety concerns will be crucial for adopting autonomous vehicles.
- **Infrastructure:** Autonomous vehicles will require changes to how our roads and infrastructure are designed and used. For example, there may be a need for more charging stations for electric vehicles, and traffic patterns may change as autonomous vehicles become more prevalent.

9.7 Conclusion

In this research, authors have gained valuable insights into the impact of autonomous vehicles on various sectors, the industry's applications, and the limitations and challenges that currently exist. As an emergent technology, the autonomous vehicle industry has enormous potential to improve people's safety, quality of life, and productivity. The industry is rapidly growing, with large-scale adoption and infrastructure build-up underway. This shift in transportation has the potential to transform how we perceive and utilize transport across sectors. However, despite the many benefits, there are still numerous issues and drawbacks that must be addressed before we see widespread adoption of autonomous vehicles. These challenges need to be tackled head-on to ensure a smooth transition to a future with autonomous vehicles.

References

1. What is ADAS (advanced driver assistance systems)? – overview of ADAS applications, Synopsys, 2022, [Online]. Available: https://www.synopsys.com/automotive/what-is-adas.html. [Accessed: 01-Nov-2022].
2. Shuttleworth, J., SAE J3016 automated-driving graphic, SAE International, 07-Jan-2019, [Online]. Available: https://www.sae.org/news/2019/01/sae-updates-j3016-automated-driving-graphic. [Accessed: 01-Nov-2022].
3. Becker, J., A brief history of automated driving - part three: Toward product development, Apex.AI, 05-Oct-2020, [Online]. Available: https://www.apex.ai/post/a-brief-history-of-automated-driving-part-three-toward-product-development. [Accessed: 01-Nov-2022].
4. Yeruva, V., Council post: Autonomous vehicles and their impact on the economy, Forbes, 14-Feb-2022, [Online]. Available: https://www.forbes.com/sites/forbestechcouncil/2022/02/14/autonomous-vehicles-and-their-impact-on-the-economy/?sh=4c1d7c2960de. [Accessed: 01-Nov-2022].
5. Autonomous vehicle market size, share, trends, report 2022-2030, Precedenceresearch.com, Retrieved 11 January 2023, from https://www.precedenceresearch.com/autonomous-vehicle-market.
6. Nahian, R.A. and Rana, M.E., ParkoSet — An intelligent and sustainable car parking management system towards solving traffic congestion problems in developing countries. *Test Eng. Manage.*, 83, 254–268, 2020.
7. Mohseni, S., Pitale, M., Singh, V., Wang, Z., Practical solutions for machine learning safety in autonomous vehicles, 2019, ArXiv, abs/1912.09630.

8. Hutson, M., Watch just a few self-driving cars stop traffic Jams, Science, 16-Nov-2018, [Online]. Available: https://www.science.org/content/article/watch-just-few-self-driving-cars-stop-traffic-jams. [Accessed: 02-Nov-2022].
9. Straits Research, HD maps for autonomous vehicles market, Straitsresearch.com, Retrieved 7 January 2023, from https://straitsresearch.com/report/hd-maps-for-autonomous-vehicles-market.
10. Trust and safety among the top priorities for autonomous vehicles, Eltis.org, Retrieved 7 January 2023, from https://www.eltis.org/in-brief/news/trust-and-safety-among-top-priorities-autonomous-vehicles.
11. Möller, P., Autonomous driving in logistics: At the future's wheel, DHL Freight Connections, 08-Feb-2022, [Online]. Available: https://dhl-freight-connections.com/en/trends/autonomous-driving-in-logistics-at-the-futures-wheel/. [Accessed: 01-Nov-2022].
12. Eddy, N., What to know about autonomous vehicles in healthcare, Technology Solutions That Drive Healthcare, 10-Dec-2020, [Online]. Available: https://healthtechmagazine.net/article/2020/12/what-know-about-autonomous-vehicles-healthcare. [Accessed: 01-Nov-2022].
13. Igini, M., Environmental pros and cons of self-driving cars, Earth.Org, 16-May-2022, [Online]. Available: https://earth.org/pros-and-cons-of-self-driving-cars/. [Accessed: 02-Nov-2022].
14. Klaver, F., The economic and social impacts of fully autonomous vehicles, Compact, [Online]. Available: https://www.compact.nl/en/articles/the-economic-and-social-impacts-of-fully-autonomous-vehicles/. [Accessed: 02-Nov-2022].
15. Benefits of self-driving vehicles, Coalition For Future Mobility, [Online]. Available: https://coalitionforfuturemobility.com/benefits-of-self-driving-vehicles/. [Accessed: 02-Nov-2022].
16. Self-driving cars: The pros & cons of automated vehicles, Valiente Mott, [Online]. Available: https://valientemott.com/auto-collisions/self-driving-cars-pros-and-cons/. [Accessed: 02-Nov-2022].
17. Autopilot and full self-driving capability, Tesla, [Online]. Available: https://www.tesla.com/support/autopilot [Accessed: 03-Nov-2022].
18. Waymo driver - Waymo, Waymo, [Online]. Available: https://waymo.com/. [Accessed: 03-Nov-2022].
19. Iqbal, A. and Rana, M.E., Adoption of IoT in automobiles for driver's safety: Key considerations and major challenges. *Int. J. Sci. Technol. Res.*, 8, 1378–1384, 2019.
20. Waymo via - Waymo, Waymo, [Online]. Available: https://waymo.com/waymo-via/. [Accessed: 03-Nov-2022].
21. Goncharov, I., Autonomous vehicle companies and their ML, W&B, 30-Sep-2022, [Online]. Available: https://wandb.ai/ivangoncharov/AVs-report/reports/Autonomous-Vehicle-Companies-And-Their-ML–VmlldzoyNTg1Mjc1. [Accessed: 03-Nov-2022].

22. Houston, J. *et al.*, One thousand and one hours: Self-driving motion prediction dataset, 25-Jun-2022, [Online]. Available: https://doi.org/10.48550/arXiv.2006.14480. [Accessed: 03-Nov-2022].
23. Yu, F., Chen, H., Wang, X., Xian, W., Chen, Y., Liu, F., Madhavan, V., Darrell, T., BDD100K: A diverse driving dataset for heterogeneous multitask learning. *2020 IEEE/CVF Conference on Computer Vision and Pattern Recognition (CVPR)*, 2633–2642, 2018.
24. Luminar (NASDAQ: LAZR): Technology, Luminar (Nasdaq: LAZR), 2022, [Online]. Available: https://www.luminartech.com/technology/. [Accessed: 04-Nov-2022].
25. Applications, Innoviz, 2022, [Online]. Available: https://innoviz.tech/applications. [Accessed: 04-Nov-2022].
26. Martinho, A., Herber, N., Kroesen, M., Chorus,C., Ethical issues in focus by the autonomous vehicles industry. *Transport Rev.*, 41, 5, 556–577, 2021.
27. Fafoutellis, P. and Mantouka, E.G., Major limitations and concerns regarding the integration of autonomous vehicles in urban transportation systems. *Data Analytics: Paving the Way to Sustainable Urban Mobility*, 1, pp. 739–747, Jan. 2019.
28. Glancy, D.J., Privacy in autonomous vehicles. *Santa Clara L. Rev., 52*, 1171, 2012.
29. Kim, K., Kim, J.S., Jeong, S., Park, J.H., Kim, H.K., Cybersecurity for autonomous vehicles: Review of attacks and defense. *Comput. Secur., 103*, 102150, 2021.
30. Milakis, D., Thomopoulos, N., Van Wee, B., *Policy implications of autonomous vehicles*, Academic Press, 5, 354, 2020.

10

Role of Smart and Predictive Healthcare in Modern Society

Muhammad Ehsan Rana[1]* and Manoj Jayabalan[2]

[1]Asia Pacific University of Technology & Innovation (APU) Bukit Jalil, Kuala Lumpur, Malaysia
[2]Liverpool John Moores University, Liverpool, UK

Abstract

The electronic sharing of health-related data between two or more healthcare organizations like hospitals, laboratories, diagnostic centers and outpatient clinics is always challenging due to their heterogeneous infrastructure. The fragmented and disjointed nature of private and public primary care services leads to data duplication and limits the capacity to perform informed decisions. For instance, the lack of readily available and relevant data to test a hypothesis in infectious disease control such as COVID-19 has led to limited evidence-based approaches. Therefore, a comprehensive health management system with a unified framework is required to improve data sharing and availability. With the advancement of the Internet of Things (IoT), many sectors have been revolutionized by utilizing its strength to connect and exchange data with other devices and systems over the Internet. Reshaping personalized healthcare by identifying the factors influencing the utilization of IoT technology in the health sector is the need of the hour. This also signifies the use of IoT for an interconnected and collaborative healthcare platform by materializing state-of-the-art data sharing and analysis techniques. Predictive analytics can help identify patterns and trends in healthcare data, allowing healthcare professionals to make more informed decisions and anticipate potential health issues before they become severe. By leveraging IoT technology and predictive analytics, healthcare providers can gain valuable insights into patient health, disease patterns, and treatment outcomes, ultimately leading to improved quality and efficiency of treatments. This chapter investigates and proposes IoT technologies in conjunction with predictive analytics for the

**Corresponding author*: muhd_ehsanrana@apu.edu.my

Hrudaya Kumar Tripathy, Sushruta Mishra, Minakhi Rout, S. Balamurugan and Samaresh Mishra (eds.) Optimized Computational Intelligence Driven Decision-Making: Theory, Application and Challenges, (195–214) © 2024 Scrivener Publishing LLC

healthcare sector to integrate isolated and scattered patient health information systems. It also specifies the possible barriers and challenges which might occur while developing IoT-based healthcare systems. Further, the study leverages IoT-based healthcare methods to realize unparalleled benefits, focusing on improving the quality and efficiency of treatments.

Keywords: Healthcare management system, Internet of things (IoT), IoT-based healthcare management, predictive analytics, healthcare management framework

10.1 Introduction

Healthcare nowadays is undergoing a radical transformation backed by emerging technology like the Internet of Things (IoT) to change the traditional landscape of the whole sector. IoT is still a novel concept for most healthcare professionals, but its use in healthcare is inevitable. Over the years, prior to the invention of modern-age digital infrastructure, patients' engagements with health practitioners were limited to physical presence. However, a longer, healthier and happier life cannot be realized without expanding modern technology and embracing a broader platform signifying the adoption of IoT in the healthcare sector. The market size of IoT in the healthcare sector worldwide is forecasted for exponential growth from existing figures of USD 127.7 billion in 2023 to staggering proportions of USD 289.2 billion by 2028 [11].

The electronic sharing of health-related data between two or more healthcare organizations like hospitals, laboratories, diagnostic centers and outpatient clinics is always challenging due to their heterogeneous infrastructure. The fragmented and disjointed nature of private and public primary care services leads to data duplication and limits the capacity to make informed decisions. For instance, the lack of readily available and relevant data required to test a hypothesis in infectious disease control such as COVID-19 has led to limited evidence-based approaches. Therefore, a comprehensive health management system with a unified framework is required to improve data sharing and data availability. With the advancement of the IoT, many sectors have been revolutionized by utilizing its strength to connect and exchange data with other devices and systems over the Internet. Reshaping personalized healthcare by identifying the factors that are influencing the utilization of IoT technology in the health sector is the need of the hour. This also signifies the use of IoT for an interconnected and collaborative healthcare platform by materializing state-of-the-art data sharing and analysis techniques. This chapter focuses on investigating and proposing the use of IoT technologies in the healthcare sector, particularly in personalized healthcare, capable of integrating isolated and scattered patient health information systems. It also specifies the possible barriers and challenges which might occur while developing IoT based healthcare systems. Further, the study proposes a detailed IoT based healthcare management framework to realize unparalleled benefits with a focus on improving the quality and efficiency of treatments. Appropriate use of IoT based techniques can promote equal access to timely, acceptable and affordable healthcare globally to realize a smart healthcare society.

The existing research on IoT based healthcare assists to extract deliverables pertaining to the digitalization of the patient's and health practitioner's management, ease of communications and consultations and optimization of the resource utilizations. However, a specific framework for bespoke healthcare, utilizing connected devices which will seamlessly coordinate patient information among distributed stakeholders and healthcare providers, has yet to be established. Moreover, privacy is of paramount importance because patients expect that their private information will remain confidential under all circumstances. IoT based healthcare systems must allow for sharing of information that is absolutely required while providing high-quality assurance for privacy. By critically reviewing the current IoT based healthcare frameworks, while looking at their strengths, weaknesses, notable features and constraints, authors have provided a comprehensive IoT based Healthcare Management framework that captures the complete view of the IoT based healthcare ecosystem.

10.2 Healthcare System

The healthcare system consists of patient medical data such as treatments, medical histories, radiology images, etc., in a digital format. It benefits the providers to streamline the workflow, providing better decisions on patient care, and timely service and sharing the patient medical data among the participating entities.

Patient behavior towards migrating to different healthcare providers is due to better quality service, not satisfied with the present medication/practitioner, relocation of job/house, and emergency [12]. The traces of medical visits are left behind at the previous hospital/clinic causing the patient to repeat the same diagnosing procedure/medical test that leads to a time-consuming and costly operation.

The government agencies in several countries initiated the process of integrating the healthcare providers to share the patient's medical treatments to improve timely decision-making, reduce cost and improve service. The adoption of Electronic Health Records (EHR) in hospitals and clinics is growing at the fastest pace in several countries, 98% in Norway and Netherlands, 97% in the United Kingdom and New Zealand, 69% in the United States, etc. [4, 6].

10.3 Role of Predictive Analytics in Healthcare

Predictive analytics has gained significant attention in healthcare due to its potential to improve patient outcomes, reduce costs, and enhance the overall efficiency of healthcare systems. Predictive analytics involves using statistical and machine learning algorithms to analyze data, identify patterns, and predict outcomes. In healthcare, predictive analytics can be used for a wide range of applications, such as disease prevention, early detection, diagnosis, treatment planning, and resource optimization. Healthcare organizations that leverage predictive analytics can gain a competitive advantage by providing more personalized and effective care while reducing costs and improving the overall patient experience [14].

Some of the specific roles of predictive analytics in healthcare are as follows:

10.3.1 Disease Prevention

Predictive analytics can help identify individuals at a higher risk of developing specific diseases. By analyzing patient data, including demographic information, medical history, and lifestyle factors, predictive models can identify patients at risk of developing certain diseases, such as diabetes, heart disease, and cancer.

10.3.2 Early Detection

Predictive analytics can help in the early detection of diseases, which can significantly improve patient outcomes. By analyzing patient data, including biomarkers, imaging data, and other clinical data, predictive models can identify patients who are at risk of developing specific diseases, such as cancer and Alzheimer's disease.

10.3.3 Diagnosis

Predictive analytics can help in improving the accuracy of diagnosis by analyzing patient data, including symptoms, medical history, and test

results. Predictive models can identify patterns in patient data that can aid in diagnosing diseases such as cancer and infectious diseases.

10.3.4 Treatment Planning

Predictive analytics can help in optimizing treatment planning by analyzing patient data, including medical history, genetics, and lifestyle factors. Predictive models can identify patterns in patient data that can help select the most effective treatment plan for a specific patient.

10.3.5 Resource Optimization

Predictive analytics can help in optimizing resource allocation in healthcare systems by predicting patient demand and resource utilization. Predictive models can analyze patient data, including hospital admission data, to predict patient demand and resource utilization, which can aid in resource allocation and staffing decisions.

10.4 Application of IoT in Healthcare

The application of IoT in healthcare can be broadly classified into home healthcare, mobile health and e-health, and Electronic Health Record (EHR).

10.4.1 Home Healthcare

Traditionally, home healthcare was provided by practitioners to patients in their homes under the direction of physicians. However, patients with limited mobility due to older age, chronic disease, and disabilities are able to be monitored through sensors and wearable devices. As such, they assist in notifying the practitioners and emergency services when the patient's health deteriorates.

10.4.2 m-Health

Smartphone applications have proven to be efficient in real-time monitoring of health status, physiological abnormalities and even as a gateway to share medical records to either cloud or devices. Moreover, the mobile phone can help in locating the position and fill the gap for lack of information about the position of the patients on GPS localization.

10.4.3 Electronic Health Record (EHR)

The scope of EHR is broader in terms of the user accessing the medical data from the ubiquitous environment and patients' information is collected from different sources such as laboratories, pharmacies, sensors, and wearable devices. This information can assist healthcare and government agencies to analyze patients' behavior, disease severity, and groundbreaking discoveries [5].

10.5 IoT Based Healthcare Management Framework

Considering the rapid growth of emerging technologies like wireless sensor networks, cloud computing, big data analytics, machine learning and mobile computing, an applicable and effective IoT based healthcare management framework must leverage the advancement of these technologies. In this section, the authors have provided a framework that captures the complete view of the IoT based healthcare ecosystem. IoT based Healthcare Management framework is divided into five layers namely the data collection layer, connectivity layer, cloud layer, application layer and consumer layer as depicted in Figure 10.1.

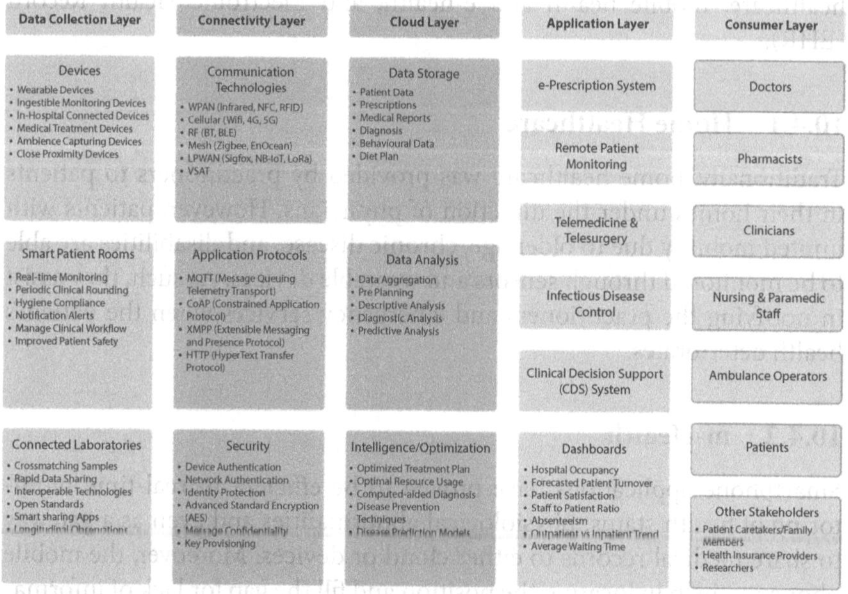

Figure 10.1 IoT-based healthcare management framework.

In the subsequent sections, the authors have discussed each of these layers to better understand their role, significance, and possible use cases from a healthcare perspective.

10.5.1 Data Collection Layer

It is now easy to reduce the risks of human errors at the time of data collection, via a way of automatic medical data collection approach. In healthcare, human error is the main source of transmission or collection of flawed information, thereby causing serious threats to patients' health. By utilizing IoT based connected devices, patient information can be seamlessly coordinated among distributed stakeholders and healthcare providers. This layer collects data from various sources like wearable gadgets, proximity sensors, smart patient rooms and connected laboratories for a smarter, scientific and more evidence-based healthcare management process.

a. **Devices**
The increased usage of wearable devices has provided an opportunity to constantly monitor the vital signs of a patient for health tracking. These wearable devices are equipped with advanced algorithms to interpret this data into meaningful health information. Patient monitoring has been dramatically improved with the help of smart watches and closed proximity devices containing specialized sensors for recording blood pressure, heart rate, electrocardiogram (ECG), pulse, body temperature and glucose levels and raising alarms in case of irregularities. There are various other sources of data collection that include ingestible monitoring devices (camera pill, pulse sensor), in-hospital connected devices (X-ray, MRI and CT scan), medical treatment devices (implantable device, infusion pump), ambient capture devices (motion sensor, light sensor, vibration sensor, gyroscope, IP camera) and closed proximity devices (proximity sensor) [2]. The technology used by these devices is getting matured with every passing day and is providing more accurate and reliable data.

b. **Smart Patient Rooms**
The use of smart patient rooms is a disruptive concept that can be materialized with the help of IoT based technologies. By using sensor-based systems, room temperature, lighting and adjusting the thermostat can be automatically

controlled. Smart patient rooms provide an ideal choice to implement real-time monitoring, hygiene compliance and notification alerts for improved patient safety. The use of technology is even more significant in critical care rooms, emergency departments and intensive units. Measures can be taken to ensure periodic clinical rounding and managing clinical workflow. Other benefits of smart rooms include instant communication, automated energy savings, integrated meal ordering and room readiness.

c. **Connected Laboratories**
With the help of IoT devices, data sharing can be made easier among the laboratories and other stakeholders of the system. Interoperable technologies in the healthcare sector using open standards will materialize the integration of heterogeneous laboratory testing equipment with smart sharing apps through the use of IoT devices. Connected laboratories will be able to compare patients' previous readings to provide a longitudinal observation of patients' conditions. Laboratories will not be collecting the data that is already available and in access to the doctor through the use of wearable and closed proximity devices.

10.5.2 Connectivity Layer

The connectivity layer is the second layer of IoT based Healthcare Management Framework. It is an important layer to determine the right technology, protocol and security mechanism required for collecting data from the first layer and channelizing it to the cloud layer for further processing. The components of this layer should be carefully selected based on the infrastructure available as well as other important factors like use case requirement, cost consideration and risk analysis to ensure the success of the system.

a. **Communication Technologies**
There are various IoT communication technologies available that are suited for different use cases. The selection of these technologies depends on multiple factors like data rate, distance, energy efficiency and cost consideration.

Wireless Personal Area Networks (WPANs) operates within short distances among a private set of member devices. It allows low cost and low power solutions to be implemented for a variety of devices.

Cellular networks like WIFI, 4G, 5G have increased capacity and access to large coverage areas. These are less prone to other signals but typically are more power conducive and might suffer from intermittent connection.

Radiofrequency (RF) technologies like Bluetooth and NFC possess low range considering the IoT application requirements but they are well suited for specialized use cases like payments and medication labeling.

Low-power WAN (LPWAN) technologies like Sigfox, LoRa and NB-IoT interconnect low-bandwidth, battery-powered devices with extremely low bit rates over long ranges. They are cost-effective and are also able to support a large number of connected devices.

Mesh technologies, such as Zigbee and Z-Wave are more suitable for smart clinics or smart hospitals. They have high data rates but consumes more battery power as compared to LPWAN technologies. Zigbee provides a better range as compared to RF technologies like BT and BLE.

VSAT uses satellite technology and is able to operate at remote places and geographically dispersed sites where other technologies are not accessible. Figure 10.2 shows the comparison of these technologies with respect to data range and rate of transmission.

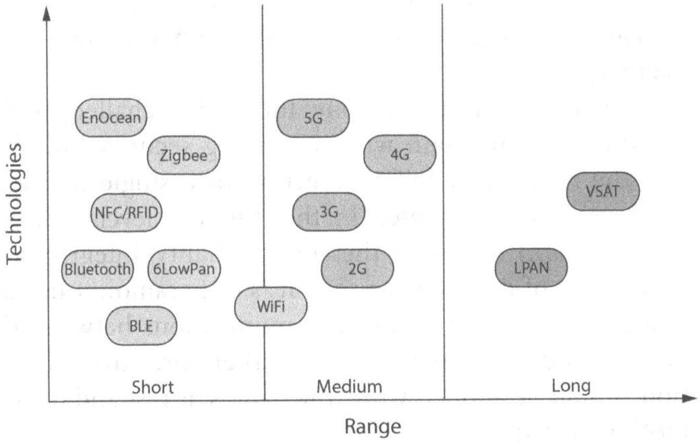

Figure 10.2 Comparison of various IoT technologies.

b. **Application Protocols**

In IoT Based Healthcare Management, communication between sensors, gateways, servers, cloud and finally to the applications and their consumers is extremely important. IoT protocols enable this communication and ensure that it happens in a structured and secure way. There are various protocols used in IoT Based Healthcare Management to provision data exchanged between connected devices. These protocols differ in power, range, and memory consumption. Application protocols allow the gathered data to communicate with various types of healthcare applications. The most used application protocols in IoT Based Healthcare Management are MQTT (Message Queue Telemetry Transport), CoAP (Constrained Application Protocol), XMPP (Extensible Messaging and Presence Protocol), and HTTP (Hypertext Transfer Protocol).

MQTT is a popular protocol for various IoT based applications. It is useful when we have limitations in terms of bandwidth, memory usage and power consumption. CoAP is a lightweight protocol suitable that requires limited power consumption. It is recommended to use for remote monitoring systems and Machine to Machine (M2M) communication. XMPP is useful for real-time communication. It is suitable for applications that are using voice and video messaging. HTTP is comparatively a more power-hungry protocol, but it can be used to easily send and receive data from the cloud layer, considering the communication with the cloud will require Internet based communication [7].

c. **Security**

Privacy and security are one the major challenges for healthcare to maintain with increasing security breaches and the number of IoT connecting to a single network. Authentication is applied in the different levels (Device and Network) to ensure the confidentiality, integrity and availability of data to the users. It is of paramount importance to implement secure communications between the devices and the medical server. Therefore, the encryption should be applied while the data is at rest and data in motion [1, 10].

10.5.3 Cloud Layer

Cloud Layer has key importance as at one side it provides a repository for the storage of healthcare data and at the other side also provides a channel to perform analysis for solving problems, finding trends, and getting meaningful data outcomes. This layer also provides intelligence and optimization to make the results more functional and effective and to unlock new insights. By applying more efficient computational methods and advanced optimization techniques, the number of iterations required can be significantly reduced [13].

a. **Data Storage**
   Data storage represents a repository containing the data gathered at the data collection layer. All stored data will be encrypted to ensure privacy and security. The non-sensitive data can be accessed by the administrators and the sensitive data are available only the users who has given consent by the patient. However, during the emergency situations the sensitive data are available to users (doctors and nurse) through overriding the access privileges.
b. **Analysis**
   Data preprocessing is the first step towards building an intelligent system that is capable of providing informed decisions to medical practitioners and patients. It includes cleaning, imputation, outlier detection and data aggregation. The processed data are used for further analysis. Descriptive analysis, predictive analysis and perspective analysis are the most widely used methods applied in healthcare. These analytics methods are applied to patient data collected through various means are analyzed for identifying the behavior, monitor the trends and patterns, detect any abnormalities, and predict the future outcome of the treatment and correlation [18].
c. **Intelligence/Optimization**
   The task of optimization is to run deterministic and rule-based policies on the patient's medical data. The traditional optimization methods heavily rely on mathematical models which require higher computation power, are complex to implement, and do not adapt to the dynamic environment. As such, the resource constrained IoT devices can be

optimized through Artificial Intelligence (AI) techniques. AI techniques provide an opportunity for healthcare providers to train auto-learning models for disease prediction and to optimize the treatment plan and resources. Specialized AI algorithms can be used to diagnose more people accurately in areas where there is a shortage of doctors and health practitioners [19]. Therefore, electronic health records become intelligent, agile and able to adapt themselves for the betterment of patient care and improve the quality of service.

10.5.4 Application Layer

The application layer has a significantly crucial role in the healthcare management framework. By utilizing this framework, it is much easier to develop cross-platform web and mobile applications that can prove to be game-changer in the coming future.

a) **e-Prescription System**
An e-Prescription (electronic prescription) system is a digital system that enables healthcare professionals to electronically create, sign, and transmit prescriptions to pharmacies. It is an alternative to the traditional paper-based prescription process, which can be time-consuming and prone to errors. With an e-Prescription system, healthcare professionals can generate prescriptions from their computer or mobile device, which are then transmitted directly to a patient's chosen pharmacy. This can help to improve the efficiency and accuracy of the prescription process, as well as reduce the risk of errors or misinterpretations. e-Prescription systems can also provide benefits such as automatic drug interaction checks, real-time access to patient medication histories, and the ability to monitor medication adherence. These systems can improve patient safety and overall quality of care by reducing medication errors and improving medication management.

b) **Remote Patient Monitoring**
Many people worldwide nowadays suffer health-related issues as a result of insufficient healthcare monitoring. As such, chronically ill people, in particular, require health examinations almost every day. Thus, the possibility of a

remote monitoring system will no doubt assist in avoiding the constant check-up visits to the hospital. This eventually will assist in overcoming the lack of doctors and medical facilities. Moreover, patients admitted to intensive care hospitals require close monitoring as well as constant attention to tackle any possible emergency in order to save precious human life. In such situations, IoT can be used for an effective health monitoring system by utilizing various sensor-based applications to gather physiological information to be analyzed and stored on the cloud and afterwards forward it to the relevant caregiver (nurse, doctor as well as family members of patients).

c) **Telemedicine and Telesurgery**

Telemedicine is the use of telecommunication and information technologies to provide remote clinical healthcare services. It allows healthcare professionals to evaluate, diagnose, and treat patients from a distance using various forms of communication technology, such as videoconferencing, remote monitoring, and secure messaging. Telemedicine can help to improve access to healthcare services, especially for patients in rural or remote areas, and can also reduce the cost of healthcare by reducing the need for in-person visits.

Telesurgery, on the other hand, is a more specific application of telemedicine that involves performing surgical procedures remotely using robotic technology. In telesurgery, the surgeon operates a robotic system from a remote location, using high-speed data connections and advanced robotic technology to control surgical instruments and view the surgical site in real-time. This allows for complex surgeries to be performed in remote locations or for expert surgeons to provide consultations and guidance from afar.

Both telemedicine and telesurgery are rapidly growing fields that have the potential to revolutionize the delivery of healthcare services, improve access to care, and enhance patient outcomes. Telemedicine and Telesurgery offers a versatile tool for improving access to care through audiovisual technologies. Telemedicine is beneficial when the patients need frequent visits to the hospital for a general checkup. Whereas, in telesurgery, surgeons perform surgery on patients in different locations [8].

d) **Infectious Disease Control**
In the case of infectious disease control, such as Covid-19, the lack of readily available and relevant data might lead to a disastrous situation. The IoT based healthcare Management framework can be used for identifying disease outbreak locations, tracking of patients, instant availability of test results and generating statistics for healthcare agencies. It can also be used to check hospital occupancy, readmission rates and to create public awareness.

e) **Clinical Decision Support (CDS) System**
Healthcare delivery is improved through the usage of a clinical decision support system to make informed medical decisions with targeted clinical knowledge and patient information. The function of CDS includes alerts and reminders, diagnostics, disease management, prescription, drug control, patient medical reports and overall clinical workflow [16]. Further, it improves the quality of service to the patients with the development of IoT based data-driven and personalized healthcare.

f) **Dashboards**
Dashboards are an effective way of real-time monitoring and visualization of data. It assists the healthcare providers, practitioners and patients to enhance understanding of data and provide a first-hand overview into the pattern of complex data [17]. The presentation of information in condensed form on an interactive dashboard that is easy to read and only displaying data to show the current status and future trends. Each visualization can interact with one another thus highlighting any impactful changes between different features.

10.5.5 Consumer Layer

The consumer layer constitutes the stakeholders of the system that involves doctors, pharmacists, clinicians, radiologists, ambulance operators, patients and their caretakers and family members. By adopting this framework, consumers will be able to leverage key benefits like easy and timely access of data, use of dashboards and smart applications for decision support, as well as integration of various processes in real-time.

a) **Doctors**
Doctors are the real beneficiaries of this framework. Timely assistance from the doctor can save precious lives. Through this system, doctors will have timely access to their respective patient data. By using smart applications, better diagnosis and medical care are possible. In most cases, doctors do not need the patients to come for a regular checkup, but a broad set of information is already accessible using wearable and proximity sensors. For patients in critical condition, they can make use of the live data available through the dashboards and make real-time decisions.

b) **Pharmacists**
Pharmacists will potentially benefit from medication distribution, real-time insurance benefits, and placing orders directly to the agents. The medication distribution is one of the tedious tasks that consume significant time for the pharmacists to verify the patient's allergies and other complications before issuing the medications [3]. With an IoT enabled pharmacy system, the process can be automated that can eventually improve the efficiency and accuracy of distribution.

c) **Clinicians**
The clinician is a healthcare professional who has a more direct connection with the patient and is more familiar with the patient's lifestyle, family history and medical records. In the IoT based healthcare system, several clinical measurements (such as temperature, blood pressure, blood glucose level, heart rate measures, etc.) can be performed without the need of a practitioner and without calling the patient at the healthcare facility. These readings can be communicated to the clinicians even from remote areas that are geographically far from the medical centers.

d) **Nursing and Paramedic Staff**
IoT based healthcare framework will assist nurses and paramedic staff in various aspects including remote patient monitoring, automated inventory management, predictive maintenance, real-time communication, managing shifts, location tracking and receiving alert messages in case of emergency. These features can improve patient outcomes, reduce the workload, and enhance the capabilities of nursing

and paramedic staff. The framework will change the healthcare perspective from a reactive treatment to a proactive treatment to ensure quality of care and patient safety.

e) **Ambulance Operators**
IoT enabled Ambulance service provides a means for the healthcare practitioners to monitor the patients' health in real-time via a remote system and assist the ambulance nurse to perform basic emergency procedures [9]. Moreover, this enables the doctors to prepare for emergency services even before the patient arrives at the hospital. This service involves outfitting ambulances with various IoT devices, sensors, and technologies that enable real-time communication and data exchange between the ambulance, the hospital, and other healthcare providers.

f) **Patients**
Patients with limited mobility, the elderly and one with chronic conditions are benefited from the IoT based healthcare framework. The real-time connectivity of sensors allows healthcare practitioners to assist patients with prescriptions, medication, monitor general wellbeing and further, reduces the need for frequent visits to hospitals. IoT technology can assist elderly patients in various ways, including remote monitoring of health and activity levels, medication management, fall detection, social interaction, and home safety and security. These devices enable healthcare professionals and caregivers to monitor patients' conditions and provide timely interventions, allowing for greater independence and reducing the burden on caregivers.

g) **Other Stakeholders**
Other stakeholders of the system involve patient family members, caretakers, health insurance providers and researchers. All these stakeholders will significantly benefit from the IoT based healthcare framework in terms of automated procedures, fast communication, cost reduction and reliable service provision.

10.6 Future Recommendations for Research

Healthcare is important to all societies because people get ill and hospitals are needed to diagnose, treat and manage different types of ailments and diseases quickly to avoid spreading to others. Therefore, views of patients, as well as medical practitioners, need to be captured to investigate the factors that influence users' acceptance of IoT technologies.

The national cultural impact on the perceived usefulness and perceived ease of use have a very important role in determining the user behavior towards the usage of new technology in the healthcare sector that can be implemented directly into healthcare operations. Therefore, for analyzing the affecting factors, particularly the technology acceptance behavior of users in the context of IoT technologies, the Technology Acceptance Model (TAM) which includes national cultures dimensions such as Individualism/Collectivism, Power Distance Index (PDI), Avoidance of Uncertainty (UA) and Masculinity-Femininity (MA) may be added for studying consumer's perceptions of usefulness, perceived ease of use, and predictive of user's intention to use IoT technology.

Considering the recent outbreak of COVID-19, an integrated and systematic healthcare plan is inevitable. Such pandemics can create emergencies at the national and international level and need to be dealt with expeditiously [15]. An applicable IoT based healthcare framework must have the capability of managing such catastrophes by enforcing all possible measures and by taking all constituents in confidence in a more coordinated, swift and efficient manner.

The potential future research directions for the role of predictive analytics in healthcare include integrating predictive analytics with electronic health records (EHR), utilizing social determinants of health data, improving the explainability and interpretability of predictive models, mitigating bias in predictive models, combining predictive analytics with precision medicine, addressing data privacy concerns, and exploring the potential of predictive analytics for public health applications.

10.7 Conclusion

In conclusion, this chapter has provided a comprehensive overview of the factors influencing the utilization of IoT technology in the healthcare sector globally. The authors have explored the various drivers and trends that have contributed to the growing adoption of IoT in healthcare, including the increasing need for personalized healthcare solutions and the potential benefits it offers in terms of improved patient outcomes. This chapter also focuses on the role of predictive analytics in healthcare. From wearable devices and remote patient monitoring to smart medication dispensers and connected healthcare systems, IoT has the potential to revolutionize the way healthcare services are delivered. By leveraging IoT technologies, healthcare providers can collect real-time patient data, enable remote monitoring and diagnostics, and facilitate personalized treatment plans tailored to individual needs. Finally, the authors have formulated a framework highlighting the functionality and significance of IoT that can be implemented to advance healthcare operations. This framework integrates various IoT devices, systems, and platforms to enable seamless data exchange, real-time monitoring, and efficient coordination among healthcare stakeholders. By implementing this framework, healthcare organizations can streamline their operations, improve resource allocation, and enhance patient care and satisfaction.

References

1. Akinkunmi, O., Rana, M.E., others, Privacy preserving data publishing anonymization methods for limiting malicious attacks in healthcare records. *J. Comput. Theor. Nanosci.*, *16*, 8, 3538–3543, 2019.
2. Albert, J., Rana, M.E., V, S., InsurFit - client health monitoring system for health insurance companies using smart wearable devices and data visualization. *Int. J. Pharm. Res.*, *12*, 02, 2020, https://doi.org/10.31838/ijpr/2020.12.02.0136.
3. Basingab, M., Distributed simulation with multi-agents for IoT in a retail pharmacy facility. *Information*, 11, 11, 2020, https://doi.org/10.3390/info11110527.
4. Conn, J., Hospitals achieve 96% EHR adoption rate; data exchange still needs work, 2016, May 31. Retrieved March 21, 2024, from Modern healthcare.com website: https://www.modernhealthcare.com/article/20160531/NEWS/160539990/hospitals-achieve-96-ehr-adoption-rate-data-exchange-still-needs-work
5. Ehsan Rana, M., Kubbo, M., Jayabalan, M., Privacy and security challenges towards cloud based access control in electronic health records. *Asian J. Inf. Technol.*, *16*, 2, 274–281, 2017, https://doi.org/10.36478/ajit.2017.274.281.
6. Top 10 countries for EHR adoption. 2013, June 27. Retrieved March 21, 2024, from Beckershospitalreview.com website: https://www.beckershospitalreview.com/healthcare-information-technology/top-10-countries-for-ehr-adoption.html
7. Jayabalan, M. and Oadaniel, T., Continuous and transparent access control framework for electronic health records: A preliminary study. *Proceedings - 2017 2nd International Conferences on Information Technology, Information Systems and Electrical Engineering, ICITISEE 2017, 2018-Janua*, pp. 165–170, 2018, https://doi.org/10.1109/ICITISEE.2017.8285487.

8. Jin, M.L., Brown, M.M., Dhir, P., Nirmalan, A., Edwards, P.A., Telemedicine, telementoring, and telesurgery for surgical practices. *Curr. Probl. Surg.*, 100987, 2021, https://doi.org/10.1016/j.cpsurg.2021.100987.
9. Lakkis, S.I., and Elshakankiri, M., IoT based emergency and operational services in medical care systems. *2017 Internet of Things Business Models, Users, and Networks*, pp. 1–5, 2017, https://doi.org/10.1109/CTTE.2017.8260983.
10. Laplante, P.A. and Laplante, N., The internet of things in healthcare: Potential applications and challenges. *IT Prof.*, *18*, 3, 2–4, 2016.
11. Mehra, A., IoT in Healthcare Market, 2023, March 21. Marketsandmarkets.com. https://www.marketsandmarkets.com/PressReleases/iot-healthcare.asp
12. Musunuru, K., Patient's choice or switching attitude towards health care services in private hospitals. *Int. J. Eng. Manage. Sci.*, *2*, 4, 172–178, 2011.
13. Naidu, A., Factors affecting patient satisfaction and healthcare quality. *Int. J. Health Care Qual. Assur.*, *22*, 4, 366–381, 2009, https://doi.org/10.1108/09526860910964834.
14. Nithya, B. and Ilango, V., Predictive analytics in health care using machine learning tools and techniques, in: *2017 International Conference on Intelligent Computing and Control Systems (ICICCS)*, pp. 492–499, IEEE, 2017, June.
15. Shah, H., *How IoT can help detect and control infectious disease outbreaks in real-time*, Chiefhealthcareexecutive, 2017, https://www.chiefhealthcare-executive.com/view/how-internet-of-things-helps-detect-and-control-infectious-disease-outbreaks-in-realtime.
16. Sutton, R.T., Pincock, D., Baumgart, D.C., Sadowski, D.C., Fedorak, R.N., Kroeker, K.I., An overview of clinical decision support systems: Benefits, risks, and strategies for success. *NPJ Digital Med.*, *3*, 1, 17, 2020, https://doi.org/10.1038/s41746-020-0221-y.
17. Wang, Y., Kung, L., Byrd, T.A., Big data analytics: Understanding its capabilities and potential benefits for healthcare organizations. *Technol. Forecasting Soc. Change*, *126*, 3–13, 2018, https://doi.org/https://doi.org/10.1016/j.techfore.2015.12.019.
18. Witten, I.H., Frank, E., Hall, M.A., *Data mining: Practical machine learning tools and techniques*, Morgan Kaufmann, 2011.
19. Yang, H., Alphones, A., Xiong, Z., Niyato, D., Zhao, J., Wu, K., Artificial-intelligence-enabled intelligent 6G networks. *IEEE Network*, *34*, 6, 272–280, 2020, https://doi.org/10.1109/MNET.011.2000195.

11

An Analytical Study on Depression Detection Using Machine Learning

Angelia Melani Adrian* **and Junaidy Budi Sanger**†

Informatics Engineering Department, Universitas Katolik De La Salle, Manado, Indonesia

Abstract

Worldwide, 264 million people suffer from depression, which is one of the main causes of inefficiency. Poor working environments can cause a variety of physical and mental problems. According to studies, people are afraid to seek help because of the societal stigma associated with mental health issues. Considering the height of machine learning, we can use various machine learning algorithms to predict depression in humans. Depression is a frequent and significant medical condition that adversely affects our thoughts, feelings, and actions. It might be fatal since a person in this mindset has given up hope and only perceives the negative sides of the circumstance. According to a 2019 WHO report, approximately 3 million people globally between the ages of 17–25 and 40–70 suffer from depression. Thus in such cases, machine learning models can be developed on a training database to determine whether or not a person is depressed. Users can interact with their loved ones and post their ideas, photos, and videos that reflect their sentiments, moods, and emotions on social media, which is a fantastic source of realistic data from online users. This opens up the possibility of analyzing social networking data to detect users' emotions and moods in order to investigate their feelings and behavior when interacting on these online platforms. Many researchers have attempted to predict depression using machine learning approaches in the past. The purpose of this paper is to review the effectiveness of ML strategies for the early prediction of depression in humans.

Keywords: Depression, machine learning, decision tree, random forest, stress

**Corresponding author*: madrian@unikadelasalle.ac.id

†Author has contributed significantly to the experimental design and data analysis.

Hrudaya Kumar Tripathy, Sushruta Mishra, Minakhi Rout, S. Balamurugan and Samaresh Mishra (eds.) Optimized Computational Intelligence Driven Decision-Making: Theory, Application and Challenges, (215–228) © 2024 Scrivener Publishing LLC

11.1 Introduction

Depression is a type of emotional illness. It's characterized as sadness, loss, or rage that interferes with one's day-to-day activities. Stress is dealt with in a variety of ways by people. It can obstruct your regular tasks, resulting in lost time and poorer production. It can also have an impact on relationships and other chronic illnesses. Emotions are really crucial in our life. Emotions come in a variety of forms that influence how we live and express ourselves. Many times in our lives, our emotions are influenced by the decisions we make, the actions we take, and the vision we have. Joy, sorrow, disgust, fear, surprise, and anger are the six traditional classifications for the various sorts of emotions we experience. Although most of us experience feelings of delight, surprise, and terror on a regular basis, we also experience sadness, loneliness, and depression from time to time as one of the most common responses to the things we've lost, life troubles, or low self-esteem. Prolonged exposure to these emotions can cause physical symptoms that make it difficult for people to live a normal and active life, while also providing insight into depression. Grief is a natural and healthy reaction to the loss of someone we care about. Traumatic, gloomy situations or challenges are common causes of sadness. Depression is an unique emotional state that lasts a long time and impacts our thoughts, ideas, and actions. A difficult or demanding scenario may not necessarily lead to depression. It manifests itself in men as weariness, irritable nature, and rage. It manifests itself in women as melancholy, remorse, and emptiness. When young children are isolated from their parents, they exhibit school rejection and anxiety. Depressed teenagers are prone to wrath and getting into mischief. Anxiety, food disorders, and substance misuse are common among the children. Depression in adults can be mild, as they are less prone to identify feelings of melancholy or bereavement, as well as health problems. To determine the presence of sadness or depression, mental health professionals follow the American Psychiatric Association Diagnostic and Statistical Manual of Mental Disruption (DSM-5 terminology). At least 5 of the following symptoms should hold well for 2 weeks. Conditions for DSM-5 include various possible symptoms of depression, which includes:

- Feeling depressed all day long or on a regular basis
- Frequent feelings of sadness, hopelessness, or anxiety
- Lack of enthusiasm and enjoyment in the things that made you happy

- Sleep problems, or excessive sleepiness
- Obesity or weight loss resulting from an eating disorder, or overeating
- Feelings of guilt or unnecessary exaggeration or unprofitable
- Inability to concentrate or make decisions
- Thinking of self-harm or suicidal thoughts or actions, coupled with excessive thoughts about fear of dying
- You experience physical problems such as headaches, fatigue, stomach aches, or sexual dysfunction.
- Having a hard time concentrating or remembering things.

Depression can be caused because of various reasons, such as

- **Brain chemistry:** It can be caused by abnormalities in brain chemical levels.
- **Drug abuse:** Prolong use of drugs and alcohol can even cause depression.
- **Childhood trauma:** Past traumas can sometimes influence how our bodies react to stressful situations.
- **Medical Problems:** Insomnia, chronic disease, ADHD, or chronic pain are among some of the health issues that can put you at risk for depression.
- **Heredity:** If you have a family member who is depressed, you are more likely to be depressed.

11.2 Literature Survey

The authors of [1] proposed a facial expression recognition computer vision software (AFER). It majorly consists of three steps: the very first step involves face detection, the second step involves facial feature extraction from facial characteristics, and the third and the final phase is the classification of these facial characteristics. They employed the open-source code library (OpenCV) for face recognition, which employs a face recognition algorithm based on Viola and Jones' characteristics. Next, for the facial feature extraction, we do the segmentation by dividing the face into 3 parts: eyes, eyebrows, nose, and mouth. Using the eyes, eyebrows, nose, and mouth FCPs extraction procedures, we perform segmentation for the facial characteristic points (FCPs) present in each face component. Then we extract features from the face image to create a feature vector, which has two sorts of features: appearance and geometric features, which are utilized to classify facial expressions. This feature vector is now fed into the radial basis function of the facial recognition artificial neural network. The AFER system was able to reliably classify facial expressions with accuracy rates of 90 to 99 percent in a person-dependent dataset and 83 to 100 percent in a person-independent dataset, according to the findings. The authors of [2] worked on depression detection in college students as they are highly ignorant about their mental state. For the paper two things have been considered, one is the student questionnaire and another is the student video. Here they tried to detect depression using video analysis for different emotions such as joy, disgust, disdain, and frustration. Depression detection has been performed by looking at the amount of negativity in the video. The various algorithms used here are Gabor filters, Viola and Jones, and SVM. The author of [3] researched the state of mind of different people and used social networks to evaluate whether a person is depressed or not. They met with 14 distinct Twitter users in person, half of whom were discouraged and the other half were not. They also looked at the future implications of social media for depressed people and advised them to fight depression by using social media to solve their issues [4]. Many studies have previously used machine learning methods such as Random Forest Tree, Support Vector Machine (SVM), and Convolution Neural Network (CNN) for data gathering and classification from blog posts to predict sadness and anxiety. Several strategies have been employed for text encoding, including title modeling, Bag-of-Words, and Term Frequency - Inverse Document Frequency (TF–IDF). Furthermore, Python programming was employed for model testing, with CNN delivering the best result of all classifiers, with an accuracy score of 78 percent and a recall score of 0.72 [5]. The authors of [6] discovered how individuals' Facebook behaviors are linked to their identities as measured by the traditional Five-Factor Model. They examined the association between Facebook users' identities and their profile attributes. For instance, the size and strength of their friends' networks, the amount of images transferred, the number of times they've gone somewhere, the number of subscribers gained, and the number of times a user has been mentioned in posts are all factors to consider. The author of [7] demonstrated a new emotional testing approach on Facebook that proposes that data about user limitations (positive, neutral, or negative), as communicated through the messages they compose; as well as indicating the limit of general user experience and identifying major enthusiastic changes, can be extracted from user-generated messages. A study has been conducted by the authors of [8] to diagnose depression using textual and audio modes. The study aims at collecting the data and improving the analysis from the text and audio features. The F1-score's mean is then examined and recorded to find the optimal machine learning techniques. As a consequence, the random forest algorithm is the top performer, with an average F1-score of 0.73. With an average F1-score of 0.50, XGBoost, also known as extreme gradient boosting, proved to be the best for audio features, which is far better as compared to other machine learning algorithms. The authors of [9] in a study proposed a framework that uses physiological signals for assessing affected areas in a depressed and anxious individual. The method uses five predetermined classes to try to anticipate an individual's affected areas: neutral, relaxed, startled, anxious, and very apprehensive. Artificial neural networks, neuro-fuzzy systems, random forest, and support vector machines are some of the machine learning algorithms used in this study. With an accuracy score of 84.3 percent, the neuro-fuzzy system came out on top, followed by random forest with an accuracy score of 80.83 percent. However, artificial neural networks and support vector machines achieved accuracies of 78.5 percent and 77.33 percent, respectively. The authors of [10] proposed a methodology for detecting early indicators of anxiety and depression. It uses psychological testing as well as ML methods like SVM, KNN, CNN, and linear discriminant analysis to categorize anxiety and depression levels across two datasets. As a consequence, the convolutional neural network (CNN) had the best accuracy, with 96 percent for anxiety and 96.8 percent for depression. The support vector machine produced excellent results as well, with an accuracy of 95 percent for anxiety and 95.8 percent for depression. The linear discriminant analysis had a score of 93 percent for anxiety and 87.9 percent for depression, but the K-nearest neighbor had the lowest accuracy of all, with a score of 70.96 percent for anxiety and 81.82 percent for depression. As a result, psychologists and counselors may find the convolutional neural network (CNN) to be a useful model for effective treatment. After giving birth to a child, mothers often experience some kind of depression known as Postpartum Depression (PDD). The authors of [11] aimed to diagnose this depression. They compared the performance of the Functional Gradient Boosting algorithm with the performance of different standard ML algorithms and deduced that the Functional Gradient Boosting method exhibited the best outcomes in diagnosing PPD using data from 173 mothers of various ethnic backgrounds. The authors of [12] also developed a mobile application to diagnose PPD in 2015. Using the psychometric, clinical, and socio-economic data of 1397 new moms, they evaluated the performance of Logistic Regression, Naive Bayes, Support Vector Machine, & Artificial Neural Networks. Later, they claimed that their research was the first to show that a Clinical Decision Support Framework could detect PPD within the first week following birth.

The authors of [13] used LIWC (Linguistic Inquiry and Word Count Tool) as a tool for classification. Among the four variants of the LIWC used in the experiment, the LIWC with ηgram model proved to be the best with an accuracy of 81.8%. The authors of [14] consider the advancement in NLP techniques & their characteristics. The authors of [15] use an NLP mechanism named MTERMS as a classification model for classifying free-text. From MTERM's knowledge-based decision tree they obtained an F-measure of 89.6%. The authors of [16] compare and analyze the various sources of data on the internet in order to collect data for NLP tools. The result shows that online support groups and blogs are the best sources of reliable data for NLP research. The method in [17] includes the following modules: Face Acquisition and Face Detection, where images are captured by the camera or obtained using some standard images database which is pre-processed automatically. Here the face is detected using the Viola-Jones algorithm. The main window is further divided into various sub-windows in which the invariant detector rescales the images repeatedly. For feature Extraction and Expression Classification: Local Binary Patterns (LBP) along with Fuzzy Rule Interpolation have been used. After which the textual image local information is extracted. Then a sparse situation in hand is dealt with where a CRI (Compositional Rule of Interference) fails to draw any conclusion. Here they have used the FIVE (Fuzzy Interpolation in Vague Environment) concept. The system uses the Cohn Kanade database and achieved an accuracy of 71%. Researchers have employed various ML algorithms to predict psychological diseases like depression, and the performance of the algorithms has varied depending on the situation; hence, no single algorithm has been shown to be totally suitable for diagnosing depression.

11.3 Proposed System

The proposed system detects depression in an individual using various machine learning techniques. Here we are analyzing the textual dataset of individuals who were clinically diagnosed with depression. This data will help our ML model to detect the signs of depression in the textual input submitted by the user. In order to get the required input, we ask the users to write a random piece of text and submit the same to the application. By doing so we try not to breach the user's privacy, as here we don't ask for access to any of their sensitive social media content. The user can access the system with full privacy which further protects the user from any social stigma and this, in turn, increases the accuracy of the input data that is fed into the system. Further our ML model trained on text classification algorithms analyzes this input text and diagnosis for depression. As we know that depression is not a sudden occurrence and it would need at least a week to appear, so we collect text pieces at a fixed interval of time and try to pass any verdict by combining all these data from the past. The conclusion can be between very depressed, mildly depressed, or not depressed. The final results consist of a short description of our conclusion, along with the graphical representation of depression levels at different times. Graphical representation helps in analyzing the depression patterns of the user. This pattern can help the user to understand their mood swings. These patterns are formed using both the past and the present readings of the user. The purpose of the survey is to analyze and predict depression at an early stage. As the symptoms get worse, the individual may lose control over the mind, which can further lead to disorders [18]. The proposed methodology is focused on the discovery of human emotion and if a feeling of sadness is detected continuous evaluation is performed to distinguish between

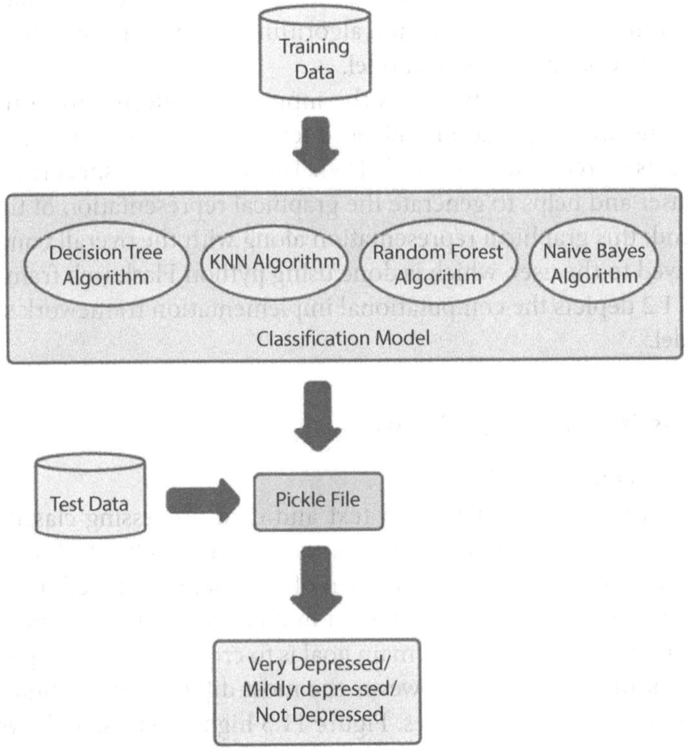

Figure 11.1 Proposed methodology workflow model.

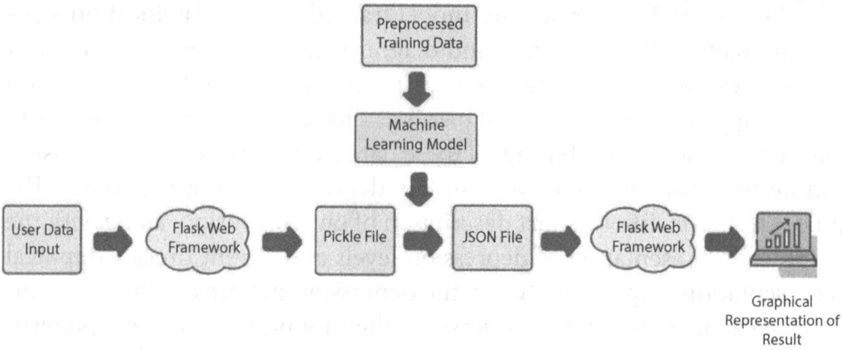

Figure 11.2 Computational implementation framework steps of the model.

sadness and depression. And if a person is not depressed but sad, we can also console and motivate the individual to fight their sadness, by proper counseling. The Machine Learning model takes the user inputted data collected from different clinically tested individuals as the dataset, then this data is preprocessed and given to the model. Then it generates a pickle file by using different text classification algorithms. Figure 11.1 shows the proposed methodology workflow model.

This generated pickle file takes the input data collected from the user through the web application and produces the depression readings that further gets stored in the JSON file [19]. This JSON file is specific to a particular user and helps to generate the graphical representation of the data. In the end, this graphical representation along with the overall conclusion is displayed to the user, which is done using python Flask web framework. Figure 11.2 depicts the computational implementation framework steps of the model.

Text Classification Algorithms

• Decision Tree Algorithm

It is one of the most widely used text and data processing classification methods. Decision tree classifiers are used in a variety of classification applications. A hierarchical breakdown of the data is included in the procedure (only train dataset). D. Morgan first proposed it as a classification algorithm [20]. The algorithm's main goal is to create trees that support the qualities of the data points, however the most difficult part is finding the parent and child level properties. Figure 11.3 highlights a sample working of decision tree.

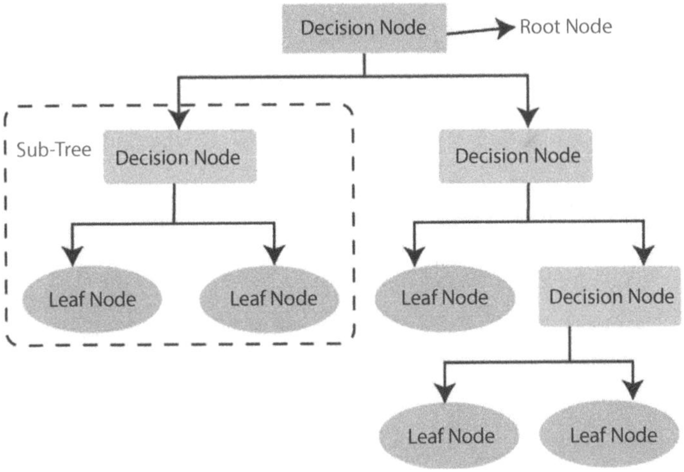

Figure 11.3 A sample demonstration of decision tree.

- **K-Nearest Neighbors Algorithm (KNN)**

It's a commonly used non-parametric classification technique. It's basically a text classification algorithm that has been used for a long time in NLP. KNN algorithm is used to classify data by finding the K nearest matches in training data and then using the label of nearest matches to predict the class [21]. Generally the Euclidean distance is calculated between two data points to find the nearest class to which they would belong. Figure 11.4 shows the working of KNN algorithm.

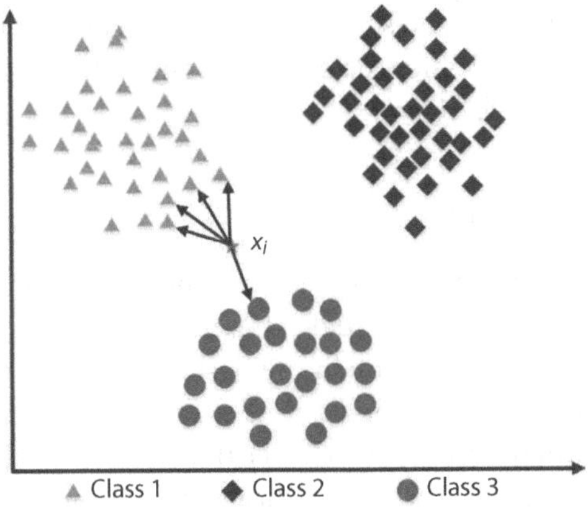

Figure 11.4 A sample demonstration of the working of KNN algorithm.

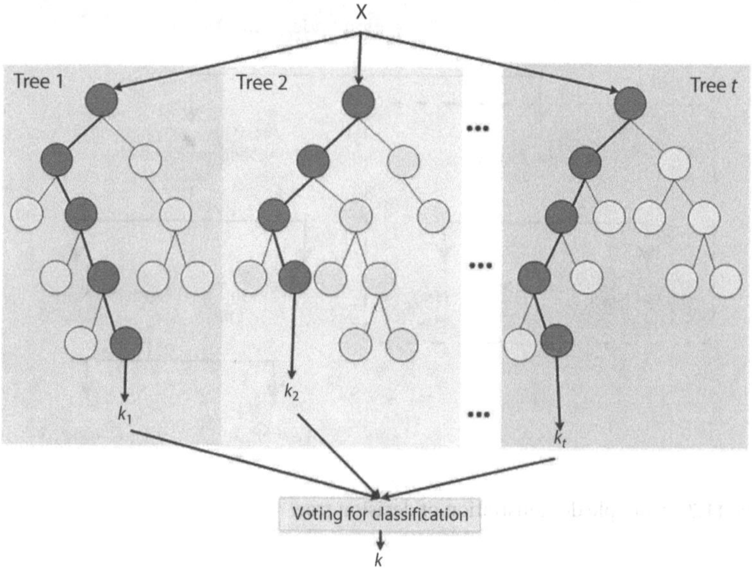

Figure 11.5 A sample demonstration of the working of random forest algorithm.

- **Random Forest Algorithm**

It's a text categorization algorithm based on ensemble learning. T. Kam Ho proposed the Supervised Machine Learning Algorithm in 1995, where he used t number of trees simultaneously. This method is generally used in various Regression and Classification problems. For regression problems it deals with continuous data, whereas for classification it mainly deals with categorical data [22]. However, it performs better for classification problems. Here we try to build decision trees on various data samples and consider the majority of votes in case of classification problems and the average of votes for regression problems. Figure 11.5 depicts the working of random forest model.

Naive Bayes Algorithm

It is a text categorization technique developed by Thomas Bayes. It has been used for document and textual categorizations since the 1950s. Many scholars have used this technique in the past for a variety of purposes [23]. Its primary application is in the field of data retrieval. It is based on the Bayes theorem, which states that we can calculate the conditional probability of two events based on the probabilities of each individual event.

$P(A|B) = [P(B|A).P(A)] / P(B)$

A, B = events
P(A|B) = probability of A given B is true
P(B|A) = probability of B given A is true
P(A), P(B) = the independent probabilities of A and B

The user must submit their input text as the initial step in the application, and this text input can be of any size, i.e. there is no limit to the input size. The front page of the application is highly user interactive, and allows the user to upload their text easily to the application. To diagnose depression, the machine learning model must be trained using a training dataset that provides the model with information on the depression pattern. The collected and cleaned dataset is subsequently sent to the model for training. The trained model is then turned to a pickle file in order to detect future depression levels. The text uploaded by the user is preprocessed, cleaned and then given to our model. Then using the generated pickle file, the input is processed in order to predict the depression levels [24]. These detected depression levels are then saved for future references which in turn saves the processing time and memory and for this purpose we record the depression levels in a local file where the number of rows is equivalent to the user entries. Then to plot a graph that represents the level of depression within a stipulated timeline we go through the depression levels of the current and previous user entries. This graph helps the user in analyzing the emotional fluctuations, which further gets us to a conclusion that whether the individual is very depressed, mildly depressed or not depressed. The final report contains the graphical representation along with the derived conclusion. In the end, this report is provided to the user, through a webpage.

11.4 Challenges of Machine Learning in Depression Detection

- Depression is not a sudden occurrence, so it might take at least a week to set in, thus one needs to wait in order to get proper reports.
- The data that we are trying to get from the questionnaires through the users might not be highly accurate as the answers can be scripted.
- Also whatever data we are collecting from social media might not show the actual emotions of an individual, as life on these platforms is mostly far away from reality.

11.5 Conclusion and Future Work

The proposed system aims to reduce the human intervention for detecting depression. Earlier, one had to depend on doctors in order to identify depression through standard methodologies. On the contrary, our application automates this process and provides a detailed report to the individual in private. This diagnostic report is highly accurate, as it is obtained from a large dataset on which our model is trained. Future developments on the proposed system will focus on making the application more accessible to people. This can be achieved by adding different diagnostic languages, especially the native languages spoken in India. Also, we can further create an Emotion Recognition System that enables real-time analysis, and interpretation of an individual's emotional states by just recording a facial video. As it is assumed that the facial expression is activated for the moment when a person is emotional that way emotion detection can be achieved by looking at the facial expressions.

References

1. Youssif, A.A.A. and Asker, W.A.A., Automatic facial expression recognition system based on geometric and appearance features. *J. Comput. Inf. Sci.*, 4, 115–124.
2. Venkataraman, D. and Parameswaran, N.S., Extraction of facial features for depression detection among students. *Int. J. Pure Appl. Math*, 118, 7, 455–463, 2018.
3. Park, M., McDonald, D.W., Cha, M., Perception differences between the depressed and non-depressed users in Twitter, in: *ICWSM*, vol. 9, pp. 217–226, 2013.
4. Wee, J. *et al.*, The influence of depression and personality on social networking. *Comput. Hum. Behav.*, 74, 45–52, 2017, DOI: 10.1016/j.chb.2017.04.003.
5. Tyshchenko, Y., *Depression and anxiety detection from blog posts data*, pp. 6-46, Nature Precis. Sci., Inst. Comput. Sci., Univ. Tartu, Tartu, Estonia, 2018.
6. Rafiqul Islam, Md., Kabir, M.A., Ahmed, A., Kamal, A.R.M., Wang, H., Ulhaq, A., Depression detection from social network data using machine learning techniques. *Health Inf. Sci. Syst.*, 6, 1, 8, 2018 Dec, Published online 2018 Aug 27. DOI: 10.1007/s13755-018-0046-0.
7. Ortigosa, A., Martín, J.M., Carro, R.M., Sentiment analysis in Facebook and its application to e-learning. *Comput. Hum. Behav.*, 31, 527–541, 2014, DOI: 10.1016/j.chb.2013.05.024.
8. Xu, A.J., Flannery, M.A., Gao, Y., Wu, Y., Machine learning for mental health detection, 2019. [online] Available: https://digitalcommons.wpi.edu/mqp-all/6732/.

9. Katsis, C.D., Katertsidis, N.S., Fotiadis, D.I., An integrated system based on physiological signals for the assessment of affective states in patients with anxiety disorders. *Biomed. Signal Process. Control*, 6, 3, 261–268, 2011.
10. Ahmed, A., Sultana, R., Ullas, M.T.R., Begom, M., Rahi, M.M.I., Alam, M.A., A machine learning approach to detect depression and anxiety using supervised learning, in: *Proceedings of the 2020 IEEE Asia-Pacific Conference on Computer Science and Data Engineering (CSDE)*, Gold Coast, Australia, 2019.
11. Natarajan, S., Prabhakar, A., Ramanan, N., Bagilone, A., Siek, K., Connelly, K., Boosting for postpartum depression prediction. *2017 IEEE/ACM International Conference on Connected Health: Applications, Systems and Engineering Technologies (CHASE)*, Philadelphia, PA, USA, pp. 232–240, 2017, doi: 10.1109/CHASE.2017.82.
12. Jiménez-Serrano, S., Tortajada, S., García-Gómez, J.M., A mobile health application to predict postpartum depression based on machine learning. *Telemed. e-Health*, 21, 7, 567–574, 2015.
13. Wolohan, J.T., Hiraga, M., Mukherjee, A., Sayyed, Z.A., Detecting linguistic traces of depression in topic-restricted text: Attending to self-stigmatized depression with NLP. *First International Workshop on Language Cognition and Computational Models*, vol. 3, Issue 5, pp. 11–21, 2018.
14. Hirschberg, J. and Manning, C.D., Advances in natural language processing. 261–266, 2015.
15. Zhou, L., Baughman, A.W., Leia, V.J., Laib, K.H., Identifying patients with depression using free-text clinical documents. *Health Technol. Inform.*, 3, 6, 629–634, 2015.
16. Srinivasan, P., Menakapriya, M., Suresh, P., *Web-based unique link recommendations using fuzzy weight ranking algorithms*, vol. 14, TAGAJOURNAL, Swansea Printing Technology Ltd, pp. 3200–3209, 2018, ISSN: 1748-0345.
17. Ofor, W.D., Nwiabu, N.D., Matthias, D., Face detection and expression recognition using fuzzy rule interpolation.
18. Sabab Zulfikar, Md., Kabir, N., Biswas, A.A., Nazneen, T., Uddin, M.S., An in-depth analysis of machine learning approaches to predict depression. *Curr. Res. Behav. Sci.*, 2, 100044, 2021, ISSN 2666-5182, https://doi.org/10.1016/j.crbeha.2021.100044. (https://www.sciencedirect.com/science/article/pii/S2666518221000310).
19. Chung, J. and Teo, J., Mental health prediction using machine learning: Taxonomy, applications, and challenge. *Appl. Comput. Intell. Soft Comput.*, 2022, 1–19, 2022, Article ID 9970363, https://doi.org/10.1155/2022/9970363.
20. Ramalingam, D., Sharma, V., Zar, P., Study of depression analysis using machine learning techniques. *Int. J. Innov. Technol. Exploring Eng. (IJITEE)*, 8, 7C2, May 2019, ISSN: 2278-3075.
21. Victor, E., Aghajan, Z.M., Sewart, A.R., Christian, R., Detecting depression using a framework combining deep multimodal neural networks with

a purpose-built automated evaluation, September 30 2018, https://doi.org/10.31234/osf.io/vqpcz.

22. Sudha, K., Sreemathi, S., Nathiya, B., RahiniPriya, D., Depression detection using machine learning. *Int. J. Res. Adv. Dev. (IJRAD)*, ISSN: 2581-4451.
23. Kowsari, K., Meimandi, K.J., Heidarysafa, M., Mendu, S., Barnes, L.E., Brown, D.E., Text classification algorithms: A survey. *Information*, 10, 4, 150, 2019.
24. Oak, S., Depression detection and analysis by using speech or text as the input. *The AAAI 2017 Spring Symposium*, Vol. 1, pp. 710–714, 2017.

12

Revolutionizing Healthcare: Empowering Faster Treatment with IoT-Powered Smart Healthcare

Prerna Kumari*, Rupali Agarwal and Shruti Kumari

School of Computer Engineering, KIIT Deemed to be University, Bhubaneswar, Odisha, India

Abstract

Smart cities around the world are adapting new forms of developmental activities because of Internet of Things (IoT) technologies. The IoT is the integration of different hardware, networks, and sensors that enables data collection and sharing from these things efficiently. There are several such devices with medical sensors in used in it. Also despite continuous monitoring for 24 hours, there may be situations when the doctor is not alerted at time of emergency. Exchanging data and information with experts, worried family members, and relatives may also be challenging, although the technology to enhance these characteristics is now accessible. But most individuals in underdeveloped countries like India cannot afford it. Therefore, the solutions to these problems may be as simple. It is by adding the functionalities to already existing IoT devices that lack them. We can create healthcare systems that are more knowledgeable, reliable, and predictable by making developments in IoT and cyber-physical devices. By fusing various computer vision, machine, and deep learning techniques, the IoT significantly supports the smart healthcare sector. It provides prompt, accurate device services that automatically detect and track patient issues. This paper will evaluate how IoT devices with AI and ML applications are applied in the healthcare sector. Also how databases help to give information about the thing which can help in saving a person's life and help them in providing medical facilities on time. This research will serve as a benchmark for understanding the IoT's impact on the health sector in smart cities.

**Corresponding author*: prernakumari2408@gmail.com

Hrudaya Kumar Tripathy, Sushruta Mishra, Minakhi Rout, S. Balamurugan and Samaresh Mishra (eds.) Optimized Computational Intelligence Driven Decision-Making: Theory, Application and Challenges, (229–250)

***Keywords*:** Wearable, IoT, smart, health, sensor, hospital, aadhar, patient

12.1 Introduction

In today's generation, technology is transforming the healthcare sector all over the world. IoT has a major impact on digital healthcare at different levels which makes the diagnosis process easier and available to most people. By implementing a real-time smart healthcare system in a hospital across the nation, we can use IoT to improve healthcare facilities. The IoT is the inter-connection of hardware, networks, and sensors that enables data collection and sharing from these things over the internet. New information technology has emerged as a result of the merging of smart healthcare. This capability to continuously monitor a patient using a variety of parameters and extrapolate a successful result from the past using regular tracking is what sets IoT in the healthcare system apart [1]. ICUs have a lot of these modern devices with medical sensors. Even with 24-hour monitoring, there is always a chance that an emergency will go unnoticed and unattended due to the lateness in information passing. The doctor was given the alarm-related data so as to build a faster monitoring model and send alert notifications to interconnected units [2]. Exchanges of data and information with experts, concerned family members, and relatives may also be challenging. Although the technology to enhance these characteristics is now accessible, most individuals in these developing countries like India cannot afford it. As a result, the solutions to these problems might be as simple as including these functionalities in already existing products. This enhances health and clinical effects, caregiver experience. It also lowers costs, and improves patient experience [3].Current developments in IoT technology enable cost effective and connected sensor networks for attaining the planned approach. The IoT-based systems collect and synthesize patient information [4], cross-check them against recorded patterns, and examine the illness indicators using data mining techniques as shown in Figure 12.1. IoT offers quick and efficient patient monitoring technologies like intelligent software systems. These are crucial for remote patient monitoring when there is no medical staff present. Nowadays, using non-intrusive devices and deep learning and computer vision algorithms is used for a useful monitoring system (based on IoT-based sensors or cameras) to be offered. Monitoring is more effective when issues with patients are identified based on visual signs, such as body posture, movements,

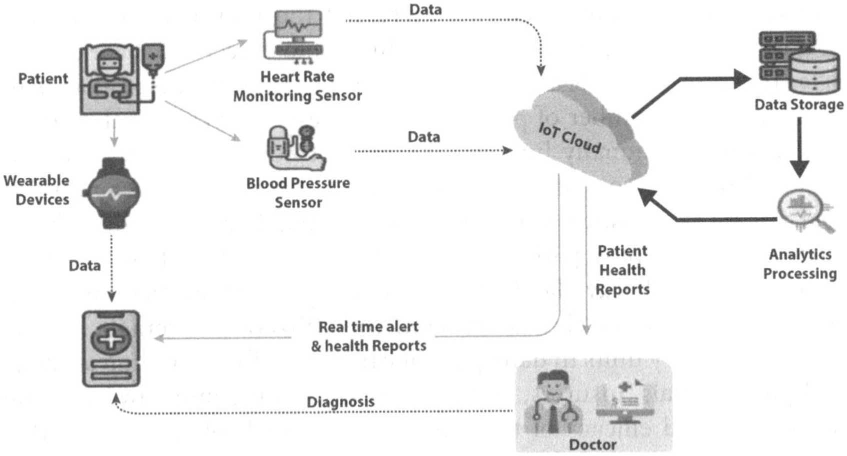

Figure 12.1 Implementation of IoT in smart healthcare.

and facial expressions. Recently, although at a higher cost, researchers have installed specialized hardware and sensors for patient monitoring systems [5]. The bulk of patient monitoring techniques that have been developed and published in the literature either rely on invasive or contact techniques or make use of many cameras or sensors. IoT is a powerful technology that is currently on the rise. It enables computing devices and objects to exchange information and move it from one place to another. Technology has been used in healthcare for a long time to improve testing, imaging, and surgical competence. Patients are cared for using more sophisticated services and improvements in healthcare delivery in the rapidly evolving technological landscape of medicine. Additionally, doctors and patients now have more diagnostic and treatment options as a result of advancements in critical medical technology. Technology has helped doctors improve the service reliability in the areas of diagnosis and treatment and also the screening and prevention. The Internet of Things sensor enhances communication without requiring the more essential human-system or person specific communication. The implementation of IoT sensor devices, as shown in Figure 12.1, has benefited medical centers, IoT based houses, configured modules, personal devices and secure enabled units. These apps have been successfully incorporated into a number of ecological models, involving mobile phones, IoT enabled speakers, and mobile devices. The health sector, which comprises several healthcare services, is one of the many industries where IoT-based sensors are being used. Global population growth and technological advancements increased the prevalence of chronic diseases, which were frequently more difficult to detect and more expensive. The ratio of people affected by chronic illnesses increased as a result of a rise in the number of people whose access to a very usual medical utility was occasionally limited. The majority of respondents thought that additional resources were needed for patient assistance programs and disease diagnostics. It was also observed that because medical functionalities were not so patient oriented, they were not available. In order to increase people's availability and generosity, smart utilities have been made available to the clinical domain. The utilization of remote healthcare and sensory units in daily patient care is rapidly expanding as medical professionals, insurers, and patients all become more aware of the advantages and efficacy of these integrated models of care [6]. In this way we can do more and serve more with less manpower. In this model we use different sensors to detect a patient's condition then by using the internet we find the location of the patient. In this way we can help the patient by providing medical help on time. Also this can help in informing the patient's family about their condition. This can be implemented using IoT and AI that can help the needy people on time.

12.1.1 Main Contribution of the Paper

12.1.1.1 Using IoT to Track Abnormalities

IoT can be utilized to monitor patient vital signs and other health metrics. By tracking this data, any abnormalities can be identified and monitored. Consider adding specific examples of the types of data that could be monitored and the specific IoT devices that could be used for this purpose.

12.1.1.2 Emergency Alerts for Patients

Patients can get immediate notifications about emergencies on their mobile devices or other connected devices. Alerts for harmful changes in vital signs or other signs that medical care is required can fall under this category. Think about going into more detail about the specific alerts that can be sent and the mechanism by which they are delivered.

12.1.1.3 Ambulance Notification

When a patient needs emergency medical attention, IoT devices can notify nearby ambulances and hospitals. This can ensure that patients receive prompt medical attention, potentially saving lives. Consider providing more details on how this notification system works, and what data is used to trigger alerts to emergency services.

12.1.1.4 Patient Medical History and Family Contacts

The use of Aadhar to access patient medical history and family contacts can be extremely valuable in an emergency. By having this information readily available, medical professionals can quickly and accurately diagnose and treat patients. Consider expanding on the specific types of medical information that can be accessed, and how the system ensures patient privacy and data security.

12.1.1.5 Early Access to Treatment

By combining all these features, Smart Healthcare can provide patients with the earliest possible access to treatment. This can lead to improved patient outcomes and overall healthcare quality. Consider elaborating on the potential benefits of this approach, and how it could be implemented in different healthcare settings.

12.2 Scope/Motivation

The developing traction of IoT within healthcare has opened opportunities to scale clinical services. By the use of IoT-powered devices, doctors are capable of observing the health of a sick person remotely, thereby making sure about the patient's health conditions and providing ongoing treatment. This has made it easier for patients to get proper treatment due to easier interaction with doctors. Non-critical patients could be treated at home rather than at a hospital by using remote health facilities. It reduces strain both on hospital resources as well as patients. Additionally, it aids in improving accessibility for senior citizens and residents of rural areas.

IoT healthcare systems have been developed in several works for a variety of purposes that are all closely related by using the same enabling technologies. Even though IoT has challenges and limitations, it won't stop it from growing. The IoT in healthcare is immensely helpful. We cannot deny that IoT has already made a huge impact. There are companies trying to make healthcare and IoT successful and soon there will be applications and technologies interacting with many devices.

12.3 Literature Survey

Using the modified particle swarm optimization method, a productive smart sensory driven medical model was produced. The earlier mentioned optimization approach in medical services was used to establish an identification system. The system made use of inertia weight elements, shrinking variables, and other information integrating features, and the effectiveness of particle swarm optimization methods was optimized using these parameters. The disease that is present in the human body was successfully detected using data that was gathered and analyzed by fuzzy rule-based neural networks classifiers. Understanding that the term "behavior analytics" is broad and includes exploratory, treatment specific, estimators, and subjective learning is important. Exploratory learning can be used to determine various clinical choices and their effects on the functionality and results of applications [7]. Diagnostic analysis may also be referred to as medical diagnosis. IoT and cloud computing are also used in a smart healthcare system that can track diabetic patients in real-time [8]. A system that uses wearable sensors and big data analytics to monitor the health of elderly people and deliver individualized care [9]. With the help of the Internet of Things, wearable sensors, and big data analytics, this improved system accurately predicted patient health data from a variety of sources [10]. A smartphone health app was created that forecasts illness in the healthcare sector using IoT sensors. At this point, the database of the UCI records was needed via smartphone features.

The development of an effective IoT-based system for diagnosing heart disease uses a hybrid ensemble learning and linear approach. In order to accurately predict cardiovascular disease, sensory units were applied to collect heart related information, which was further analyzed using the best predictive intelligence models. Making a well-informed choice was made easier through the utilization of a smart module driven validation method, which also reduced the difficulties of the diagnostic process. The developed system's accuracy for detecting cardiovascular disease using IoT sensor data was 88.7%. IoT-based real-time health monitoring systems were used together with deep learning techniques. The system was entirely dedicated to studying real-time ECG data to forecast changes in heart functionalities and assess heart risks level. Additionally, the approach could increase illness detection rates at a lower cost. The IoT-based ECG tracking system reduces treatment process complexity and improves communication between the healthcare industry and patients.

12.4 Smart Technology

ICT (information and communication technology) refers to an enormous variety of technologies, including wearable and smart home technology. It is also frequently used to denote smart healthcare technology and health surveillance systems. Computers, smartphones, voice-controlled systems, touchpad control devices, Independent Digital Assistants, and other gadgets like remote controls that can interact with and control the machinery in a person's workplace are all examples of smart technology. Radio frequency, sound, and infrared extension units are potential system control methods. Smartphones with added features are the main application for smart technologies. Investigating the use of mobile devices for data sharing, like contacting a doctor, is essential to assist people with their health care requirements, such as therapies for sound sleeping control issues or sugar status control in diabetes disorder.

Advanced methodology has been used in healthcare for a long time to improve testing, imaging, and surgical competence. In the rapidly evolving technology environment of medicine, increasingly sophisticated services and innovations in healthcare delivery are deployed to offer patient care. These smart healthcare technologies have helped medical professionals provide better service in the domain of processing and assessment of disease treatment.

Growing these achievements to patients as well as carers remotely as well as at society level is a crucial and ongoing aspect of health interference. Clinical experts, insurance companies, and patients are all becoming more aware of the merits and effectiveness of these hybrid frameworks of care, which is why the use of remote applications and IoT based modules in regular patient care is growing quickly.

12.4.1 IoT-Enabled Healthcare

IoT services, IoT apps, and other solutions are now available in the healthcare sector, making connections within the sector more creative. The cost-effectiveness of therapy is one of the core aims of digital health. Wearable medical technology enables remote patients to obtain diagnosis and treatment. Before being transmitted to the carer via the patient's smartphone, smart wearables gather information on the individual's blood glucose, oxygen saturation, and pulse rate. It is possible to quickly determine the patient's condition using IoT-based sensors-equipped devices. Medical professionals can speak with patients via online medical care rather than having them physically visit a clinic or hospital. By changing bad behavior and embracing better lifestyles, behavior modification is a technique that may aid people in enhancing their health.

12.4.2 Importance of Smart Healthcare System

A few of the participants in the healthcare system are hospitals, patients, physicians, and research organizations. The cornerstones of the smart healthcare sector are the dynamic online service, smart computing, 5G method, big data, cloud service, machine intelligence, and biotechnology. To provide intelligent healthcare, the health care models use many approaches. Patients use wearable technology to continuously monitor their health, to act as virtual assistants in their medical care, and to facilitate telehealth operations; medical staff utilize many complicated medical functionalities to aid systems to improve diagnosis. Doctors' forms address integrated information platforms, which are composed of devices like messaging apps, picture archives, lab data coordination models, and electrical health samples. The usage of mixed reality technology by the surgical robots is helpful.

The supply chain of the hospital and the tracking of individuals and their possessions are both done via radio frequency identification (RFID). RFID helps in collecting data and supports decision-making in combination with integrated management systems. Mobile medical platforms are frequently used to enhance other mobile medical platforms. In scientific research, machine learning is routinely utilized in place of manual drug testing, and big data may be used to identify the right candidates. The cost of medical procedures is reduced, and there may be significantly less risk involved, thanks to smart healthcare technology. The IoT can be incorporated into a smart healthcare system to provide specialized healthcare services like self-service healthcare, resource efficiency, and regional collaboration and exchange.

12.5 Methods and Materials

The model and sensors employed in this study support technological advancements to efficiently analyze and administer lead to treatment to patients through smart healthcare devices in an efficient way. This model is sensor based and its performance is totally dependent upon the sensor's performance. Sensors are an essential element of this IoT based medical model. This medical model provides IoT-enabled healthcare services.

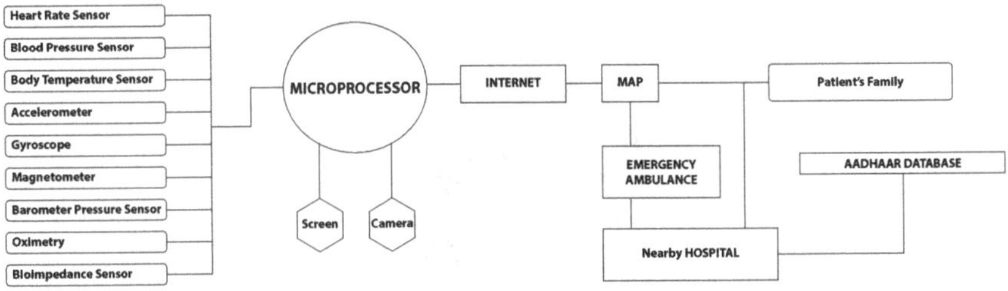

Figure 12.2 Proposed model of smart healthcare.

The healthcare model described in Figure 12.2 in this paper provides the capability of making decisions and facilitating patients to make an emergency call based on data collected related to the detection of the patient's heart rate, heartbeat, body temperature, blood pressure and oxygen level. Better architecture algorithms to provide best longevity of the device along with the sensors used and can also manage cost. In this healthcare model communication channels and sensors are there which are utilized to monitor the IoT based tracking service system with a patient control module. IoT has helped in the implementation of smart devices in many rural areas. The smart device with the help of sensors collects all the medical data of the patient and starts evaluating the current situation and suggests the user make an emergency call to an ambulance helping the patient to reach the appropriate hospital where the doctor provides proper treatment.

12.5.1 Smart Sensors Deployed in Model

Here we need a variety of sensors which will make the devices smart and they are as follows:

12.5.1.1 Heart Rate Sensor

A human's heart rate is measured in Beats per Minute using an optical LED light source and an LED light sensor by a heart rate sensor shown in Figure 12.3. The light shines through skin, and then the sensor helps in computing the light quantity which reflects as shown in Figure 12.4. By analyzing the variations in light reflections caused by the pulsating blood under the skin, heartbeats are detected.

There is an emission of infrared signals into the surface of the body/skin and it estimates the variation in flow of blood when heart beats as a variation in the quantity of light propagated within the body/skin. Penetration of light is feasible where this approach is restricted to limited regions.

Figure 12.3 Heart rate sensor.

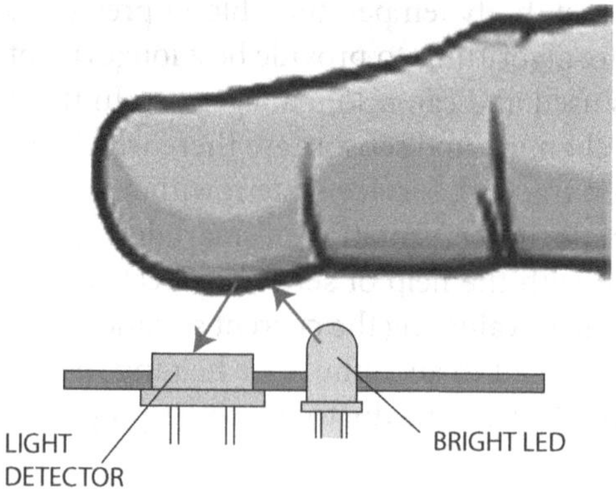

Figure 12.4 Light detector.

12.5.1.2 Blood Pressure Sensor

Smart Devices use the optical heart rate sensor shown in Figure 12.5 or ECG (if available) are used for measuring the blood pressure. The time taken for a pulse to travel from the heart to the wrist is detected by it. This time is also known as Pulse Transit Time (PTT), based on which it also interprets the blood pressure readings. When the pulse transit time is faster then blood pressure is high and if the pulse transit time is slow then the blood pressure is low.

12.5.1.3 Body Temperature Sensor

Figure 12.6 presents a sensor which is designed specifically for measuring the temperature of the human body by focusing on the skin temperature.

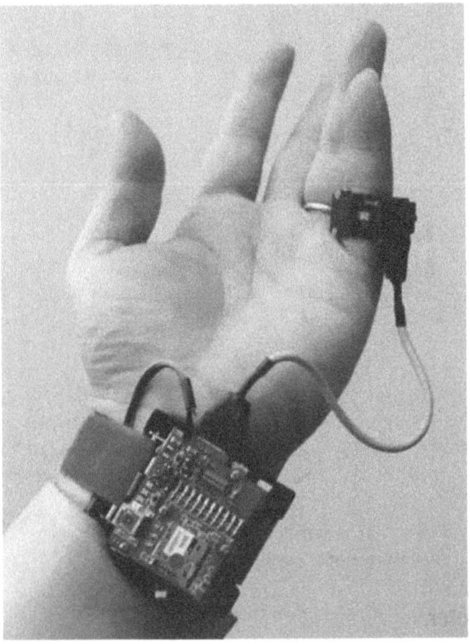

Figure 12.5 Blood pressure sensor.

Figure 12.6 Body temperature sensor.

Even if there is a slight change in the body temperature this sensor is capable of detecting it accurately. It also provides insights into potential fever conditions. These sensors also play an important role in detecting and tracking menstrual cycles.

12.5.1.4 Accelerometer

In Figure 12.7 there is a mechanism for tracking of body movement along the X, Y, and Z axes by counting the number of steps taken by an individual. Person's sleep patterns and distinguishing between periods of sleep and wakefulness are also determined by this tracking system.

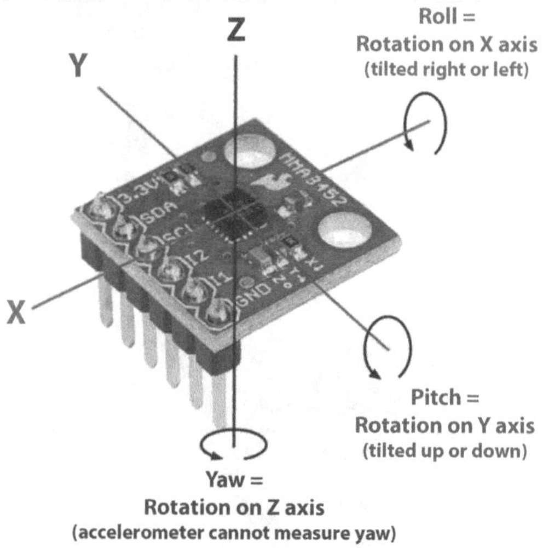

Figure 12.7 Accelerometer.

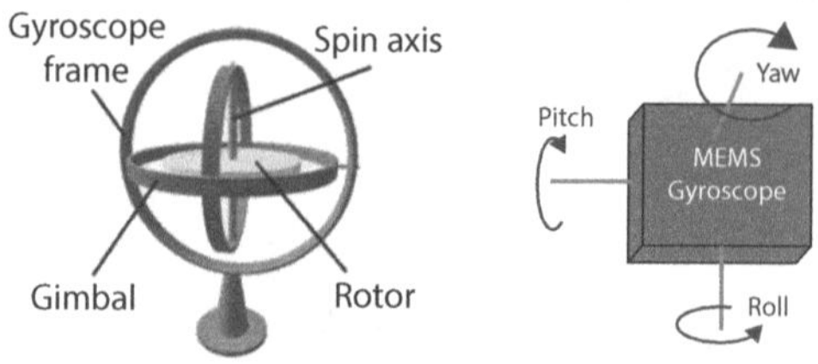

Figure 12.8 Gyroscope.

12.5.1.5 Gyroscope

It measures the angular velocity. It focuses on measuring all other movements other than linear movement shown in Figure 12.8. By combining the 3-axis gyroscope and 3-axis accelerometer sensors, a 6-degree motion tracking system is achieved. To detect body movements, gravity, and speed two sensors are used in order to accurately detect and find motion, gestures, and track them accordingly. This functionality helps in determining various activities performed by the user.

12.5.1.6 Magnetometer

The sensor in Figure 12.9 helps in determining your exact location where you are standing by working with the GPS coordinates.

Figure 12.9 Magnetometer.

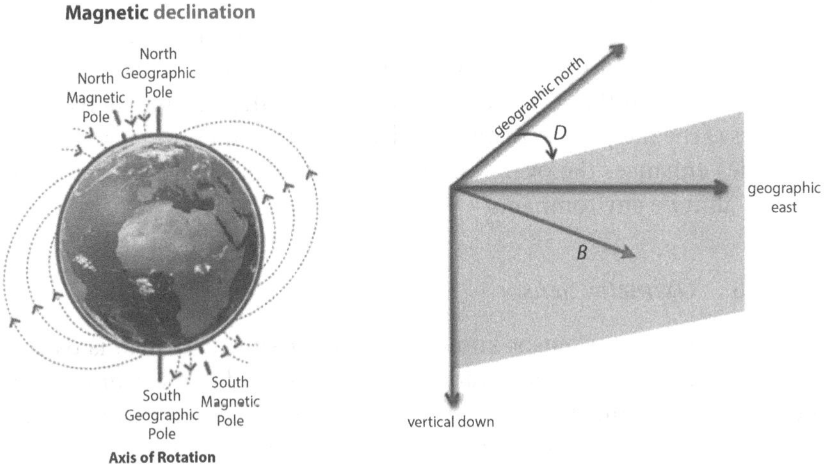

Figure 12.10 Magnetic declination.

Figure 12.10 illustrates a sensor which by sensing the magnetic fields is capable of detecting the precise position in the coordinate space. Both the magnitude (strength) and direction of the magnetic field are measured by this sensor and it provides information necessary for determining the exact position.

12.5.1.7 Barometric Pressure Sensor

Figure 12.11 showcases a sensor which is incorporated in smart wearables and helps in detecting altitude. This data plays a valuable role during activities such as hiking and mountain climbing. The altitude data, combined

Figure 12.11 Barometric pressure sensor.

with other measurements obtained by the smart wearable devices, provides crucial insights into the environmental conditions and gives necessary information to the user. For instance, the sensor can be used to know if the oxygen saturation (SpO2) levels are declining as the altitude increases or vice versa. The integration of altitude data with the other measurements of wearables enhances the overall understanding of physiological responses in high-altitude environments.

12.5.1.8 Oximetry Sensor

Figure 12.12 shows a sensor known as a pulse oximeter, which is used to determine the oxygen concentration in the blood and this sensor relies on the presence of hemoglobin in red blood cells, which carries oxygen. There are two forms of hemoglobin which can exist: saturated (containing oxygen) and unsaturated (lacking oxygen). Varying amounts of red light get

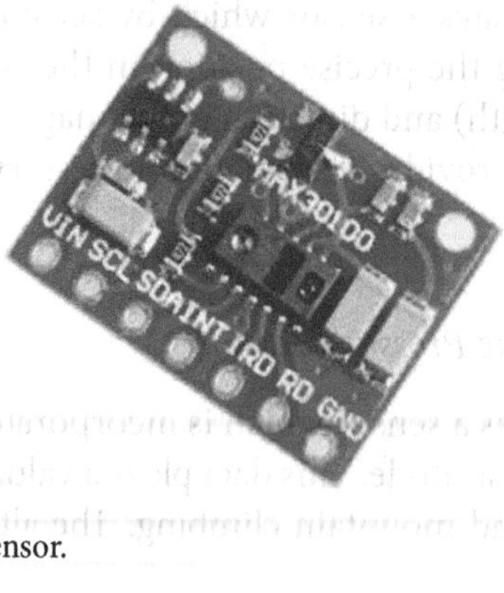

Figure 12.12 Oximetry sensor.

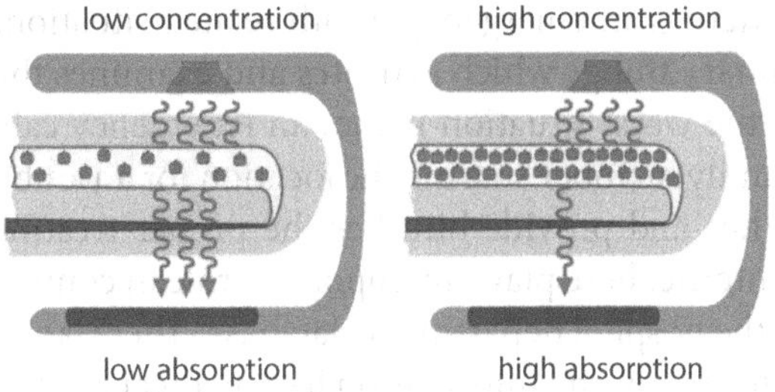

Figure 12.13 Low absorption and high absorption of infrared lights.

absorbed by these two forms of hemoglobin. When a wristwatch is worn, it emits visible red light and infrared light through the skin. Figure 12.13 illustrates the measurement of the ratio of light absorption between the two types of hemoglobin and also provides insights into oxygen levels in the blood.

12.5.1.9 Bioimpedance Sensor

Figure 12.14 illustrates the measurement of skin resistance to varying levels of electric current. At the bottom of the smart wearable there are electrodes which get charged and deliver small amounts of currents . Sleep, heart rate, and respiration data can be measured with the help of these currents.

12.5.2 Working of Model

If every minute is precious when it comes to human life and requires treatment in appropriate time then this healthcare model with these components is going to provide the proper treatment and efficient care in the best possible way. Let's explain how they'll help.

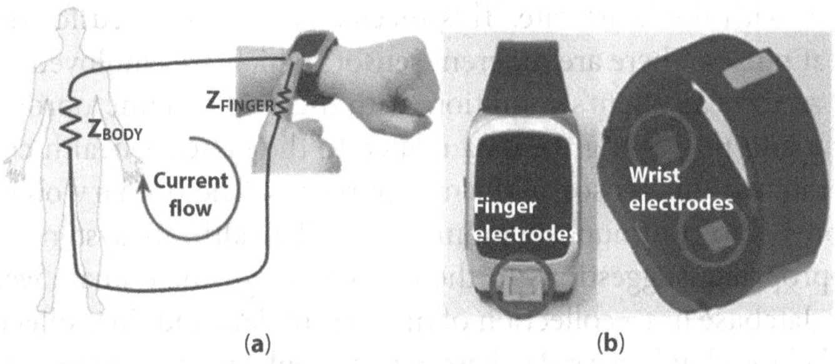

Figure 12.14 Working of bioimpedance sensor.

Whenever there is an emergency health related situation, these sensors are the preliminary things which activates and examines their health conditions and at the worst situation makes an emergency call to the nearby hospital and family member sharing the location for an ambulance to come and take him/her and provide him/her the proper treatment as soon as possible. The Internet here plays an important role in connecting the smart wearable with the hospital department/database. There is an Aadhaar database which will contain the information like medical history, blood group, any allergic problem, contact number, family info and many more of each and every person staying in that city and all these details are available to hospitals. Hospitals will have access to the Aadhaar database which will help the doctors in quick analyzing and providing patients a proper treatment as per the past medical details of the patient. Here maps will play an important role in helping the ambulance department providing live location of the person and guiding the ambulance to reach the sufferer and take him/her to the hospital within the time to make avail the best possible treatment. By using a variety of devices, resources and sensors this smart healthcare system is made. The smart device is developed by using sensors like body temperature sensor, heart rate sensor, pulse rate sensor, accelerometer, blood pressure sensor, oximetry, bioimpedance sensor, barometric pressure sensor, gyroscope, magnetometer. The sensitivity of the sensor can be easily improved by the potentiometer. The detection and quantification of smoking, nicotine, benzene, CO, NH3 and CO2 present in air can be analyzed by sensors. There are digital pins which are there in the sensor module that allow it to operate independently and eliminate the need for a microcontroller. This feature of eliminating the need for microcontrollers proves highly beneficial in the detection of various gases. The amount of light passing through the skin also tells the changes in the blood volume. Devices which focus on heart rate and do the assessment heavily are generally relying on precise pulse timings. The light signal pulse and distribution of blood volume is affected by the expansion of blood, which directly correlates with the heart rate. This mechanism is also used for detecting heartbeat pulses. There are different sensors which are employed to monitor and analyze a patient's condition, with the help of a microprocessor by receiving and processing the data related to the patient's health condition to generate various sensor performance results. The primary objective of this model is to continuously monitor an IoT healthcare system and provide appropriate suggestions to the users through their smart wearables. Aadhar database has a collection of millions of data and this collected data is stored in the cloud server database, and can only be accessed by authorized users and hospital departments via the internet or a designed application.

While the concept of microprocessor is well-known, this system represents a complete and economical solution by using a microprocessor. Users can easily update their health details online and its aim is to provide alerts during any kind of health emergency, and deliver immediate proper treatment during emergencies. This smart healthcare device has the ability to autonomously trigger emergency healthcare facilities if an elderly person falls sick/down. By quickly measuring, analyzing, and transmitting the patient's health information to a healthcare system cloud platform so that medical practitioners/doctors can significantly reduce errors and provide better treatment in the right time. Moreover, on these IoT devices there are several algorithms running that can offer suggestions related to the proper treatment based on the available details of user's data which further results in enhancing the healthcare process and making this healthcare system an important part for every human life.

12.6 Result

The implementation of a smart healthcare system facilitates doctors in studying a patient's behavior, leading to more accurate disease diagnosis. Patients can easily monitor their health condition regardless of time and location through the help of the smart health management system. This system enables the provision of health information on a digital platform that can be accessed globally. By using real-time health data, the smart healthcare system demonstrates its practical application in the real world. It efficiently calculates the health status of patients.

We obtained experimental results for a developed system that collects data from various sensors used in the device. Figure 12.15 shows the heart rate sensor results. A typical adult heart rate is 60–100 bpm. In most cases, a lower heart rate indicates a more efficient heart and better cardiovascular health. If a person's heart rate exceeds 100 bpm, the person is considered to have a rapid heart rate, also known as tachycardia. If one's pulse is somewhere between 60 and 100 beats per minute, the heartbeat is normal at that point.

Figure 12.16 shows the pressure rate of a particular person over time. The average blood pressure for adults is 120/80 mmHg. A person's pressure rate is measured over an hour.

The data is stored after detection. If the data exceeds the normal range, it alarms the patient's smart device as well as the devices connected to it. And if the need occurs, the patient can even activate the emergency call which notifies the nearby emergency service as well as one of the family members

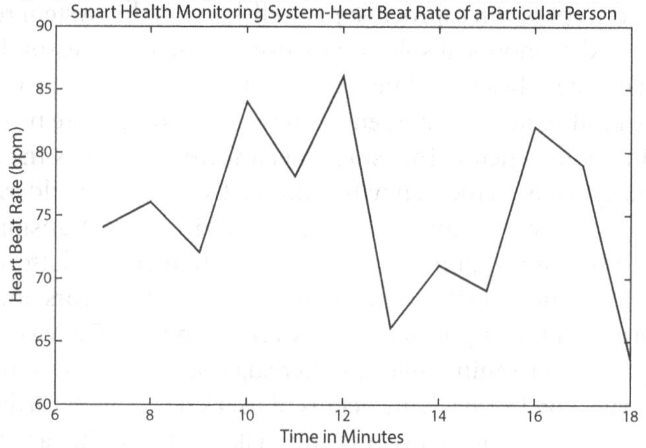

Figure 12.15 Rate of the heartbeat.

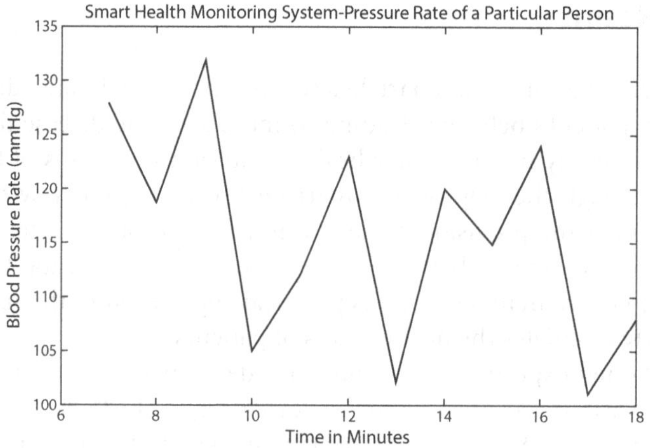

Figure 12.16 Pressure rate.

saved in the emergency contact. The patient can be taken to the nearby hospitals for the treatment as soon as possible. The details of the previous medical records can also be accessed by using the Aadhaar database for providing proper treatment. Doctors can utilize these records to treat the patient accordingly.

Figure 12.17 shows the data collected from the Accelerometer and Gyroscope. These data represents whether a person is actively changing the position and moving. Some part of the data also shows that the person is not moving actively.

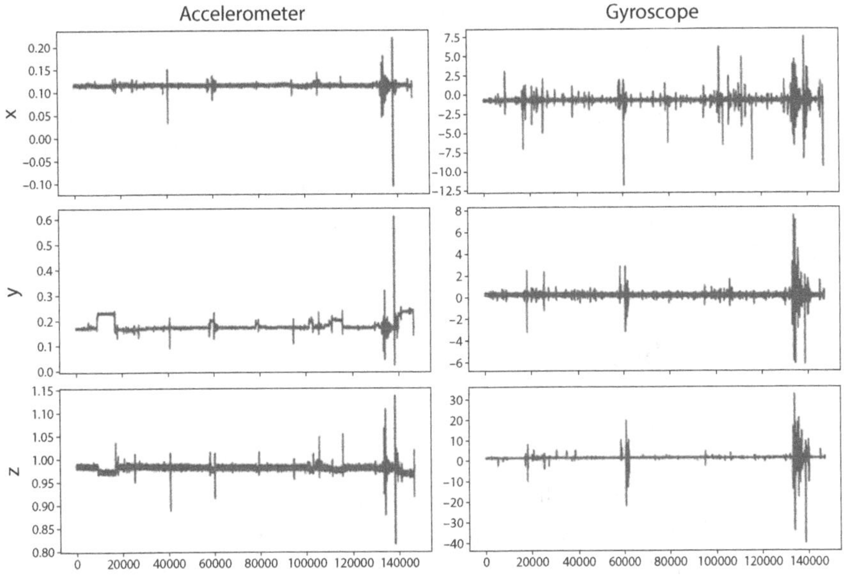

Figure 12.17 Raw data collected from the accelerometer and gyroscope.

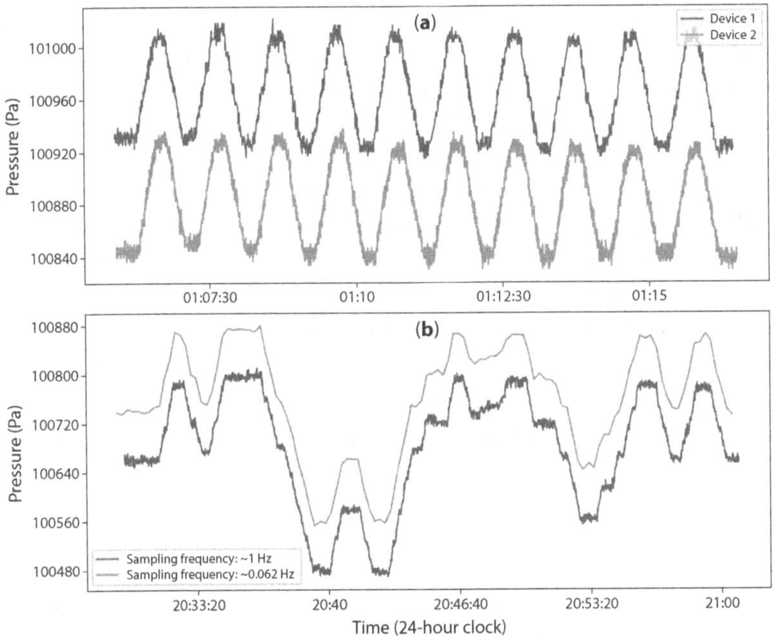

Figure 12.18 Data collected by barometric sensors.

Figure 12.18 shows the time series of barometric pressure recorded by two devices with two different sampling rates i.e.1 Hz and 0.062 Hz. Although the overall patterns of the two time series were parallel, some details in the device with the sampling rate of ~0.062 Hz (orange line) were missing in comparison to the barometric pressure recorded with the ~1 Hz device. This indicated that altitude or human activity might not be accurately estimated or recognized during certain periods if the sampling rate is not high enough.

12.7 Conclusion

Nowadays, continuous health monitoring has become very important, and the healthcare field is benefiting from the help of IoT technology. The utilization of IoT in healthcare has enhanced communication between patients and the doctors, resulting in more effective treatment planning. Furthermore, IoT-based methods of disease diagnosis have revolutionized the healthcare industry. The advantages of sensor-based IoT healthcare services include real-time reporting and monitoring, cost-effectiveness, seamless connectivity, telemedicine support, data analysis, tracking, and timely notifications. These advancements have improved treatment outcomes, facilitated coherent communication between patients and healthcare providers, and enabled accurate disease diagnosis through IoT-enabled devices. By using data-driven analytics tools, these devices generate graphs based on variations in health data, leading to correct decision-making. In case of potential risks, the medical devices right away alert both patients and physicians, creating a real-time tracking system and triggering alerts across connected systems. This enables healthcare professionals to make timely decisions, provide appropriate treatment, and ultimately improve patients' quality of life. IoT sensor devices have proven successful in monitoring, alerting, and tracking patient behavior, thereby enhancing effective treatment and overall healthcare. These devices are particularly useful for remote monitoring, ensuring personalized medication plans customized to individual patients' conditions, and optimizing treatment effectiveness.

References

1. Seeger, C., Van Laerhoven, K., Buchmann, A., Myhealth Assistant: An event-driven middleware for diverse medical applications on a

smartphone-mediated body sensor network. *IEEE J. Biomed. Health Inform.*, 19, 2, 752–760, 2014.

2. Jagadeeswari, V., Subramaniyaswamy, V., Logesh, R., Vijayakumar, V., A research on medical Internet of Things and Big Data in customised healthcare systems. *Health Inf. Sci. Syst.*, 6, 1, 14–20, 2018.
3. Larratt, C., Williams, E.S., McCoffee, D., Ishqaidef, G., Patient perceptions of patient-centred care: An empirical test of theoretical models. *Health Expectations*, 18, 2, 199–209, 2015.
4. Petrakis, E.G.M., Sotiriadis, S., Soultanopoulos, T., Renta, P.T., Buyya, R., Bessis, N., Internet of Things as a Service (iTaaS): Challenges and solutions for managing sensor data in the cloud and fog. *Internet Things*, 3–4, 156–174, 2018.
5. Al Rizq, M., Solangi, S.A., Alghamdi, A., Nizamani, M.A., Memon, M.A., Hamdi, M., Architectures supporting intelligent mobile healthcare using HCI principles of human-computer interaction. *Comput. Syst. Sci. Eng.*, 40, 558, 2022.
6. Al-Shargabi, B. and Abuarqoub, S., IoT-enabled healthcare: Benefits, problems, and challenges. *Proceedings of the Fourth International Conference on Future Networks and Distributed Systems (ICFNDS)*, St. Petersburg, Russia, pp. 1–5, November 2020.
7. Mohammed, B.G., & Hasan, D.S., Smart Healthcare Monitoring System Using IoT. *Int. J. Interact. Mob. Technol.*, 17, 141–152, 2023.
8. Khan, S.M. and Khan, S.I., A smart healthcare system for diabetic patients using IoT and cloud computing, 6, 3004, 2023.
9. Goyal, M. and Sharma, A.K., Smart healthcare system for elderly people using wearable sensors and big data analytics, 43, 365–370, 2020.
10. Reddy, A.R. and Kumar, P.S., Predictive big data analytics in healthcare, in: *Computational Intelligence & Communication Technology (CICT), 2016 Second International Conference on*, IEEE, pp. 623–626, February 2016.

13

Machine Learning Algorithms for Initial Diagnosis of Parkinson's Disease

Udayan Das*, Manish Jena and Manish Roy

Kalinga Institute of Industrial Technology, Bhubaneswar, India

Abstract
Parkinson's disease (PD) is a variety of physical and mental signs associated with Parkinson's disease (PD), a cognitive condition, are caused by the gradual death of dopaminergic cells in the substantia nigra. The etiology of PD is multifactorial, involving a complex interplay of genetic, environmental, and age-related aspects. The bradykinesia, stiffness, tremor, and instability of posture are the four main motor symptoms of disease, while non-motor symptoms encompass cognitive impairments, mood disorders, sleep disturbances, and autonomic dysfunction. Diagnosis relies on clinical assessment and may be aided by neuroimaging techniques. Current treatment options primarily focus on alleviating symptoms through pharmacotherapy, deep brain stimulation, and rehabilitative therapies. However, no cure for PD exists, highlighting the need for continued research to unravel the disease's underlying mechanisms and develop effective disease-modifying interventions.

***Keywords*:** Parkinson's disease, logistic regression model, K-nearest neighbors, machine learning

13.1 Overview of Parkinson's Disease

Parkinson's disease (PD) is a well-known nervous sickness that disturbs loads of individuals across the world. The gradual loss of dopaminergic nerve cell in the substantia nigra, a particular area of the brain, is a trademark of the disease. Tremors, muscular stiffness, bradykinesia, and

**Corresponding author*: 2005280@kiit.ac.in

Hrudaya Kumar Tripathy, Sushruta Mishra, Minakhi Rout, S. Balamurugan and Samaresh Mishra (eds.) Optimized Computational Intelligence Driven Decision-Making: Theory, Application and Challenges, (251–274) © 2024 Scrivener Publishing LLC

instability of posture are the key motor-powered signs of PD. Disruption of the dopamine-mediated signaling pathways involved in motor control causes these motorized symptoms. The sickness is not just a movement problem, though. Additionally, it exhibits a range of non-motor symptoms, such as mental deterioration, depressive disorders, insomnia, and autonomic dysfunctions. The quality of life of those with PD is substantially impacted by these non-motor-powered indicators. Parkinson's disease is thought to be caused by a mix of hereditary and environmental factors, while its specific etiology is yet unclear. Diagnosing Parkinson's disease can be challenging, as there are no specific tests or biomarkers available. Parkinson's condition currently has no known cure, despite there are a variety of therapies that can help individuals manage their symptoms and live better lives. Figure 13.1 shows an illustration of Parkinson's disease-affected portion of the brain. Cranial trauma, family history, age, and even interaction with specific pesticides can all contribute to dopamine deficits caused by neuron death in the black material.

Non-motor and motor disorders are the two basic categories for PD signs. Loads of individuals are mindful of motor-powered signs since they may be seen with the unaided eye. The primary symptoms are often a condition known as stiffness, balance issues, and sleeping tremors. It is now understood that there is a window of opportunity for identifying non-motor symptoms, also known as dopamine-nonresponsive symptoms. Constipation, cognitive decline, sleep disorders, loss of smell, inexplicable

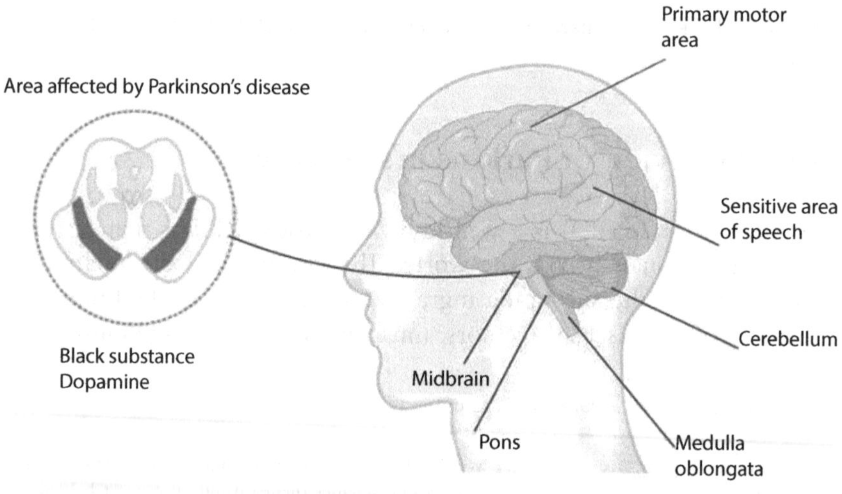

Figure 13.1 A schematic depicting an area of the brain damaged by the disease.

pains, difficulties speaking and swallowing, drooling, and low blood pressure upon standing are some of these symptoms. It is crucial to emphasize that nobody of these non-motor-powered symptoms are absolute proof that a person has Parkinson's disease. However, when paired with additional indicators like dopamine transporter imaging and cerebrospinal fluid measurement (CSF), they can aid in the prediction of the illness.

Symptoms normally appear gradually and progress over time. Parkinson's disease first appears inthe olfactory tracts, lower brain stem, and enteric nervous system. The disease extends from these areas to higher sections of the brain, specifically the brain shell, and the substantia nigra. The disease is believed to develop years before motor symptoms including tremor and slowness of movement, loss or decrease of smell, trouble falling asleep and staying constipated. A definitive diagnosis cannot be established through a single test. Instead, medical professionals must conduct a comprehensive clinical assessment, taking into account the patient's medical history. However, this diagnostic approach has limitations. Nevertheless, early diagnosis remains crucial for effective treatment. The global prevalence of PD has significantly increased, with approximately 10 million individuals affected worldwide. In India, it affects around one million people annually, while in the United States, it impacts an estimated 500,000 to one million individuals, accounting for about 1% of those aged 60 and above. Although PD is more commonly observed in older individuals, approximately 4% of cases are diagnosed before the age of 50. Men are affected by PD 1.5 times more frequently than women [1]. At any given time, PD affects one to two individuals per 1,000, and its commonness surges with age, affecting approximately 1% of individuals over the age of 60. The development of PD is influenced by a combination of genetic and environmental factors [1].

At this time, there is no solo definitive method for spotting Parkinson's disease (PD) [2]. Instead, a combination of symptoms and diagnostic tests are utilized in the diagnostic process. In order to detect Parkinson's disease early and halt its development, scientists have investigated a variety of biomarkers. Currently, no PD medications reduce or stop disease progression. While PD does not currently have a cure, various treatment options are available to assist individuals in managing their symptoms. Medications may assist patients with Parkinson's disease manage problems with walking, movement and tremor. In some circumstances, surgeries might be recommended. Despite extensive research on Parkinson's disease, we're still not sure what's causingit. We even have difficulty diagnosing it at times [3]. In the works, numerous ML constructed classification algorithms have been suggested to differentiate between people with the disease and vigorous individuals using speech signals, as mentioned in the paper. Due to the luxury of use and excellent exactness, Machine Learning is being embraced more and more for the diagnosis of medical conditions.

According to published research, machine learning has also been secondhand to treat the disease. ML practices have been used in the supervision of PD, particularly predicting of clinical progression, the personalization of therapy, and the creation of assistive technology to enhance cognitive function [4]. Tsanas *et al.* [4] utilized 23 speech samples obtained from the 43 individuals, with 76.7% of the dataset coming from persons with Parkinson's and the rest being healthy. They used a revised version of the data set from the previous study [4]. Little *et al.* [4] present a review of methods for identifying dysphonia in PD patients to differentiate them from healthy individuals. They discovered 8 healthy individuals and 23 Parkinson's disease sufferers, and their data set contained vowels. They classified using SVM, and the accuracy of their classification was 91.4%. In order to identify between people with PD and healthy subjects, classifiers such as SVM, Nave Bayes (NB), KNN and MLP were also used. They revealed that the most effective technique for identifying Parkinson's disease is phonation. K-NN, MLP, SVM were the classifiers evaluated in the research. For the ML-based diagnostic, artificial neural networks were utilized to reduce voice characteristics.

13.2 Scope

While Parkinson's disease has no known cure, new research has resulted in better therapies. Scientists and clinicians are collaborating to develop a therapy or preventative method. Researchers are also trying to figure out who is more likely to have the condition. Furthermore, experts are investigating the environmental and genetic elements that increase the probability of a diagnosis.

The most recent treatments for this degenerative neurological condition are:

1. **Biomarker** – Medical professionals have limited resources for evaluating the progression of the disease. Although staging is helpful, it primarily focuses on monitoring the advancement of motor symptoms linked to the condition. Other grading scales are available, but they are not extensively used enough to be recommended as a general guideline. A promising field of research, on the other hand, may

make assessing Parkinson's disease easier and more precise. The goal of research is to identify a biomarker—a cell or gene—that will enable the development of more effective treatments.

2. **Gene Therapy** – Researchers have yet to discover a sure way to treat Parkinson's, decrease its course, or reverse the brain damage caused by the disease. Gene therapy has the ability to accomplish all three of these goals. Gene therapy can be an efficient and secure treatment for Parkinson's disease, according to several researches.
3. **Deep Brain Stimulation** – In 2002, the FDA authorized it as a Parkinson's disease therapy. Still, improvement has been hindered due to the fact that only one company has obtained clearance to manufacture the necessary equipment for this therapy. In June 2015, the FDA gave the Brio Neurostimulation System approval. By sending brief electrical pulses throughout the body, an implantable device helped alleviate discomfort.
4. **Neural Transplantation** – A promising avenue of treatment in the future is replacing the brain cells lost to PD. This technique replaces damaged and dying brain cells with new, growing and multiplying cells. However, research into brain transplanting has yielded conflicting outcomes. Some individuals improved as a result of the treatment, while others saw no change and even acquired new issues. Parkinson's illness patients can lead healthier lives with the use of medications, treatments, and lifestyle changes as we search for a cure.

13.3 Related Works

Researchers used a variety of features and data to detect Parkinson's disease. When related to expression and eloquence, it was determined in 1998 that the speech is the major impairment utmost commonly impacted the primary phases of Parkinson's disease. The scientists created a model that analyses patients' voices to predict whether or not they have Parkinson's disease. The dataset analysis involved the utilization of FCM cluster and pattern recognition algorithms, resulting in an accuracy rate of 68.04%. The sensitivity was measured at 75.34%, indicating the ability to correctly identify positive cases, while the specificity was recorded at

45.83%, representing the accuracy in identifying negative cases. Statistical approaches were employed to assess the lingering indication auto-interaction in patients having the disease, substantial changes in pitch fullness amongst patients and strong collection were discovered. Some modern ML procedures were elected for prophecy, besides their performance was compared using exactness, range underneath the ROC curvature, and additional system of measurement. Other ways of detecting Parkinson's disease have recently been researched. Various human traits, such as phonation, or speech, and other aspects of speech, were extracted for analysis of voice recordings.

In countless investigations [5, 6] led by scholars, for the sake of simplicity, acceleration signals for PD detection have been examined. Subsequently, applying machine learning (ML) methods, various classification systems were created to discern between different types of tremors for, e.g., PD and ET [7, 8]. The information was gathered via a smart watch, and a cataloguing method was developed to distinguish essential tremor from Parkinson's tremor [9]. A dysphonia measure was created to differentiate between individuals with Parkinson's ailment and individuals without the condition by analyzing variations in voice frequency. Continual vowel sound, phrases, and judgments after a series of dialogue challenges were discovered to contain PD-discriminatory data in 2013, which may be used with machinelearning methods. Arrangement was accomplished by SVM. Unlike everyone else, an unconfirmed technique was correspondingly cast-off for PD. [10] used phonation data to train XGBoost replicas to envisage fluctuations in medical ratings of 50 Parkinson's disease patients. [11] employed speech-based evidence to parameterize anarchic bacterium finding optimization (CBFO) using an upgraded fuzzy k-nearest neighbor (FKNN) for the first identification of Parkinson's illness. For the identification of PD, various statistical features and ML techniques have been employed. Researchers demonstrated that the easiest job for PD detection is vocalization.

The study evaluated four classification algorithms, namely KNN, MLP, Optimum Path Forest, and SVM. The authors of reference [12] proposed the utilization of Modified Grey Wolf Optimization (MGWO) for feature selection exploration, along with Random Forest, k-nearest neighbor classifiers, and decision trees for classification purposes. The fundamental technique is based on separating distinct indicator components and choosing the most relevant parts to analyze interlimb difference in a certain value region. As indicated in Table 13.1, an assembly of 6 distinct models, including a variety of cataloguing categories, was deployed. Tests of each model's detection accuracy on the member datasets guided the selection

Table 13.1 ML models employed in the cataloguing group.

Model	References and remarks
1. Support Vector Machine (SVM)	[12]
2. Multi-Level Perceptron (MLP)	Subtle to feature grading, hence it was necessary to normalize the data [13]
3. Logistic Regression Model (LRM)	[14]
4. Random Forest (RFC)	[15]
5. Nu-Support Vector Classification (NSVC)	Management of the quantity of support vectors and retraining mistakes in an improved SVC
6. Decision Tree Classifier (DTC)	[16]

from each model. The use of a large ensemble was justified by the observation that, despite the fact that observational classification accuracy varied significantly amongst systems, reliability rose as the methods were pooled together to form a meta-classifier.

The work with speech signals sustained in the ages that followed, the inclusion of more functionalities separators and statistics. The furthermost prominent leanings remain the use ofsmartphone expertise in everyday life to gather voice samples and sense indicators of speech loss in people, as well as the growth of package for scrutinizing compulsive language signs, pleasing vocalization, enunciation, metrics, and lucidness into account. In reference [13], it was suggested to employ weights obtained from a MLP for feature selection and a LSVM for classification. The objective of this approach is to enable early detection of Parkinson's disease (PD) in a medical setting. [14] achieved an accuracy of 86.47% by using functionality choice approaches based on Pearson's and Kendall's association constants, principal component analysis (PCA), self-organizing maps (SOM), and feed-forward Artificial Neural Networks in abundance. In alternative technique, the Cuttlefish technique was enhanced for disease detection, but is done with an effectiveness of about 94% [15]. CNN is still employed to identify illness into speech signals, however the results do not approach 90% accuracy [16].

Current imaging techniques offer great spatial and color resolution, which has led to their being the primary tool for diagnosing neurological disorders. In order to assess the deficiencies of dopamine transporters of the pvn, DaTSCAN SPECT scanning is frequently utilized aimed at the uncovering of PD. Nevertheless, even for experienced clinicians, visual analysis of these pictures to arrive at a definitive diagnosis is a time-consuming and difficult process that necessitates taking into consideration several aspects. This issue could be solved using pattern recognition that enables mixing several inputs to generate a result. Notwithstanding, the enormous quantity of data included in DatSCAN pictures necessitates the employment of machine techniques to be fully used, enabling the discovery of multifaceted, ailment-associated forms to improve diagnostic precision. Primary software methods suggested in this architecture are then reviewed. The Unified PD Rating Scale (UPDRS) is the utmost often used approach for diagnosing PD indicators, and motorized exams are performed on the scale of 0 to 4 that is given in Table 13.2 for both tremor and speech.

Palumbo and Towey are two of the first research to assess the potential of deep learning systems with DaTSCAN (2011). To distinguish among PD and movement disorders, the prior contrasted a PNN with a classification

Table 13.2 Conveying UPDRS tags constructed on tremor and speech features.

UPDRS	Tremor	Speech
0	Vague	No issue
1	Trivial and rarely present	Forfeiture of variation yet comprehensible
2	Insignificant and adamant	Forfeiture of intonation with uncertain words
3	The majority of the time, modest and current	Speech is not well comprehended
4	Noticeable and extant majority of the period	Most of the conversation is incoherent.

tree CIT. Three slices' worth of the cingulate and putamina's striatal binding ratios were used as picture defining qualities. Instead it used PCAcessation of the striatal area's vertices along with Naïve bayes. Those are employed in a number of studies where SVMs served as the primary classification tool, along by rectilinear or Along with different picture properties and RBF kernels. Oliveira and Castelo-Branco used voxel-as-structures, or the direct use of picture voxel concentrations as structures, for the first time in 2015 after Illán *et al.* DaT pictures were divided into scores and loaded using a Partial Least Square (PLS) method by Segovia. The greatest Linear Discriminate Proportion ratings also were employed as a parameter in the SVM. PLS was also employed by Khedher. The idea of using 2D empirical phase segmentation to divide DaTSCAN pictures into several intrinsic mode functions, accounting for various frequency subsets, was proposed by Rojas. The elements were utilized to choose PD-related characteristics that distinctly set them apart from NC, enabling a quick visual evaluation. The decomposition of the DaTSCAN pictures into quantitatively independent components by Martnez-Murcia identified patterns connected to PD. This method also prioritized picture clusters according to their empirical importance for class classification.

Machine learning techniques, in particular, have lately made it possible to find complex behavior and, as a result, surpass the accuracy rate attained by traditional methodology. It is possible to calculate simulations with a high range of parameters by using models with heaps of blocks made up of several units that each execute basic actions. Furthermore, by using a process of learning that is structured as an optimizer, those enormously multithread systems are capable of identifying extremely complicated patterns in the information [16]. A predictor that uses a single layer perceptron using voxel-as-features from several slices is proposed by Zhang and Kagen. Larger networks must be used to increase accuracy, but doing so increases the likelihood that the network may overfit and strain the capabilities of the training methods. In order to efficiently grow the number of layers, designs incorporating increasingly complex building components, such those in He, have also been suggested. Through this study, authors present a predictor thatlearns from isosurfaces computed from the problematic regions and is based on the well-known CNNs LeNet-5 and AlexNet. The intention of isocontours, which considerably lessens the involvedness of the data, allows for the creation of substantial training sets with minimal computational effort.

13.4 Comparative Analysis of Parkinson's Disease

Parkinson's disease research encompasses various disciplines, including biology, chemistry, and heredity. A few of these eventually evolved by utilizing machine learning and artificial intelligence, which aided in the early detection of this ailment greatly [17–19]. Although Parkinson's disease is a diagnostic condition with no confirmed symptom or assured detection procedure, many people havegiven it their all in the diagnosis and inhibition of this ailment. Because ML is based on extrapolative norms, It is often suggested for such investigations. Parkinson's disease has various symptoms, which leads to a variety of discoveries, such as identification by analyzing a single symptom or identification by reviewing multiple symptoms [20–22]. This review effort prioritizes detection utilizing one ailment at a moment. Figure 13.2 depicts common Parkinson's sickness signs.

Various sensors are currently employed to identify motorized symptoms, however, because devices are not always obtainable in all part of the earth and data inputs are inadequate, the results' accuracy may degrade. The detection method is the primary topic of this review article, which is extensively explored [23]. A variety of research articles are presented along with their implementations, as well as some of the strategies that have been implemented and compared to the original ones. It should be emphasized that not all articles are included, particularly those that incorporate MRI images.

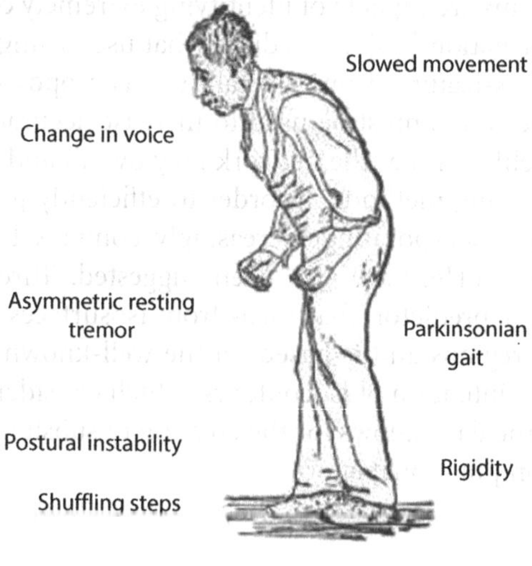

Figure 13.2 Signs of Parkinson's sickness.

A. Bradykinesia

Among the greatest common signs of the illness is bradykinesia. It is brought on by a deficiency in basal ganglia output, which impairs the processes in the brain that produce and carry out motor system commands, resulting in motility. Dopamine receptors deficiency in the cerebellummay cause increased neural activity, inhibiting striatal output [24, 25]. Because of this insufficiency, the person begins to have atypical movement activity, resulting in difficulties with self-paced motions.

Individuals experiencing dyskinesia, characterized by uncontrolled involuntary muscle movements and convulsions, face challenges in reacting effectively. Bradykinesia, akinesia, and hypokinesia are terms used to describe different aspects of reduced voluntary movement in the disease. Akinesia refers to a lack of voluntary movements, while hypokinesia pertains to movements that are slower and weaker than intended, such as impaired handwriting. These terms highlight the motor clumsiness observed in Parkinson's disease patients. Research indicates that finger tapping frequency decreases with age, and there are variations between genders and dominant/non-dominant fingers [26]. The Finger Tapping Test (FTT), which utilizes altimeter and tactile sensors, demonstrated significant differences between disease patients and hale and hearty persons. Various machine learning techniques have been developed to evaluate signs and extract useful information from the data. Figure 13.3 presents the results of all algorithms, with the combination of Random Forest (RF) and Hard Voting (HV) achieving a maximum accuracy of 97.5%.

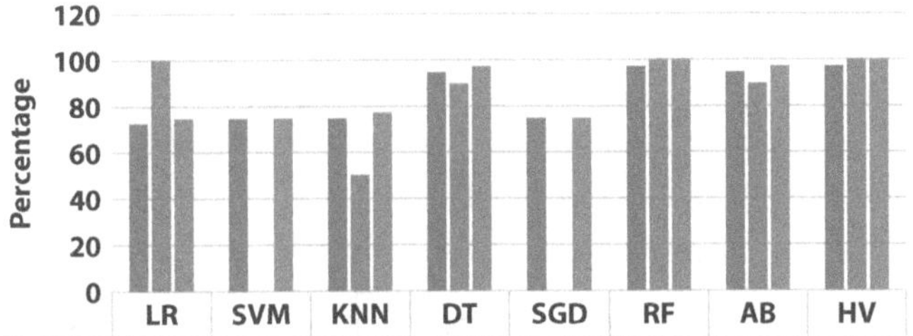

	LR	SVM	KNN	DT	SGD	RF	AB	HV
Accuracy (%)	72.5	75	75	95	75	97.5	95	97.5
Specificity (%)	100	0	50	90	0	100	90	100
Sensitivity (%)	75.67	75	77.78	96.67	75	100	96.67	100

Figure 13.3 Results of bradykinesia.

B. Tremor at Rest

Tremor refers to the rhythmic and uncontrollable oscillation of body parts and is observed as a symptom in various medical conditions [27]. However, there are distinct characteristics that differentiate tremors associated with the ailment Tremors primarily affect the hands in PD patients compared to other body parts. PD patients often experience difficulties in activities like handwriting when compared to healthy individuals. Small hesitations and pauses during tasks can be considered as tremor manifestations. Since vibration at leisure is frequently noted in preliminary diagnosis, tremor detection offers important insights into the motor disturbances associated with Parkinson's disease. Figure 13.4 shows the results analysis of SST analysis.

The spiral test was predominantly employed to evaluate the impact of movement issues in this study, which was divided into eight challenges [28]. The initial challenge involved drawing Archimedean spirals, followed by writing letters and sentences in cursive. Handwriting signals in the x-y plane were captured using a Wacom digitizing tablet, with a third axis measuring pressure. The feature vectors underwent preprocessing before being classified using a Support Vector Machine (SVM) classifier. After considering all variables, an accuracy of 79.4% was achieved, with similar specificity and sensitivity values [29]. This research presents a multifunction strategy combining the validity of two different tests. Using the spiral tremor test, 77 participants—62 people with Parkinson's disease and 15 healthy controls—were screened for tremors. The spiral test was further divided into three tests: SST, DST, and STCP. During the preprocessing step, feature selection was performed using min-max scaling, and correlations were assessed to

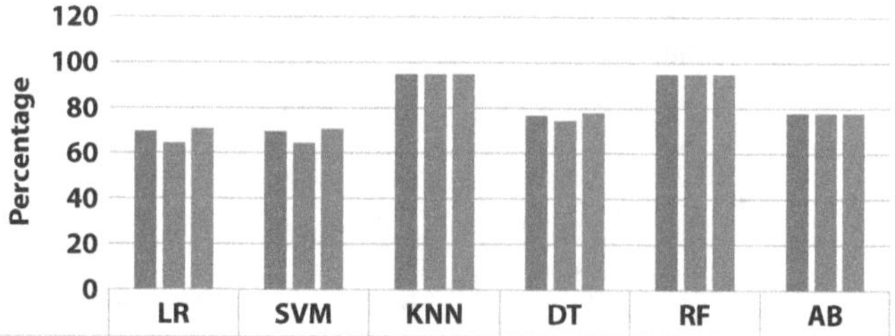

	LR	SVM	KNN	DT	RF	AB
Accuracy (%)	72.13	72.13	99.36	80.59	99.79	81.87
Specificity (%)	67.86	67.86	99.08	78.66	99.61	81.23
Sensitivity (%)	74.22	74.22	99.55	81.61	99.61	82.21

Figure 13.4 Results of SST.

eliminate redundant features. The information was then partitioned into the three trials for scrutiny: Figure 13.5 shows the results analysis of DST analysis and figure 13.6 highlights the outcome of STCP analysis.

C. Voice Impairment

The inclusion of voice impairment as an early indicator for identifying Parkinson's disease has been a subject of ongoing debate [30, 31]. Some renowned neurologists do not consider vocal symptoms to be a reliable early sign of the disease. Others contend, however, that voice problems are more pronounced in the early stages of Parkinson's disease combined with

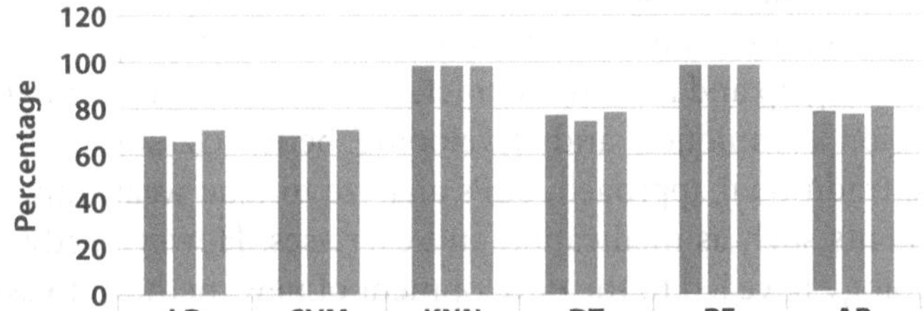

	LR	SVM	KNN	DT	RF	AB
Accuracy (%)	69.42	69.42	99.28	77.7	99.76	79.51
Specificity (%)	65.28	65.28	99.26	74.95	99.58	77.42
Sensitivity (%)	71.31	71.37	99.3	79.97	99.88	80.72

Algorithms

Accuracy (%) Specificity (%) Sensitivity (%)

Figure 13.5 Results of DST.

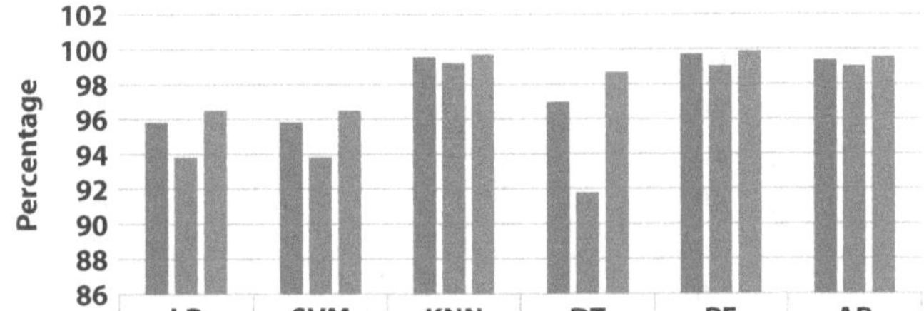

	LR	SVM	KNN	DT	RF	AB
Accuracy (%)	95.88	95.88	99.58	97.05	99.71	99.4
Specificity (%)	93.75	93.75	99.25	91.74	99.17	99.14
Sensitivity (%)	96.43	96.42	99.68	98.65	99.87	99.49

Algorithms

Accuracy (%) Specificity (%) Sensitivity (%)

Figure 13.6 Results of STCP.

tremors and stiffness. There is inconsistent evidence that voice impairment can be seen in people with PD, and there is unambiguous evidence that dysarthria, a speech issue, often develops later in people with PD. Dysarthria is a speech problem caused by poor functioning of the muscles used for speaking. Other forms of speech impairments connected with Parkinson's disease include hypophonia and tachyphemia. This has the effect of restricting the Dopamine secretion, which aids in the production of smooth muscle movement, so influencing the face musclesand the afflicted person's expression. In recent years, there has been intense competition among researchers to develop cost-effective and dependable methods for acoustically diagnosing Parkinson's disease.

The introduction of Artificial Intelligence (AI) has further fueled this competition and continues to advance the field [19]. Researchers have been setting up and refining Artificial Neural Networks, enhancing feature extraction techniques, and developing better classification approaches. In the traditional approach, SLPs analyze tongue weakening and provide treatments such as medication and exercises. However, advancements in technology have led to the development of new methods to assist patients. One example is EchoWear, a wristwatch device that enables remote monitoring of patients' activity, providing benefits and overcoming the limitations of in-person SLP treatments. issue of Parkinson's disease, notably K-Nearest Neighbors (KNN). To improve PD detection, a variety of acoustic, prosodic, and vocal cues have been retrieved. The dataset used in this study came from the UCI repository and included information on 31 people, including 8 healthy volunteers and 23 people with Parkinson's disease. The results of using several machine learning approaches for illness

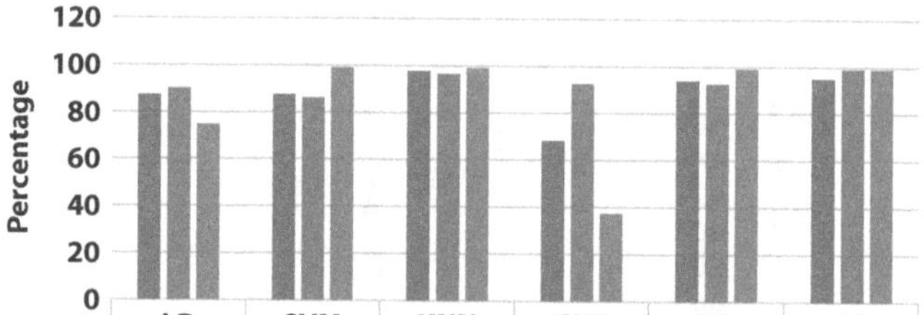

	LR	SVM	KNN	GNB	RF	AB
Accuracy (%)	87.76	87.76	97.96	69.39	93.88	95
Specificity (%)	90.24	86.67	97.5	92.8	92.85	100
Sensitivity (%)	75	100	100	38.09	100	100

Figure 13.7 Results of voice impairment.

classification are shown in Figure 13.7, where KNN attained an ideal precision of 97.96% on the identical sample.

D. Rigidity

One of the four key signs for diagnosing the illness is this. It refers to the significant difficulty experienced in the limbs or other body parts, which hinders the proper expansion or relaxation of muscles. This symptom can manifest on either one side or both sides of the body. Stiffness is defined by muscular rigidity, such as that found in the facial muscles. The inability to present a visage and articulate fluently cleared the path for speech treatment. Body-worn sensors have emerged as a successful approach for diagnosing balance and gait issues in a novel manner. Individuals were requested to participate in activities, and their metrics were recorded. The parameter postural sway was used to assess postural instability. It was monitored using a sensorimotor control loop, and numerous data were recovered, including velocity, jerk, sway, and frequency [20]. The postural assessment was graphed with the x and y axes being mediolateral and anteroposterior. The freezing ratio's proposal for gait freezing. The report analyzed studies on various bodily components that influenced the patient's gait. Physio net's Gait Database is used to comprehend the relevance of gait analysis for stiffness identification. Pre-processed the data, retrieved the features, and used cutting-edge machine learning methods. As demonstrated in Figure 13.8, the Random Forest Classifier with Adaptive Boosting had the greatest accuracy of 83.12%.

To develop the PD diagnosis technique, numerous methodologies such as clustering, dimensionality reduction, and erudition of estimate tactics

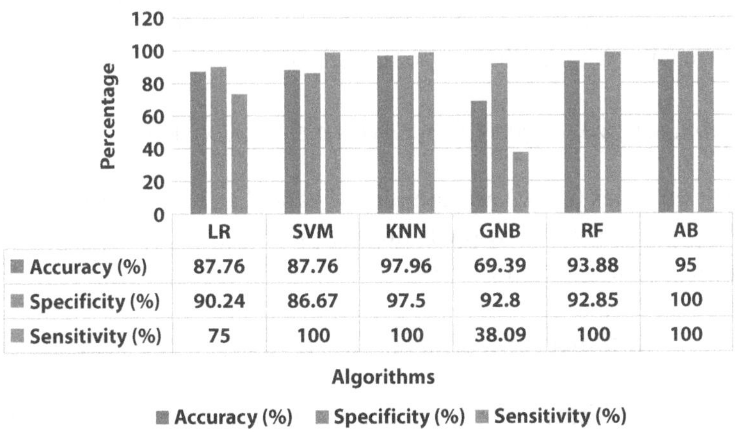

	LR	SVM	KNN	GNB	RF	AB
Accuracy (%)	87.76	87.76	97.96	69.39	93.88	95
Specificity (%)	90.24	86.67	97.5	92.8	92.85	100
Sensitivity (%)	75	100	100	38.09	100	100

Figure 13.8 Result of voice impairment.

are applied. Preprocessing stage, linear regression with PCA, gathering with EM troupes, and estimation with SVR troupes are wholly steps of the strategy used to envisage UPDRS based on a collection of real-world PD data. SVM was created by Cortes and Vapnik [21] as a machine learning approach towards predicting issues with the possibility to be widely utilized as a baseline. SVM's two primary divisions are SVC and SVR. SVR predicts the quality of a specimen by retraining it with predicted input sets. By employing a Φ(x) function to map the information into a level interplanetary, the process involves using Support Vector Regression (SVR) to solve complex nonlinear equations complete linear and nonlinear regression [22]. To transform the data into a feature with a lot of dimensions space, kernel functions are used.

The following formula is used to represent a learning data with N dimensions: T = (x1, y1), (x2, y2),..., (xN, yN), where y ∈ R and xk ∈ Rn, for k=1,2,..., N. In the original SVR, a single output (y) is described as follows using a linear model formulation:

$$y \wedge x = w, x + b, \tag{13.1}$$

where y stands for the anticipated output, b for bias, w for weight, and ⟨., .⟩ denotes the inner product of vectors. SVR, nonlinear issues may be dealt with using nonlinear kernel functions like Φ(x).

$$y \wedge x = w, \Phi x + b. \tag{13.2}$$

Radial basis functions are one of the most commonly used nonlinear SVR kernels. To get optimal generalization performance in SVR, the weight vector w must be as flat as feasible. As a result, we must minimize the norm (.) of w for each data point in the dataset as,

$$minimize12w2, subject\,to\,yiw, \Phi xi - b \leq \varepsilon \text{w}, \Phi xi + b - yi \leq \varepsilon. \tag{13.3}$$

Slack variables ($\xi i, \xi i^*$) are announced as well as ε-insensitive loss function $L\varepsilon, y, y\wedge$. As a result, by predicting the parameters w and b, the following cost function must be minimized,

$$minimize12w2C\Sigma ilN\xi i{+}\xi i{*}yi{-}w,\ xi - b \leq \varepsilon + \xi i,$$

$$subject\,to\,yiw, \Phi xi - b \leq \varepsilon \leq \varepsilon{+}\ \xi\partial i{*}\xi i, \xi i{*}{\geq}0, \tag{13.4}$$

In order to address the above optimization problem, the regularization parameter C is introduced. The constraints can be handled in the dual form by incorporating Lagrange multipliers η, η∗, β, and β∗. As a result, the Lagrangian (L) can be expressed as follows:

$$L=12w2+C\Sigma i = 1N\xi i+\xi i*-\Sigma i = 1N\beta i\varepsilon + \xi i - yi + w, \Phi xi + b-$$

$$\Sigma ilN\beta i*\varepsilon + \xi i* + yi - w, \Phi xi - b - \Sigma ilN\eta i\xi i - \eta i*\xi i*. \quad (13.5)$$

The following quadratic programming final problem produces the least value. Karush-Kuhn-Tucker's limitations include,

$$minimize 12\Sigma i - 1N\Sigma i - 1NKij\beta i-\beta i*\beta j-\beta j*+\varepsilon\Sigma i-1N\beta i+\beta i*-$$

$$\Sigma i-1Nyi\beta i+\beta i*, subject\ to \Sigma i-1N\beta i+\beta i*=00\leq\beta i, \beta i*\leq C. \quad (13.6)$$

Kij stands for the kernel function. The dual space model form may be expressed as follows in SVR:

$$y \wedge x = \Sigma i -1N\beta i - \beta i * Kx, xi + b. \quad (13.7)$$

13.5 Pros and Cons Using ML Algorithms

The pros for detecting PD using Machine Learning algorithms:

I. **Easily recognizes trends and patterns:** Hefty volumes of information might be evaluated by Machine Learning, which can also recognize precise inclinations and outlines that individuals would slip [23]. For example, an E-commerce website like Amazon, it helps give operators with the appropriate goods, promotions, and prompts that are pertinent to them by understanding their browsing habits and past purchases. It uses the outcomes to show them relevant adverts.

II. **No human intervention required:** Thanks to machine learning, one can now complete the project without constantly monitoring it. Providing systems the ability to acquire knowledge permits them to create algorithms and make predictions by themselves. Windows firewall is a common example of this. As new threats are discovered, they adapt to retrieve them out. ML is adept at detecting junk.

III. **Continuous Improvement:** Machine Learning procedures continue to improve in accuracy and efficacy as they gain expertise. They can subsequently make shrewder choices. Consider developing a framework for weather predictions. As the data collection grows, these approaches get more efficient and precise at generating insights.

IV. **Handling data in several dimensions:** Reinforcement learning is considered reliable for various reasons, and one of them is its ability to handle data. Machine learning (ML) plays a crucial role in data processing and analysis. ML can handle various types of data, including multidimensional data, which makes it versatile compared to traditional systems. Data is a vital component of any machine learning model, and the ability to process and analyze data is a distinct area of expertise.

V. **Wide Applications:** As an online retailer or a supplier of healthcare, one can benefit from using machine learning. Since it does, it possesses the ability to help in giving customers a much more affecting perceived also while focusing on the right individuals. The utilization of machine learning can expedite the process of disease discovery and contribute to the growth of businesses. Therefore, investing in machine learning technology is considered valuable.

VI. **Capital asset competence:** Businesses may find it stimulating to envisage when capital assets will necessitate maintenance or developments, and these tasks may be expensive By employing predictive machine learning models, businesses have the capability to automate the collection of performance data from equipment and components. They can effectively monitor the status of these assets and make informed decisions regarding their lifespan.

VII. **Faster decision making:** Machine learning empowers organizations to make prompt, and even instant, decisions by enabling them to process and analyze data at a much faster pace than ever before. Forexample, package based on machine learning that has been taught to spot incongruities in a company's security atmosphere may quickly identify a data breach and alert that organization's tech operate [25].

These solutions help businesses protect client data, uphold their brand reputations, and evade high-priced corrective actions by making quick verdicts regarding successful restoration.

VIII. **Personalizing customer engagement:** Moreover, personalization has arisen as a critical course for challenging in the modern economy. Online traders may occupy customers more personally and increase sales by using machine learning platforms that monitor user activity and offer product endorsements based on past procurements. With its use of machine learning to compile lists of suggested items and provide ideas to customers, global powerhouse Amazon is a great example.

Due to the lack of an accurate sickness test the identification of the condition currently relies on medical and behavioral criteria [26]. However, many of the symptoms associated with PD are non-specific and can also be present in other neurodegenerative diseases. The Integrated Parkinson's Rating Scale is a tool used for assessment, but it has limitations in terms of objectivity, reproducibility, and sensitivity [27].

The premotor stages of PD can occur for several years or even decades before the characteristic motor symptoms of the disease manifest [28, 29]. Tremor is the most common premotor symptom observed within the first two years after diagnosis, with a significantly higher prevalence compared to healthy individuals. Premotor signs show promise in the first detection of PD, despite the traditional reliance on motorized indicators for diagnosis [30]. Recent advancements have been made in identifying premotor indications as a means of diagnosing PD at an earlier stage. Imaging techniques and biomarkers hold potential for detecting early PD with tall levels of understanding and specificity, offering the possibility of a robust early diagnosis [31].

Some other cons of detecting PD using Machine Learning algorithms:

I. **Moral apprehensions:** Humanity has benefited from the use of computational models to organize tasks, analyze vast amounts of data, and reach complex decisions. Even though, putting your reliance in machine learning techniques has its drawbacks [32]. Procedures can be prejudiced anywhere at phase of progress. Furthermore, eliminating bias is very impossible because people both design and teach algorithms.

II. **Deterministic problems:** ML is a powerful approach that can be used to many different disciplines, including weather prediction and climate and environmental study. ML replicas can be used to support in the correction of sensors, permitting one to tweak the presentation of instruments that display ecological predictors such as temperature, heaviness, and moistness. Representations may be built so to mimic climate and discharges to the sky, for example, to forestall contamination [33]. Based on the amount of data and the model's complexity, this might be extremely costly and can take a month or longer.

III. **Insufficient Information:** Because of their intricate structures, complex systems need a substantial amount of training information to deliver useful results. As neural networks increase in size, they require more data to effectively train and optimize their performance. While some may consider reusing existing data to compensate for the increased information needs, this approach is unlikely to produce satisfactory results. For instance, presume the information used to train an algorithm to detect Parkinson's Disease uses skull radiography. If so, the model developed using this set of data may be biased, leading it to make unfounded assumptions when interpreting radiographic images of the Brain.

IV. **Absence of interpretability:** Machine learning algorithms' interpretability is a decisive matter. Consider that one is assigned the duty of creating a model to identify fraudulent transactions by your employer, a financial company. The model must be able to clarify its contract categorization in these circumstances. For this task, a machine learning method might be accurate and quick, but its conclusions might not be verified.

V. **Lack of duplicability:** Lack of code transparency and model testing processes make the lack of predictability in deep learning a challenging and growing issue. As neural networks increase in size, they require more data to effectively train and optimize their performance. While some may consider reusing existing data to compensate for the increased information needs, this approach is unlikely to produce satisfactory results. Larger neural networks with complex architectures and numerous parameters require diverse and representative datasets to capture their expanded capacity.

13.6 Conclusion and Future Works

Artificial intelligence and therapeutic disciplines have established a connection that contributes to the management and treatment of prevalent conditions like Parkinson's disease.

To facilitate the early detection of Parkinson's disease, various symptoms including reduced body movement (Bradykinesia), involuntary shaking at rest (Tremor), stiffness in muscles (Rigidity), and difficulties with voice can be recognized. There is no specific medical technique or diagnostic to cure parkinsonism, which also applies to bioinformatics. However, powerful methods like as Machine Learning has reduced the method of detecting Parkinson's disease by making it commercially feasible and efficient. Machine learning, according to the experiments cited in this paper, can assist physicians in diagnosingParkinson's disease. For detection, simple technological equipment such as a smart phone intended for audio footage, software for detecting slowness in movement, and many more can be used. The diagnosis of bradykinesia and tremor, as illustrated in section V, leads to tangible findings for the early detection of this condition. Furthermore, it was discovered that the detection accuracy may be boosted in two ways: by incorporating collective algorithms such as trapping, improving, and voting, and by growing the dataset extent. We will attempt to adapt new methods and mathematical structures in the future to help with and resolve the premature convergence issue. Setting technique examination of fertility rate, the search range of every target individual, and other topics are further potential subjects for future research. These topics will be covered in our next publications.

13.7 Bibliography

Acronyms used in this paper:

- **PD** - Parkinson's Disease
- **ML** - Machine Learning
- **RFC** - Random Forest
- **CSF** - Cerebrospinal Fluid Measurement
- **SVM** - Support Vector Machines
- **MLP** - Multilayer Perceptron
- **DBS** - Deep Brain Stimulation
- **FCM** - Fuzzy m-means

- **FKNN** - Fuzzy k-nearest Neighbor
- **CBFO** - Chaotic Bacterial Foraging Optimization
- **MGWO** - Modified Grey Wolf Optimization
- **CNN** - Convolutional Neural Network
- **PLS** - Partial Least Square
- **FTT** - Finger Tapping Tests
- **SLPs** - Speech-Language Pathologists
- **LRM** - Logistic Regression Model
- **DTC** - Decision Tree Classifier
- **NSVC** - Nu-Support Vector Classification
- **EMG** - Electromyographic
- **UPDRS** - Unified PD Rating Scale
- **SVC** - Support Vector Classification

References

1. Alves, G., Forsaa, E.B., Pedersen, K.F., Dreetz Gjerstad, M., Larsen, J.P., The epidemiology of Parkinson's disease. *J. Neurol.*, 255, Suppl 5, 1832, 2008.
2. Sinha, K., Miranda, A.O., Mishra, S., Real-time translator for sign language, in: *Cognitive Informatics and Soft Computing*, 1, pp. 477–489, Springer, Singapore, 2022.
3. Sahoo, P.K., Mishra, S., Panigrahi, R., Bhoi, A.K., Barsocchi, P., Improved detection of laryngeal cancer using CT images and a deep learning model. *Sensors*, 22, 22, 8834, 2022.
4. Tsanas, A., Little, M.A., McSharry, P.E., Ramig, L.O., Accurate telemonitoring of Parkinson's disease progression using noninvasive speech tests. *IEEE Trans. Biomed. Eng.*, 57, 4, 884–893, 2010.
5. Lahmiri, S. and Shmuel, A., Detection of Parkinson's disease based on ranking voice patterns and optimized support vector machine. *Biomed. Signal Process. Control*, 49, 427–433, 2019.
6. Harrou, F., Sun, Y., Hering, A.S., Madakyaru, M., *Statistical process monitoring using advanced data-driven and deep learning approaches: Theory and practical applications*, Elsevier, 2020.
7. Little, M.A., McSharry, P.E., Hunter, E.J., Spielman, J., Ramig, L.O., Suitability of dysphonia measurements for telemonitoring of Parkinson's disease. *IEEE Trans. Biomed. Eng.*, 56, 4, 1015–1022, 2009.
8. Das, R., A comparison of multiple classification methods for diagnosis of Parkinson's disease. *Expert Syst. Appl.*, 37, 2, 1568–1572, 2010.
9. Ashour, A.S., El-Attar, A., Dey, N., El-Kader, H.A., Abd El-Naby, M.M., Long short-term memory based patient-dependent model for freezing of gait detection in Parkinson's disease. *Pattern Recognit. Lett.*, 131, 23–29, 2020.

10. Sharma, S., Moon, C.S., Khogali, A., Haidous, A., Chabenne, A., Ojo, C., Ebadi, M., Biomarkers in Parkinson's disease (recent update). *Neurochem. Int.*, 63, 3, 201–229, 2013.
11. Galaz, Z., Mekyska, J., Zvoncak, V., Mucha, J., Kiska, T., Smekal, Z., Gomez-Vilda, P., Changes in phonation and their relations with the progress of Parkinson's disease. *Appl. Sci.*, 8, 12, 2339, 2018.
12. Cai, Z., Gu, J., Wen, C., Zhao, D., Huang, C., Huang, H., Chen, H., An intelligent Parkinson's disease diagnostic system based on a chaotic bacterial foraging optimization enhanced fuzzy KNN approach. *Comput. Math. Methods Med.*, 2018, 1–25, 2018.
13. Sharma, P., Sundaram, S., Sharma, M., Gupta, D., Diagnosis of Parkinson's disease using modified grey wolf optimization. *Cognit. Syst. Res.*, 54, 100–115, 2019.
14. Parisi, L., Ravi Chandran, N., Manaog, M.L., Feature-driven machine learning to improve early diagnosis of Parkinson's disease. *Expert Syst. Appl.*, 110, 182–190, 2018.
15. Berus, L., Klancnik, S., Brezocnik, M., Ficko, M., Classifying Parkinson's disease based on acoustic measures using artificial neural networks. *Sensors*, 19, 1, 16, 2018.
16. Li, G., Chen, L., Liu, J., Fang, X., Comparative study on deep transfer learning strategies for fault diagnosis in building energy systems under different conditions. *Energy*, 125943, 2022.
17. Afonso, L.C., Rosa, G.H., Pereira, C.R., Weber, S.A., Hook, C., Albuquerque, V.H.C., Papa, J.P., Recurrence plot-based approach for identification of Parkinson's disease. *Future Gener. Comput. Syst.*, 94, 282–292, 2019.
18. Mishra, S., Thakkar, H.K., Singh, P., Sharma, G., Combined unsupervised-supervised model for chronic disease risk assessment using metaheuristic attribute selection. *Comput. Intell. Neurosci.*, 2022, 2022.
19. National Institute for Health and Care Excellence (NICE), Parkinson's disease: Diagnosis and management in primary and secondary care. NICE Clinical Guidelines 35, 2006, Retrieved from http://www.nice.org.uk/guidance/cg35/resources/guidance-parkinsons-disease-pdf.
20. Chakraborty, S., Sahoo, K.S., Mishra, S., Islam, S.M., AI-driven cough voice-based COVID detection framework using spectrographic imaging, in: *2022 IEEE 7th International Conference for Convergence in Technology (I2CT)*, IEEE, 1, pp. 1–7, 2022.
21. Suman, S., Mishra, S., Sahoo, K.S., Nayyar, A., Vision navigator: A smart and intelligent obstacle recognition model for visually impaired users. *Mobile Inf. Syst.*, 2022, 1–15, 2022.
22. Karapinar Senturk, Z., *Early diagnosis of Parkinson's disease using machine learning algorithms*, Duzce University, Engineering Faculty, Department of Computer Engineering, 138, 109603, 2022.

23. Tsanas, A. *et al.*, Novel speech signal processing algorithms for high-accuracy classification of Parkinson's disease. *IEEE Trans. Biomed. Eng.*, 59, 124–141, 2012.
24. Abhishek, Tripathy, H.K., Mishra, S., Usability of encryption methods in healthcare data security: An analytical study, in: *Next Generation Healthcare Informatics*, 1, pp. 87–96, Springer, Singapore, 2022.
25. Mishra, S., Anand, K., Jhanjhi, N.Z., Smart personal assistant for visually challenged individuals, in: *Proceedings of the 2nd International Conference on Mathematical Modeling and Computational Science*, pp. 505–515, Springer, Singapore, 2022.
26. Little, M.A. *et al.*, Suitability of dysphonia measurements for telemonitoring of Parkinson's disease. *IEEE Trans. Biomed. Eng.*, 56, 4, 1015–1022, 2009.
27. Giancardo, L. *et al.*, Computer keyboard interaction as an indicator of early Parkinson's disease. *Sci. Rep.*, 6, 34468, 2016.
28. Patnaik, M. and Mishra, S., Indoor positioning system assisted big data analytics in smart healthcare, in: *Connected e-Health*, 1, pp. 393–415, Springer, Cham, 2022.
29. The National Health and Medical Research Council (Australia), National statement on ethical conduct in human research, The National Health and Medical Research Council, The Australian Research Council, The Australian Vice-Chancellors' Committee, 2007.
30. Yu, H.F., Huang, F.L., Lin, C.J., Dual coordinate descent methods for logistic regression and maximum entropy models. *Mach. Learn.*, 85, 1-2, 41–75, 2011.
31. Sivani, T. and Mishra, S., Wearable devices: Evolution and usage in remote patient monitoring systems, in: *Connected e-Health*, 1, pp. 311–332, Springer, Cham, 2022.

14

Towards a Sustainable Future: Harnessing the Power of Computational Intelligence to Track Climate Change

Satyam Sinha*, Shreyash Kumar Agnihotri and Oshmita Sarkar

Kalinga Institute of Industrial Technology, Deemed to be University, Bhubaneswar, Odisha, India

Abstract

The change in climate is a pressing global challenge, with significant impacts on the environment, economy, and society. To mitigate these impacts, emerging technologies such as artificial intelligence (AI) are being used to develop new strategies for climate change adaptation and mitigation. This paper provides a comparative analysis of two case studies that leverage AI for these purposes: drought prediction and carbon dioxide capture. AI can also be used in climate change modeling by predicting future climate trends and simulating the effects caused by climate change on ecosystems and society. However, the use of AI in climate change modeling poses several challenges and limitations, such as the need for accurate data and concerns related to ethics and responsibility. The paper presents a comparative analysis of two case studies: drought prediction and carbon dioxide capture. The study demonstrates the potential of AI to support climate change adaptation efforts by predicting drought conditions and informing water management strategies. The study also highlights the challenges associated with the implementation of AI-based carbon dioxide capture systems, such as high costs and technical limitations. To effectively use AI in modeling of climatic change, it is essential to build partnerships and collaborations, ensure the availability of data, and address ethical and social issues. The paper concludes by outlining the emerging trends in AI and adaptability to climate change, highlighting the opportunities and challenges relating to the application of AI to climate change adaption and mitigation, and suggesting future research directions in this area. The paper provides

**Corresponding author*: 2005689@kiit.ac.in

Hrudaya Kumar Tripathy, Sushruta Mishra, Minakhi Rout, S. Balamurugan and Samaresh Mishra (eds.) Optimized Computational Intelligence Driven Decision-Making: Theory, Application and Challenges, (275–304) © 2024 Scrivener Publishing LLC

a comprehensive overview of the role of AI in addressing the climate crisis and inspires further research in this critical area.

Keywords: Artificial intelligence, climate change, climate change mitigation, climate change adaptation, drought prediction, carbon dioxide capture, comparative analysis

14.1 Introduction

Climate crisis is arguably one of the biggest challenges faced by humanity, especially in the 21st Century. The globe would experience catastrophic impacts if greenhouse gas emissions are not controlled within the next 30 years, according to the 2018 global report on climate change [1]. The development of machine learning (ML) algorithms has led to advancements in many areas of research and has been suggested as a tool for climate studies [2]. The use of artificial intelligence models to address the issue of climate change has been studied and researched.

Weather trade is one of the most urgent problems dealing with the arena these days. The results of weather change, collectively with developing sea levels, excessive weather activities, and converting precipitation styles, are having an exquisite effect on the surroundings, monetary machine, and society. Mitigating climate trade calls for collective efforts to reduce greenhouse gasoline emissions and adopt sustainable practices. But, irrespective of such efforts, some modifications inside the climate are inevitable, and models are important to lessen the horrific impacts of weather exchange.

Artificial intelligence (AI) has emerged as an effective device in climate trade research, allowing us to count on and reply to modifications within the weather. AI can help in climate change mitigation via ways of identifying patterns and developments in statistics and presenting insights into the effectiveness of mitigation techniques. By predicting future weather trends and simulating the effects of weather change on ecosystems and society, AI can also help with the climate change model. In Lee *et al.* (2013)'s research, a Gaussian process was used to simulate the effects of parameter uncertainty on climate model output and to explore the reasons for uncertainty in cloud-aerosol processes [3].

Climate change is currently one of the world's most urgent problems. The outcomes of weather change, which includes developing sea tiers, immoderate weather activities, and changing precipitation styles, are having a big effect on the surroundings, financial tools, and society. Mitigating weather change calls for collective efforts to reduce greenhouse gas emissions and adopt sustainable practices. However, despite such efforts, a few adjustments inside the climate are inevitable, and variability is critical to limit the damaging influences of climate trade.

Artificial intelligence (AI) has emerged as an effective device in weather alternate studies, permitting us to expect and respond to modifications within the climate. AI can help in weather alternate mitigation through identifying patterns and traits in statistics and supplying insights into the effectiveness of mitigation strategies. AI also can assist in climate alternate models via ways of predicting future climate tendencies and simulating the results of climate change on ecosystems and society.

In this paper, we explore the role of AI in addressing the climate crisis, focusing on its potential applications, ethical implications, challenges, and limitations. We first discuss an overview of the background and related works in this area, highlighting the opportunities and challenges associated with the use of AI in this context. Next, we provide a comparative analysis of the two case studies, starting with drought prediction. We describe the abstract, methodology, and outcomes of this study, which demonstrates the potential of AI to support climate change adaptation efforts by predicting drought conditions and informing water management strategies. We then turn to carbon dioxide capture and present the abstract, methodology, and outcomes of our analysis. Then the various algorithms applied in these two cases are discussed along with the importance of interdisciplinary collaboration, and outline of the future directions of AI research in this field is provided. We conclude by discussing the advantages and disadvantages of the case studies and highlighting the emerging trends in AI and climate change adaptation. We hope this paper provides a comprehensive overview of the role of AI in addressing the climate crisis and inspires further research in this critical area.

14.2 Artificial Intelligence and Climate Change Adaptation

AI can assist in climate change mitigation by means of analyzing enormous quantities of facts from diverse sources, which includes climate sensors, satellite pix, and power consumption facts. AI algorithms can become aware of patterns and trends on this data, supplying insights into the effectiveness of various mitigation strategies. As an example, AI may be used to optimize renewable electricity structures, reducing the fee and enhancing the performance of energy production. AI also can assist to reveal and manipulate greenhouse fuel emissions, identifying regions wherein emissions may be reduced.

By anticipating future climatic patterns and modeling the effects of climate change on ecosystems and society, AI can help with climate change adaptation. Policymakers and communities can use this information to help them prepare for and deal with the effects of climate change. For instance, farmers can modify their planting and irrigation strategies by using AI to predict changes in agricultural yields and water availability. Richard's (2018) work indicates their capacity to enhance historical climate data estimates by leveraging nearby stations to account for inhomogeneities non data collection procedures [4]. Deep learning is also used to aid climate crisis by improving the accuracy of climate models, predicting extreme weather events, and analyzing satellite imagery to monitor changes in the environment. It have been recently applied to climate sciences in various ways, such as Buckland's (2019) study of detecting dryland disturbance [5].

AI can also assist in disaster management by predicting the likelihood and severity of natural disasters and supporting emergency response efforts, providing real-time information and analysis during emergencies. For example, AI can analyze satellite images and social media posts to identify areas that are most affected by a disaster, enabling first responders to prioritize their efforts. AI can also help to coordinate rescue efforts by analyzing data on the location and status of people in affected areas. Additionally, AI can support disaster preparedness efforts by simulating the effects of different disaster scenarios and identifying strategies to mitigate their impacts.

14.3 Related Works

There are numerous studies and researches that discuss about AI and its usage against climate change along with the pros and cons. Machine Learning and Deep Learning models can substitute expensive components of climate models as explored in Gentine 2018 investigation [6]. It is true that improving forecasts is essential for better preparation and management of extreme weather events. McGovern (2017) conducted a study to evaluate the performance of artificial intelligence (AI) methods in predicting high impact weather events, such as storms and their duration, using historical data. They utilized the Gradient Boosted Regression Trees (GBRT) algorithm to operationalize the prediction model. The results of the study showed that the GBRT algorithm was able to outperform traditional statistical methods in predicting the duration of severe weather events [7]. Droughts are severe weather events that have resulted in a global cost of around $1.5 billion between 1998 to 2017, accounting for 33% of the total cost of weather hazards during this period. Utilizing recent advancements in drought forecasting accuracy, AI has the potential to enhance decision-making processes and mitigate the impact of droughts, according to Belayneh (2016) [8]. Chris Hunting ford (2019) in "Machine learning and artificial intelligence to aid climate change research and preparedness" suggested various machine learning techniques and identified three scenarios where the interconnectivity of systems is expected to be intricate, namely a UK extreme event, the phenomenon of 'warming hiatus', and the construction of equations for terrestrial ecosystems where AI models can be effectively used [9]. Mark Coeckelbergh in "AI and Ethics (2021) 1:67–72" discussed that though AI can be used to build a cleaner, greener and sustainable environment by tackling climate changes, it also raises privacy, political and ethical issues like influencing people to behave in a particular way to be more climate friendly thus changing their natural behavior. He suggested that the AI project for climate can be a success only if these issues are taken care and every individual takes responsibility and support the actions taken to resolve these problems [10]. However, since both processing and storage of data demand power, they have an effect on the nature and climate. Deep learning model training also requires a large amount of data. According to a study by Strubell E, Ganesh A, and McCallum A (2019), training a single model of NLP can result in the emission of about three hundred thousand kilograms of the carbon dioxide equivalent is equivalent to five times the emissions generated by an average car throughout its lifespan [11]. Anthony (2020) has suggested that in addition to the standard performance metrics, reporting the energy consumption and carbon footprint associated with deep learning model development and training is also necessary. This will allow AI practitioners to track the energy use to create a climate-friendly ecosystem and take into account less energy-intensive models as another metric of success in addition to performance and accuracy [12]. In "Tackling climate change with Machine Learning" (2019), it has been discussed that ML has proven to be highly beneficial in the battle against climate change, with its significant contributions being evident across multiple domains. These contributions include the ability to remotely sense and monitor environmental changes such as deforestation, gather building data, and assess disaster damage. Optimizing systems to improve efficiency is another key area where ML can make a difference, by consolidating freight, designing carbon markets, and reducing food waste [13]. Schmidt, Victor & Luccioni (2019). proposed that ML can also be used to spread education and promote climate awareness, Utilizing AI, it is feasible to produce individualized educational materials such as visual representations of the projected consequences of extreme weather conditions at the learner's address [14] or by simulating their real-world location in a digital environment and enable them to investigate how climate change will affect that particular area as suggested by. This approach involves anchoring the learning process in a familiar setting and providing an interactive platform for exploration [15]. Every year, electricity systems contribute to approximately 25% of the greenhouse gas emissions that are caused by human activities [16]. ML models also aid in reducing emission from generators and provide a carbon free system by techniques like forecasting, scheduling and control as AI has been called the new electricity as it has the potential to transform the electrical industry for example smart grids which requires data driven by AI and ML [17]. In "Task-based end-to-end model learning in stochastic optimization (2017)" the authors used a deep learning techniques to generate demand predictions that emphasize electricity scheduling costs over forecast precision as a priority [18]. This concept could be expanded to create predictions that minimize greenhouse gas emissions. Climate crises can have a huge impact on our planet, the AI community reduces the damage caused by it to the ecosystem and the harm it poses to the people. For instance climate change can lead to potential epidemic by increasing the amount of water borne and vector disease, the tools provided by ML can be very beneficial to health workers where specialized lab equipment is not accessible like John A Quinn's (2014) work on diagnosis of malaria using photographs captured with a mobile phone, which depict prepared pathology slides [19]. In "Tackling Climate Change with Machine Learning (2019)", it has been discussed that ML has significant contribution across several domains that helps to manage climate crisis by monitoring climatic changes like deforestation and optimizing existing systems to balance the carbon cycle. It also recommended that climate relevant dataset should be released by concerned entities to foster involvement from ML community [20]. We shall further discuss about this topic in the next sections.

14.4 Comparative Analysis of Technological Frameworks to Handle Climate Crisis

In this section we discuss two specific climate crises - droughts and excessive carbon dioxide in the atmosphere, we will also present a comparative study of how different AI models are used to control this crisis respectively.

The objective of this analysis is to identify key success factors and challenges of using AI in climate change adaptation.

14.4.1 Artificial Intelligence for Drought Assessment and Forecasting

Drought is a climate crisis that occurs when a region receives lower than average precipitation for an extended period, causing shortage of water resources. Droughts can cause significant consequences on agriculture, natural ecosystems, and human populations, including food and water scarcity, economic losses, and social and political conflicts [21].

AI models can help mitigate the impacts of drought by predicting and monitoring weather patterns, precipitation levels, soil moisture, and water availability in real-time. These models can provide valuable insights for policymakers, farmers, and other stakeholders to make informed decisions about water management, irrigation strategies, crop selection, and drought-resistant farming practices. Figure 14.1 shows the intelligent approach for drought analysis.

Park (2016) investigated the relationship between factors and conditions of droughts (like yield of crop and SPI), and to create drought indicators through the assessment of variable significance using three ML techniques—random forest (RF), boosted regression trees (BRT), and Cubist—have shown adaptability and resilience in multiple regression scenarios. The following subsections provide an explanation of the three techniques.

Random Forest:
Random Forest is a machine learning algorithm that combines multiple decision trees to make predictions by averaging their results. It is known for its versatility, robustness, and ability to handle complex data.

Boosted Regression Tree:
Boosted Regression Trees (BRT) is a powerful machine learning algorithm that combines the strength of decision trees and boosting techniques to build predictive models for both regression and classification problems. BRT iteratively builds decision trees to predict the target variable, and each subsequent tree is trained in such a way that error of previous tree is reduced [22].

Cubist: Cubist, a rule-based software that uses ML methodology, utilizes a modified regression tree and has been applied in drought studies. It produces more easily interpretable results compared to Random Forest (RF) and Boosted Regression (BR) by generating rules and equations for forecasting goal values. The Cubist approach removes the complexity associated with RF and BR [23, 24].

Remote sensing data:

1) MODIS data were used to identify the drought-related components. MODIS is a component of the Terra platform, which was unveiled in December 1999. Various outputs, including drought variables, are available through MODIS.
2) TRMM rainfall data: TRMM provides rainfall data with a 0.25 degree resolution.

For a total of 16 input variables were used.

Precipitation rate, expressed in mm/hr and then converted to inches/month, is included in the data.

In the machine learning models, a total of 16 metrics related to droughts and SPIs were employed as independent and dependent variables at varied

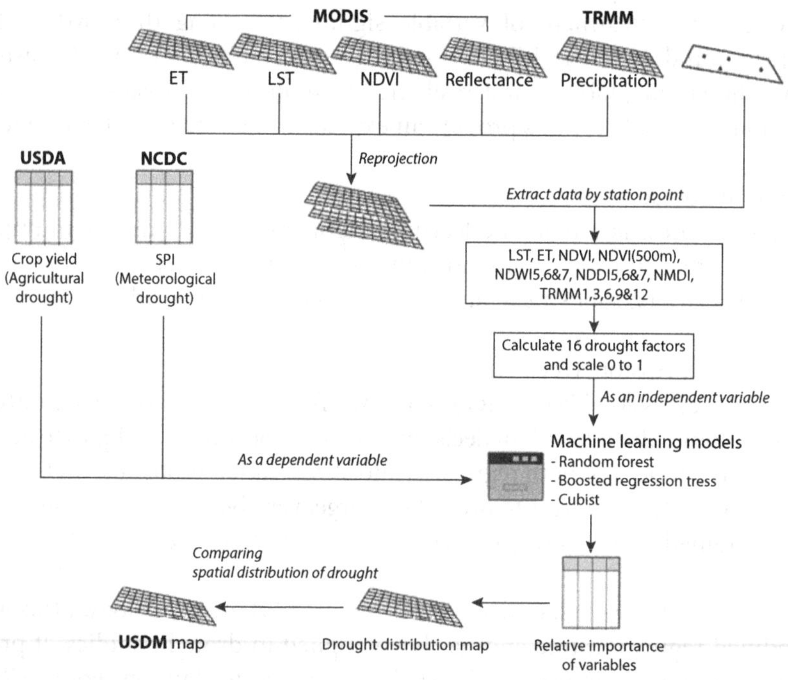

Figure 14.1 The workflow diagram.

time intervals of 1, 3, and 12 months. This modeling technique was applied separately to the three regions of varied climatic condition. Each machine learning model's efficiency was assessed using the coefficient of determination (R square) and the root-mean-square error (RMSE).

Due to the considerable interdependence of some elements, the study only chose a small fraction of the factors with the maximum relative relevance from each group of linked factors for further analysis. The study then used three machine learning techniques to reevaluate the relative importance of these six variables for both agricultural and meteorological drought in both arid and humid regions, as determined by yield of crop and SPIs, respectively. Through the use of weights that represented the relative importance of the elements, the study generated drought maps for both regions.

In order to study the spatial patterns of drought conditions, a variety of drought maps were employed, and their yearly changes were compared to maps of USDM that are often used in the United States.

With an R2 value more than 0.89 and Root Mean Square Error values ranging from 0.21 to 0.39, the study discovered that RF outperformed BRT and Cubist among the three ML models (Figure 14.2). Similar R2 values of 0.3 to 0.6 and RMSE values of 0.6 to 0.8 were achieved by the other two models. In the arid, humid, and combination environments, Figure 14.3 shows the link between the actual 12-month SPI values and the anticipated values from the three models. In contrast to BRT and Cubist, RF's anticipated range of 12-month SPI values was substantially wider than the actual range, showing that RF was more effectively trained than the other two models. In contrast to the models that employs an ensemble strategy to prevent the issues overfitting.

All three ML techniques were capable of modeling drought conditions on longer time intervals, including the 9- and 12-month SPIs, more effectively than on shorter time scales. This is due to the fact that markers of dryness, including vegetation indices, frequently indicate the impacts of a prolonged absence of precipitation, as noted by Gessner (2013) and Nezlin (2014), (2005) [25, 26]

The relative significance of variables is instead assessed by BRT by counting the frequency of variable selection for splitting.

Table 14.1 summarizes the relative importance of the 16 drought factors at different time scales when using RF, BRT, and Cubist in the arid (Table 14.1a), humid (Table 14.1b), and combination (Table 14.1c) regions.

Comparison with the US drought monitor map: Drought conditions in arid and humid regions were mapped using the derived weights from Random Forest of six selected drought indicators to establish their

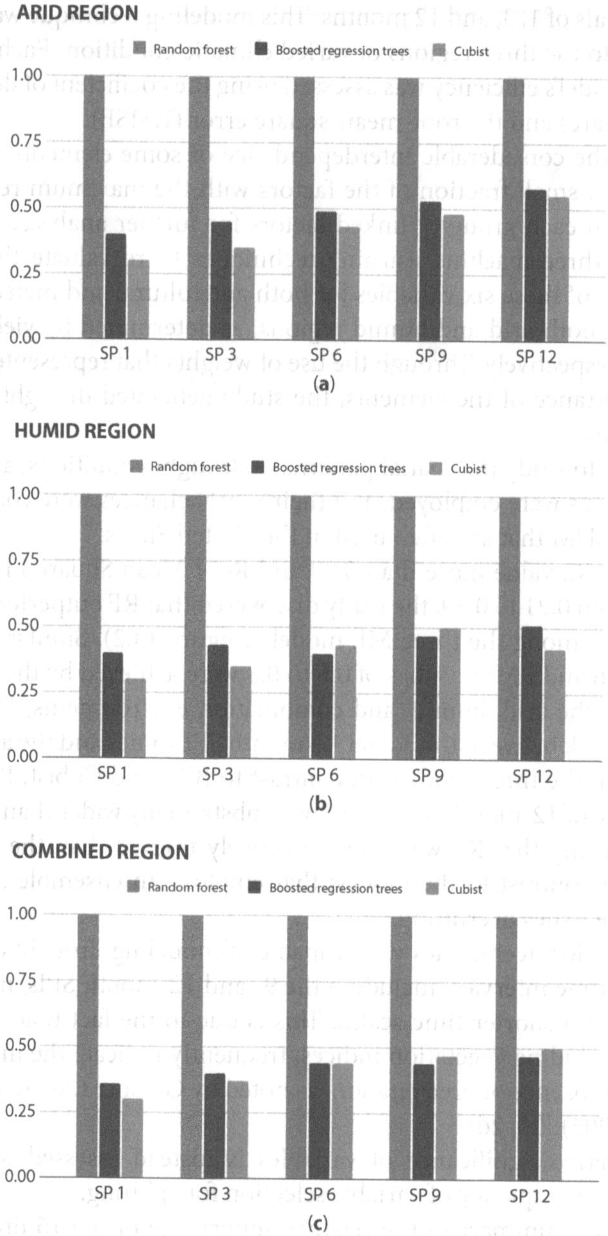

Figure 14.2 (a) Model performance of three ML techniques for 1-, 3-, 6-, 9-, and 12-month SPIs in the dry zone. (b) Model effectiveness of three ML methods for 1-, 3-, 6-, 9-, and 12-month SPIs in humid region. (c) shows the model results of three ML techniques for 1-, 3-, 6-, 9-, and 12-month SPIs in the combined region.

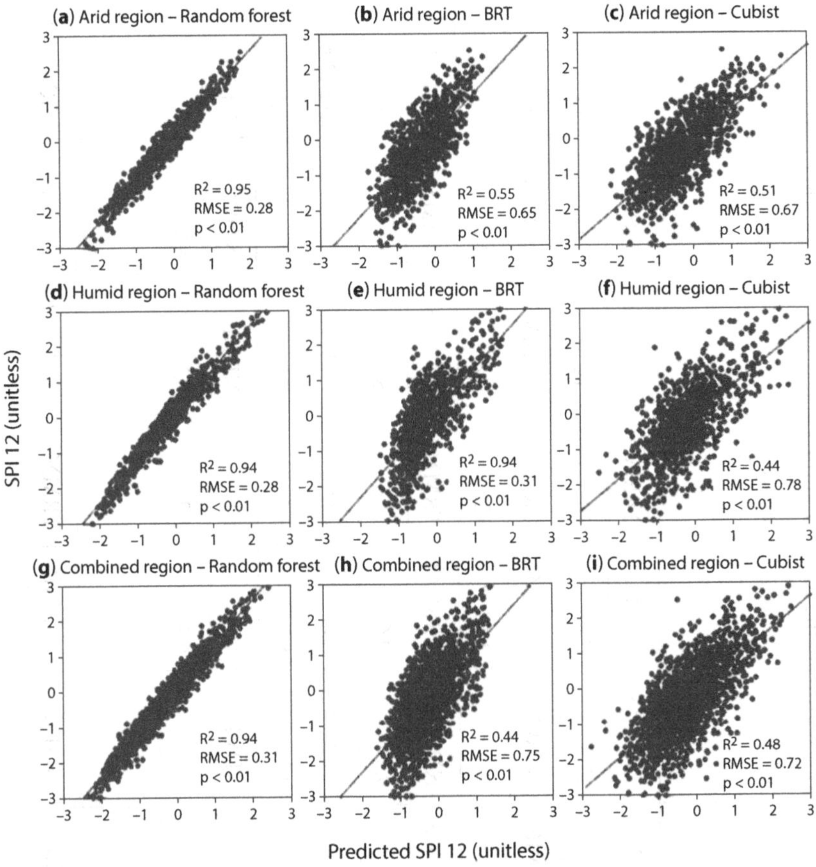

Figure 14.3 Shows the correlation between actual 12-month SPI values and those predicted by the three models across various region.

spatial distribution. Thirteen years of data (2000–2012) were used to establish the relative importance of the six variables, and weight factors were allocated in accordance. The data from 2013 and 2014 were then analyzed using these weight factors and compared to the USDM.

The USMD data revealed that the maps with remote sensors were able to detect the extreme drought in the dry region in 2013 and 2014. Although the remote sensing maps in the humid zone showed a more severe drought situation, the USDM map and the drought distribution in 2014 were in agreement. Contrary to the SPI maps of 3-months, the maps of crop yields showed a comparable drought state to the USDM data. The proposed importance-based weightings are therefore thought to be appropriate and useful for drought monitoring.

Table 14.1 Shows data on meteorological droughts in various regions using Random forests and Boosted regression trees.

	SPI1		SPI3		SPI6		SP19		SPI12	
(a)	Arid region									
1	TRMM1	TRMM1	TRMM6	TRMM6	TRMM6	TRMM1	TRMM1	TRMM6	TRMM6	TRMM6
	21.3	38	12.3	19.8	17.8	21.3	38	12.3	19.8	17.8
2	TRMM6	TRMM3	TRMM3	NDWI6	NDWI6	TRMM6	TRMM3	TRMM3	NDWI6	NDWI6
	10.3	12	9.8	13	9.6	10.3	12	9.8	13	9.6
3	TRMM3	TRMM6	NDWI6	TRMM3	NDWI7	TRMM3	TRMM6	NDWI6	TRMM3	NDWI7
	9.6	11.6	8.1	12.6	9.3	9.6	11.6	8.1	12.6	9.3
4	ET	LST	NDW17	ET	NDDI6	ET	LST	NDW17	ET	NDDI6
	8.7	8.5	7.2	9	9.1	8.7	8.5	7.2	9	9.1
5	LST	ET	NDDI6	NDVI	NDVI	LST	ET	NDDI6	NDVI	NDVI
	6.4	6.3	7.1	8.7	7.8	6.4	6.3	7.1	8.7	7.8
(b)	Humid region									
1	TRMM1	TRMM1	TRMM3	TRMM3	TRMM6	TRMM6	TRMM9	TRMM9	TRMM12	TRMM12

(*Continued*)

Table 14.1 Shows data on meteorological droughts in various regions using Random forests and Boosted regression trees. *(Continued)*

	SPI1		SPI3		SPI6		SP19		SPI12	
	25.4	50	13.2	27.2	15	40.7	15.2	47.8	21.3	55.4
2	LST	LST	TRMM1	TRMM1	TRMM9	TRMM9	TRMM12	TRMM12	TRMM9	TRMM1
	10.1	14.1	11.5	16.9	10.6	11.9	11.9	10.8	8.6	7.8
3	TRMM3	TRMM6	TRMM6	LST	TRMM3	TRMM1	TRMM6	TRMM1	TRMM6	LST
	7.4	4.8	9.4	11	8.4	9.8	9.6	9.3	7.1	5.7
4	TRMM6	NDV1500	TRMM9	TRMM9	TRMM12	LST	TRMM1	LST	NDDI6	NDVI
	7	4	8.8	9.1	8.1	8.8	8.2	5.9	6.8	4.3
5	TRMM9	TRMM3	TRMM12	NDWI7	TRMM1	NDVI	TRMM3	NDVI	TRMM1	NMDI
	6.4	2.9	8.4	5.3	8	3.8	6.7	3.8	6.8	3.9
(c)	Combined region									
1	TRMM1	TRMM1	TRMM6	TRMM3	TRMM6	TRMM6	TRMM9	TRMM9	TRMM12	TRMM12
	13.2	52.7	14.1	28.2	12.2	45.9	15.2	47.1	13.2	54.3

(Continued)

Table 14.1 Shows data on meteorological droughts in various regions using Random forests and Boosted regression trees. (*Continued*)

	SPI1		**SPI3**		**SPI6**		**SP19**		**SPI12**	
2	TRMM3	LST	TRMM3	TRMM6	TRMM9	TRMM9	TRMM12	TRMM12	ET	TRMM6
	12.3	13.1	10.3	14.8	10	9.7	11.9	9.1	12.1	7.4
3	TRMM6	TRMM3	NDWI7	NDWI6	TRMM12	LST	TRMM6	TRMM6	TRMM6	NDWI7
	9.5	7.9	8.1	9.4	7.6	7	9.6	8.1	7.6	5.8
4	LST	TRMM6	NDWI6	TRMM1	NDWI7	NDWI7	TRMM1	NDWI6	NDVI	NDVI
	8.5	5.4	7.4	9.3	7.4	6.4	8.2	4.7	6.6	5.7
5	ET	ET	ET	LST	ET	NDWI6	NDDI7	NDVI	NDWI6	LST
	6.9	2.8	6.9	9.3	6.9	5.4	5.5	4.7	6.4	4.2

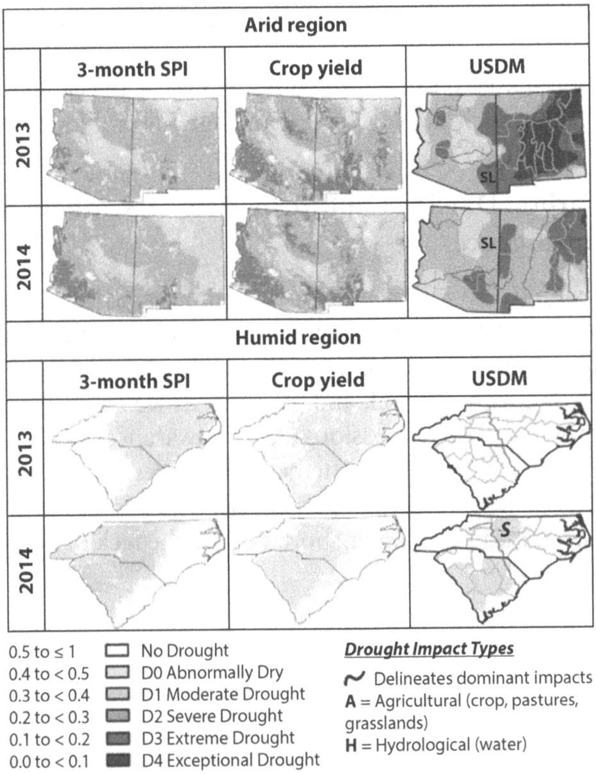

Figure 14.4 The evaluation of drought monitoring maps (2013 and 2014) took place in both arid regions with irrigation and humid regions without irrigation.

The features of sixteen drought parameters for both meteorological (as measured by SPI) and agricultural (as assessed by crop production) droughts were examined in this study using three ML techniques. The six most important variables were chosen based on their relative importance, taking into account their interaction. The RF strategy performed better than the other approaches. In order to compare the drought's spatial distribution to USDM maps, the modeling outcomes based on the six factors were displayed. Compared to the previous displayed maps, the crop yield-based maps were more accurate with the USDM maps. Despite the lack of readily available in situ data, this method can be used to vegetated areas utilizing remote sensing data. A sample drought monitoring map is shown in Figure 14.4.

In order to make the suggested approach more generalizable, clear rules and guidelines on thresholds should be set, as the current thresholds are arbitrary. Yet, there are some shortcomings in the study that can be improved by more studies and research. The creation of drought indicators should also take into account soil moisture and evapotranspiration, and remote sensing-based optimum methods should be developed for various forms of drought, such as hydrological and agricultural droughts.

14.4.2 Carbon Dioxide Capture by Help of Synthetic Intelligence

Carbon dioxide (CO_2) is a prominent greenhouse gas responsible for the retention of heat in the Earth's atmosphere, resulting in the escalation of global temperatures and consequential climate changes. The burning of fossil fuels along with coal, oil, and gasoline, as well as deforestation, are the primary assets of CO_2 emissions. As the awareness of CO_2 within the ecosystem increases, the planet's temperature rises, inflicting a range of devastating outcomes such as sea level rise, more frequent and extreme herbal disasters, meals and water shortage, and the lack of biodiversity.

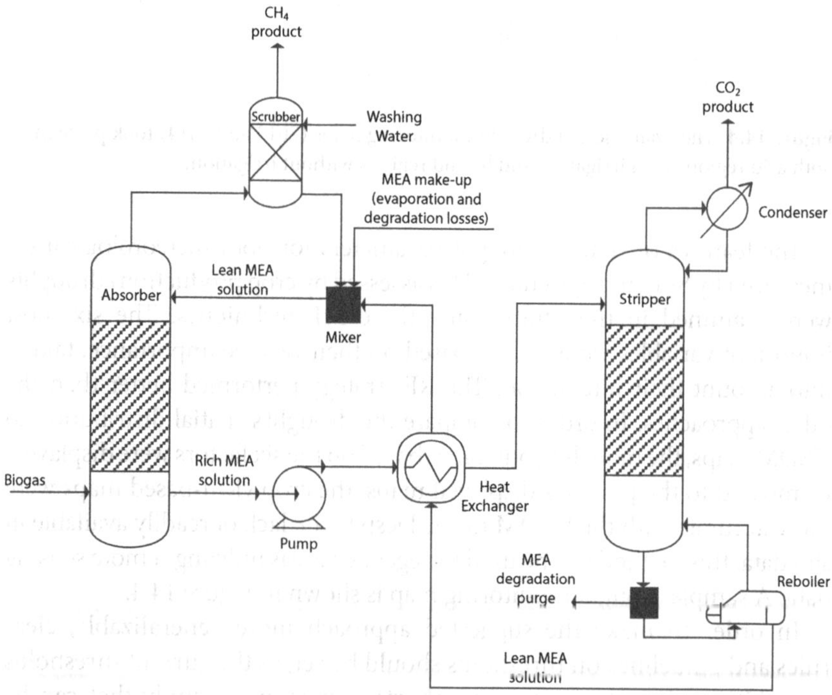

Figure 14.5 Flow chart of the process used in the study.

Synthetic intelligence (AI) can play a massive position in mitigating the outcomes of CO_2 emissions and fighting weather exchange. AI algorithms can assist optimize the layout and operation of carbon seize structures, which could boost their performance and decrease their charges. AI also can be used to predict and monitor the behavior of carbon seize systems, assisting to discover capacity troubles and enhance their overall performance. Moreover, system learning techniques may be applied to carbon capture facts to perceive patterns and optimize tactics. Selecting the maximum suitable AI-based total method to implement CO_2 tactics is essential because it appreciably impacts the accuracy of the effects acquired.

In the study by Cao, L.(2021), various AI-based models have been outlined and evaluated to assist users in selecting the optimal strategy based on their application and desired precision. Depending on the specific application, each model has the potential to outperform the others, making it critical to investigate their benefits and drawbacks through additional use cases [27]. The models have been developed accordingly.

Artificial Neural Network (ANN):
The ANN, composed of neurons interconnected to form a mathematical model, is a nonparametric technique that possesses outstanding interpolation capability for nonlinear systems. For modeling of large dimensions, this technique works fairly well. With this mathematical approach, the incoming data are distributed and processed. This method can be used to describe either linear or nonlinear systems without any prior knowledge of the relationship between input and output parameters.

ANFIS:
ANFIS stands for Adaptive Neuro-Fuzzy Inference machine, which is a form of synthetic neural network that mixes fuzzy common experience and neural networks to construct a hybrid version capable of handling complex structures with obscure or unsure records. ANFIS has been utilized in carbon dioxide seizure to expand accurate models for predicting several parameters which include solubility, mass switch, and absorption fee. The hybrid nature of ANFIS lets it capture the strengths of each fuzzy commonplace feel and neural networks, making it a powerful tool for modeling complicated carbon dioxide seizure systems.

LSSVM:
LSSVM stands for Least Squares help Vector gadget, that's a device mastering set of rules that has been utilized in carbon dioxide capture to expand accurate models for predicting diverse parameters consisting of

solubility, mass transfer, and absorption charge. LSSVM has been shown to be powerful in coping with noisy and incomplete facts, making it properly-applicable for modeling carbon dioxide seize systems, in which there are many variables that can impact the capture procedure.

AdBoost-CART:

AdaBoost-CART, additionally known as Adaptive Boosting with category and Regression bushes, is a machine studying algorithm that combines a couple of decision timber to create a stronger, greater correct predictive model. In the context of carbon capture, AdaBoost-CART can be used to expect the performance of carbon seize technologies, together with amine-based absorption or membrane separation, primarily based on enter variables consisting of gas composition, temperature, and pressure.

The CCS (Carbon Capture and Storage) approach was developed to attract, transport, and store CO_2 in a location other than the atmosphere. This is done by capturing CO_2 from large sources such as biomass power plants or cement production facilities and depositing it underground in geological formations to prevent the release of large amounts of CO_2 into the environment, which can lead to ocean acidification and global warming. The CCS approach can help mitigate the effects of CO_2 release caused by heating. It should be noted that there are also other objectives for injecting CO_2 into geological formations, such as improved oil recovery. A mathematical model is proposed as the objective.

As seen in Figure 14.5, four alternative techniques were used to examine the CO_2 equilibrium absorption in aqueous solution of Piperazine[28]. The findings show that compared to the other models, the AdaBoost-CART model has a substantially greater level of agreement with the empirical data. In comparison to the other models, this implies that the ANN model is not effective at forecasting CO_2 solubility in PZ. The study also shows that for this particular application, LSSVM and ANFIS produce results that are fairly close.

The following equation connects R2, ARD%, and AARD%, which stand for coefficient of determination, average relative deviation, and average absolute relative deviation percentage, respectively. These equations are utilized in the models CART that simulate balanced CO_2 capture in PZ solution. The letters "n," "," and "" in this sentence stand for the overall number of points, the "data," and the "assessed," respectively.

CO2 solubility obtained towards predictions of developed AI methods is shown in Figure 14.6.

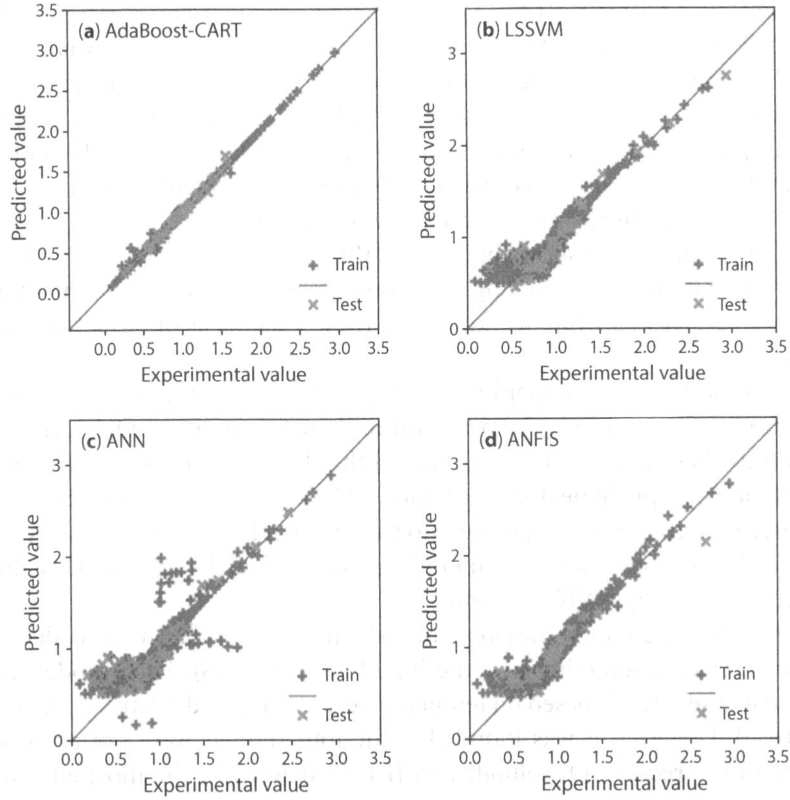

Figure 14.6 CO_2 solubility obtained experimentally towards predictions of developed AI methods.

$$R^2 = 1 - \frac{\Sigma_i^n (exp_i - pred_i)^2}{\Sigma_i^n (pred_i - average(\exp_i))^2} \tag{14.1}$$

$$ARD\% = \frac{100}{n} \Sigma_i^n \frac{(exp_i - pred_i)}{\exp_i} \tag{14.2}$$

$$AARD\% = \frac{100}{n} \Sigma_i^n \frac{|exp_i - pred_i|}{exp_i} \tag{14.3}$$

ANFIS, LSSVM, AdaBoost-CART, ANN, and ANFIS models were used to assess an amine solution made up of water, monoethanolamine, and carbon dioxide. Figure 14.7 shows the outcomes of a comparison of the models based on variables like R2, ARD%, and AARD%. The AdaBoost-CART model obviously outperforms the other models by a large margin, as evidenced by the AARD% numbers, which indicate accuracy, and ARD% values, which indicate error distribution. The AdaBoost-CART model's advantage as compared to other models examined is further supported by the R2 values, which show the accuracy of the repeated results.

Figure 14.8 and 9 show that the AdaBoost-CART model has the lowest error rate when employing additional amine solutions. It is therefore strongly suggested while working with these systems as opposed to other models.

AdaBoost-CART outperforms other methods as the most efficient method for CO_2 capture from the atmosphere using dry NaOH sorbents. On the other hand, it has been claimed that LSSVM can forecast the precipitation of asphaltene during a variety of industrial processes with a high degree of accuracy. Gaussian function has shown better results than ANN with SA in terms of heat duty modeling, even though large scale modeling necessitates using ANN technique.

In order to compare different models, five sets of data were collected to predict CO_2 solubility in ionic liquid systems (ILS). Each model was evaluated separately based on temperature, pressure, and relative deviation % (RD). However, it was found that the four models were not accurate enough to predict CO_2 solubility in ILs. To determine the most effective

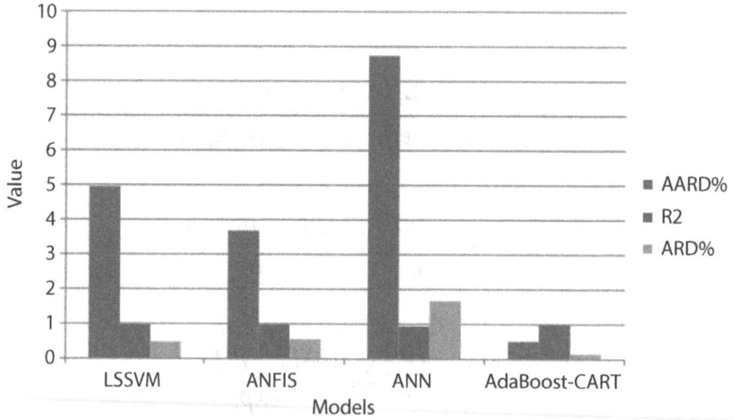

Figure 14.7 Bar graph showing a comparison between the four models - CO_2,$H_2$0 and MEA.

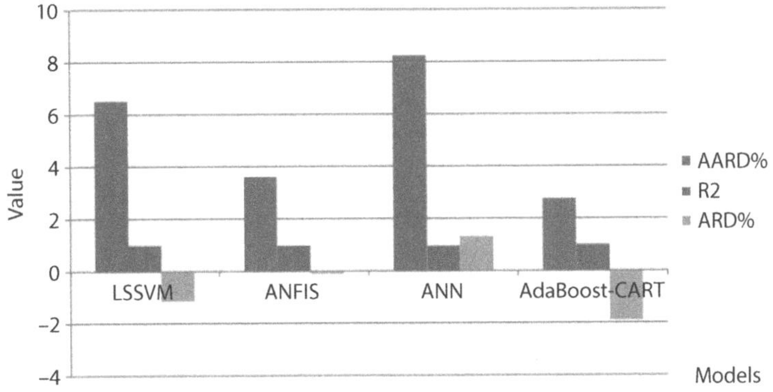

Figure 14.8 Bar graph showing a comparison between the four models - CO_2,H20 and DEA.

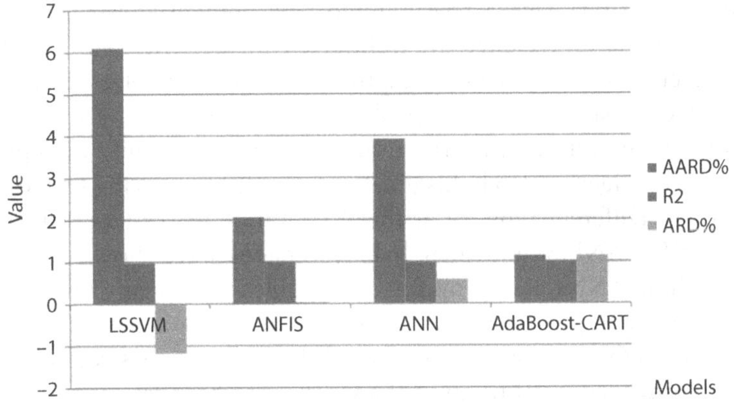

Figure 14.9 Bar graph showing a comparison between the four models - CO_2,H20 and TEA.

model, all four models were used to evaluate systems containing carbon dioxide, water, and PZ. The results are presented in Figure 14.10, where the AdaBoostCART model was found to be superior to the other models, while the ANN model was deemed the worst for the specified system.

Table 14.2 summarizes and lists a set of techniques that are effective for CO_2 capture in various applications and identifies the suitable technique for various models.

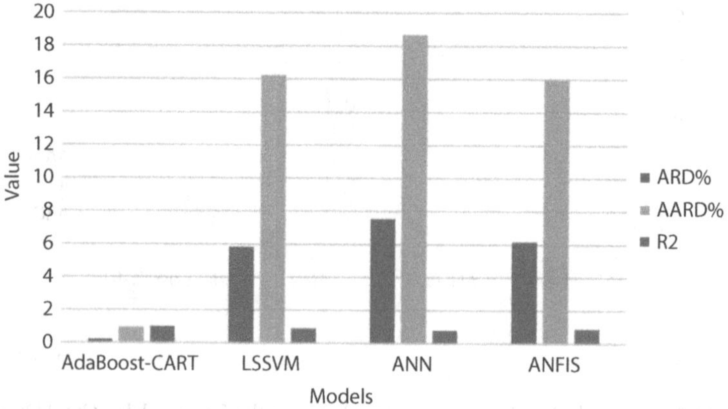

Figure 14.10 Bar graph showing a comparison between the four models - CO_2+water+PZ.

Carbon dioxide capture is a technology that has been developed for commercial use. However, research and development efforts continue to improve and commercialize the technology further. The selection of the most effective and feasible CO_2 capture technology involves considering various factors such as the achieved level of CO_2 removal, economic and environmental concerns, energy usage, and the availability of suppliers. Artificial intelligence also holds great promise for improving CO_2 capture techniques. Different types of artificial neural networks (ANNs) such as BPNN, MLP, ANFIS, and their hybrid versions have shown impressive results. These ANNs have proven to be reliable in managing uncertainties and data loss and providing accurate results in estimating events and variables. Although many studies have used these approaches recently, more research is needed to refine and enhance these techniques.

14.4.3 Discussion of the Case Studies

Now we will discuss the pros and cons of the above mentioned case studies:

A) In the first analysis we discussed about usage of artificial intelligence in assessment and forecasting of drought. The study helped to achieve several objectives, including

- Possibility of larger application and specialized strategy: utilizing sixteen drought-related parameters to predict drought conditions using machine learning techniques.

Table 14.2 Well-received methods for a variety of real world usage.

No.	Technique	Application	Remark
1	Absorption with MEA solvent	after burning	excellent potential for absorbing low concentration flue gases
2	Adsorption process	Solid solids, such as molecular sieves and activated carbon	The replicated adsorbent could be utilized in a subsequent cycle.
2.1	Adsorption (PSA)	gases with reduced absorbency	being quick when getting ready for another round
2.2	Adsorption (TSA)	use cases with high temperatures	relative pricey
2.3	Adsorption (ESA)	use cases requiring low electrical voltage	boilers that burn coal and natural gas cheaply (not economical for the plant)
3	Cryogenic process	streams with elevated CO2 levels	ability to differentiate between drugs using various processes
4	Membrane technology	processing of natural gas, a pre-combustion system, and a post-combustion system	being suitable in small places
5	Enzyme based separation	a crew room or a plant chamber	Efficiency increases as CO density increases,
6	Facilitated transport membrane	CO2 is being excluded in various ways from different gases.	Process that is simple to apply and is suited for separating inorganic materials
7	Mixed matrix membrane	organic substance	When it's hot in power plants
8	Calcium looping CO2 capture	Organic metal structures	excellent potential for absorbing low concentration flue gases

- Combining information from various sources: assessing the possibility of developing the most accurate drought indicators.
- Practical Application: The study highlights the effectiveness of the approach in accurately assessing and monitoring drought conditions in different regions by providing a case study of the United States.

However there were some limitations of the techniques used in the study:

- Limited scope: The limited crop yield samples were a result of selecting a specific study area. To improve agricultural drought modeling, future studies will test the proposed approach in regions with higher yield with different dominant crops, such as wheat. This will allow for a more generalized parameterization of the approach.
- Lack of consideration of uncertainty: The study does not explicitly address the uncertainty associated with the remote sensing data and machine learning techniques used, which could affect the accuracy of the drought assessment.
- Data availability: The study relies on the availability of multiple remote sensing datasets, which may not be available in all regions. This could limit the applicability of the approach in certain regions.

B) In the second case study analysis, we discussed about the problems caused by carbon dioxide which is a greenhouse gas to the environment at a global scale and also at an individual level. The following were its benefits:

- Improved efficiency: The use of AI in CO_2 capture can improve the efficiency of the capture process. By analyzing data from various sources, AI models can identify optimal operating conditions and process parameters to reduce energy consumption and increase the capture rate.
- Reduced costs: The improved efficiency of the CO_2 capture process through AI can lead to cost savings. AI models can help optimize the use of resources, reducing the need for expensive equipment and processes.

- Enhanced accuracy: AI models can provide more accurate predictions of CO_2 capture rates and process outcomes than traditional methods.
- Faster development of new technologies: AI can speed up the development of new CO_2 capture technologies by predicting their performance and identifying potential areas for improvement. This can help researchers and engineers to focus their efforts on the most promising solutions.

However it also had scope of improvement for future research work:

- Lack of Commercial and Practical Model: The development of a commercial application of the methods discussed above in the case study is still under process which requires a suitable and practical model to be chosen.
- Limited data availability: One of the major limitations of using AI in CO_2 capture is the availability of data. Although significant research has been done in this area, there is still a limited amount of data available, especially in terms of real-world applications.
- Uncertainty: Finally, there is often a high level of uncertainty associated with CO_2 capture, which can limit the effectiveness of AI models. This uncertainty can arise from factors such as fluctuations in feed gas composition, variations in temperature and pressure, and changes in operating conditions.

14.5 Future Scope of Climatic Crisis Handling with AI

There are numerous emerging tendencies in AI and weather exchange adaptation which have the capacity to deal with some of the restrictions and demanding situations mentioned earlier. One of the most promising traits is the use of gadget mastering (ML) and deep mastering (DL) techniques. ML and DL can analyze large amounts of statistics and become aware of styles that won't be apparent to humans. this will assist in perceiving dangers and develop effective version measures.

Another emerging fashion is the usage of AI for decision-making. AI systems can provide real-time data and guidelines that may help policymakers and stakeholders make knowledgeable decisions. This can assist in boosting the adoption of effective climate change model measures.

Finally, there is a developing fashion toward the usage of AI for weather exchange mitigation efforts. AI can be used to optimize energy systems, enhance transportation efficiency, and decrease greenhouse fuel emissions. This may assist mitigate the outcomes of climate change and make contributions to a greater sustainable destiny.

The use of AI in climate change adaptation presents several opportunities and challenges. Some of the opportunities include Improving the accuracy and speed of risk assessments, real-time information and recommendations for decision-making, more effective and efficient adaptation measures, better allocation of resources and an improved understanding of the impacts of climate change.

However, there are also several challenges that need to be addressed, including: ensuring data privacy and security, addressing bias and discrimination, ensuring transparency and interpretability of AI systems, ensuring human-machine interaction is optimized and addressing limited resources and applicability. To fully realize the opportunities presented by AI in climate change adaptation, it is essential to address these challenges.

There are several future research directions that can help address the challenges and limitations of AI in climate change adaptation. These include: Developing transparent and interpretable AI systems like ensuring the ethical collection and management of data, addressing bias and discrimination in AI systems, developing AI systems that can work with limited resources, ensuring effective human-machine interaction, assessing the applicability of AI systems on a case-by-case basis and involving stakeholders in the development and deployment of AI systems.

Furthermore, there is a need for interdisciplinary research to fully realize the potential of AI in climate change adaptation. This includes collaboration between experts in AI, climate science, social sciences, and policy-making.

14.6 Conclusion

In this paper we mentioned approximately climate disaster and AI and the way artificial Intelligence modeling have huge capacity to assist weather exchange variation efforts. We additionally mentioned about the prevailing paintings in this domain taking examples of two key factors which are due to climatic exchange (drought and excessive carbon dioxide in environment) and future scope of the subject.

In conclusion we can say that AI can assist in expecting destiny climate traits and simulate the influences of climate exchange on ecosystems and society, enabling communities and policymakers to prepare for and respond to those affects. but, the usage of AI in version also poses numerous challenges and obstacles, inclusive of the want for correct data and concerns associated with ethics and duty.

To efficiently use AI in a model, it's crucial to construct partnerships and collaboration, make certain the availability of statistics, and address moral and social issues. There are numerous regions for future studies and improvement in this area, and continued innovation and collaboration could be critical to effectively use AI in the weather exchange version.

References

1. Chen, Y., Connors, S., Gomis, M., Lonnoy, E., Matthews, J.B.R., Moufouma-Okia, W., Pean, C., Pidc, R., IPCC. 1.5 degrees of global warming. An IPCC special report on the impacts of global warming of 1.5 degrees Celsius above pre-industrial levels and related global greenhouse gas emission pathways, in the context of strengthening the global response to the threat of climate change, sustainable development, and efforts to eradicate poverty, 1, 1–3, 2018, https://www.ipcc.ch/site/assets/uploads/sites/2/2019/06/SR15_Full_Report_Low_Res.pdf.
2. (Reichstein 2019 Nature 566 195–204, Schneider 2017, *Geophys. Res. Lett.* 44 12396–417) doi: 10.1088/1748-9326/ab4e55.
3. Lee, L.A., Pringle, K.J., Reddington, C.L., Mann, G.W., Stier, P., Spracklen, D.V., Pierce, J.R., Carslaw, K.S., The magnitude and causes of uncertainty in global model simulations of cloud condensation nuclei. *Atmos. Chem. Phys.*, 13, 8879–8914, 2013, https://doi.org/10.5194/acp-13-8879-2013.
4. Rischard, M., McKinnon, K.A., Pillai, N., 2018 daily maximum and lowest temperature observations with bias correction Gaussian process modelling in 13. *Environ. Res. Lett.*, 14, 124007, 2019, [statistics] arXiv:1805.10214v2. AP] https://doi.org/10.48550/arXiv.1805.10214.
5. Bailey, R.M., Thomas, D.S.G., Buckland, C.E., Forecasting future dryland reactions to anthropogenic and climatic changes using artificial neural networks. *Research*, 9, 3855, 2019, https://doi.org/10.1038/s41598-019-40429-5.
6. Gentine, P., Pritchard, M., Rasp, S., Reinaudi, G., Might, Y.G., A 2018 study by convection parameterization be solved via machine learning? *Geophys. Res. Lett.*, 45, 5742–51, https://doi.org/10.1029/2018GL078202.
7. McGovern, A., Elmore, K.L., Gagne, D.J., II, Haupt, S.E., Karstens, C.D., Lagerquist, R., Smith, T., Williams, J.K., Using artificial intelligence to improve real-time decision-making for high-impact weather. *Bull. Am.*

Meteorol. Soc., 98, 10, 2073–2090, 2017, Retrieved Mar 5, 2023, from https://journals.ametsoc.org/view/journals/bams/98/10/bams-d-16-0123.1.xml.
8. Belayneh, A., Adamowski, J., Khalil, B., Quilty, J., Combining machine learning techniques with wavelet transformations and the bootstrap and boosting ensemble approaches for drought prediction. *Res. Atmos.*, 172, 37–47, 2016, https://doi.org/10.1016/j.atmosres.2015.12.017.
9. World Meteorological Organization, Greenhouse gas concentrations surge to new record, October 2017, available at: ttps://public.wmo.int/en/media/press-release/greenhouse-gas-concentrations-surge-new-record.
10. Coeckelbergh, M., AI for climate: Freedom, justice, and other ethical and political challenges. *AI Ethics*, 1, 67–72, 2021, https://doi.org/10.1007/s43681-020-00007-2.
11. Strubell, E., Ganesh, A., McCallum, A., Energy and policy considerations for deep learning in NLP, 57, 1–3, 2019, https://arxiv.org/abs/1906.02243.
12. Anthony, L.F., Kanding, B., Selvan, R., Carbontracker: tracking and predicting the carbon footprint of training deep learning models. *ICML Workshop on "Challenges in Deploying and Monitoring Machine Learning Systems"*, 1, 1–11, 2020, https://arxiv.org/abs/1910.09700.
13. Rolnick, D., Donti, P., Kaack, L., Kochanski, K., Lacoste, A., Sankaran, K., Ross, A., Milojevic-Dupont, N., Jaques, N., Waldman-Brown, A., Luccioni, A., Maharaj, T., Sherwin, E., Karthik Mukkavilli, S., also contributed to this article, Using machine learning to combat climate change, 55, 1–96, 2019, https://doi.org/10.48550/arXiv.1906.05433.
14. Schmidt, V., Luccioni, A., Bengio, Y., Balasooriya, N., Karthik Mukkavilli, S., Sankaran, K., Chayes, J., Utilising cycle-consistent adversarial networks to visualise the effects of climate change, 1, 1–7, 2019, Preprint arXiv:1905.03709, https://doi.org/10.48550/arXiv.1905.03709.
15. Dulic, A., Sheppard, S., Iype, D.M., Angel, J., LaValle, A., Future delta 2.0 is a setting for experiential learning for a game that addresses local climate change. Page 12 of *SIGGRAPH Asia 2015 Symposium on Education*, ACM, 2015, https://doi.org/10.1145/2818498.2818512.
16. Mitigation of Climate Change, IPCC, Climate Change 2014. Working Group III's contribution to the Intergovernmental Panel on Climate Change's Fifth Assessment Report, Y. Sokona, E. Farahani, S. Kadner, K. Seyboth, A. Adler, I. Baum, S. Brunner, P. Eickemeier, B. Kriemann, J. Savolainen, S. Schlomer, C. von Stechow, T. Zwickel, J.C. Minx (Eds.), https://www.ipcc.ch/site/assets/uploads/2018/03/WGIIIAR5_SPM_TS_Volume-3.pdf.

17. Ramchurn, S., Vytelingum, P., Rogers, A., Jennings, N.R., Putting the "smarts"into the smart grid: A grand challenge for artificial intelligence. *Commun. ACM*, 55, 4, 86–97, 2012, https://doi.org/10.1145/2133806.2133825.
18. Donti, P., Amos, B., Zico Kolter, J., Task-based end-to-end model learning in stochastic optimization, in: *Advances in Neural Information Processing Systems*, pp. 5484–5494, 2017, https://doi.org/10.48550/arXiv.1703.04529.
19. Quinn, J.A., Andama, A., Munabi, I., Kiwanuka, F.N., Automated blood smear analysis for mobile malaria diagnosis, in: *Mobile Point-of-Care Monitors and Diagnostic Device Design*, vol. 31, p. 115, 2014, http://cit.mak.ac.ug/staff/jquinn/papers/AutomatedMalariaDiagnosisChapter.pdf.
20. Jaques, N., Waldman-Brown, A., Luccioni, A., Rolnick, D., Donti, P.L., Kaack, L.H., Kochanski, K., Ross, A.S., Milojevic-Dupont, N., Maharaj, T., Sherwin, E.D., Karthik Mukkavilli, S., Kording, K.P., Gomes, C., Ng, A.Y., Hassabis, D., John, C., Using machine learning to combat climate change, 2019, https://doi.org/10.48550/arXiv.1906.05433.
21. Park, S., Im, J., Jang, E., Rhee, J., Drought assessment and monitoring through blending of multi-sensor indices using machine learning approaches for different climate regions. *Agric. For. Meteorol.*, 2016, https://doi.org/10.1016/j.agrformet.2015.10.011.
22. United Nations Environment Programme, The emissions gap report 2017, November 2017, available at: https://wedocs.unep.org/bitstream/handle/20.500.11822/22070/EGR_2017.pdf?sequence=1&isAllowed=y.
23. Brown, J.F., Wardlow, B.D., Tadesse, T., Hayes, M.J., Reed, B.C., The Vegetation Drought Response Index (VegDRI): A new integrated approach for monitoring drought stress in vegetation. *GI Science Remote Sens.*, 45, 1, 16–46, 2008, http://refhub.elsevier.com/S0168-1923(15)00746-7/sbref0030.
24. Tadesse, T., Wardlow, B.D., Hayes, M.J., Svoboda, M.D., Brown, J.F., The vegetation outlook (VegOut): A new method for predicting vegetation seasonal greenness. *GI Science Remote Sens.*, 47, 1, 25–52, 2010, https://www.tandfonline.com/doi/abs/10.2747/1548-1603.47.1.25.
25. Gessner, U., Naeimi, V., Klein, I., Kuenzer, C., Klein, D., Dech, S., The relationship between precipitation anomalies and satellite-derived vegetation activity in Central Asia. *Glob. Planet. Change*, 110, Part A, 2013, https://doi.org/10.1016/j.gloplacha.2012.09.007.
26. Nezlin, N.P., Kostianoy, A.G., Li, B.–L., Inter-annual variability and interaction between remote-sensed vegetation index and atmospheric precipitation in the Aral Sea region. *J. Arid Environ.*, 62, 4, 677–700, 2005, ISSN 0140–1963, https://doi.org/10.1016/j.jaridenv.2005.01.015.

27. Cao, L., Application of artificial intelligence on CO_2 capture: A review. *J. Therm. Anal. Calorim.*, 145, 1751–1768, 2021, https://doi.org/10.1007/s10973-021-10777-4.
28. Yarveicy, H., Ghiasi, M.M., Mohammadi, A.H., Performance evaluation of the machine learning approaches in modeling of CO_2 equilibrium absorption in Piperazine aqueous solution. *J. Mol. Liq.*, 255, 375–83, 2018, https://doi.org/10.1016/j.molliq.2017.11.156.

15

Impact of Computational Intelligence and Modeling in Tackling Weather Fluctuation

Rohan Karn*, Aniket Rouniyar, Ranjit Kumar Das and Amit Gupta

Kalinga Institute of Industrial Technology, Deemed to be University, Bhubaneswar, Odisha, India

Abstract

Humanity is going through a pressing worldwide hassle right now because of the climate disaster. knowledge and forecasting climate trade's complicated strategies is essential for minimizing its dangerous effects. Modeling techniques and artificial intelligence (AI) have turned out to be beneficial gear for fixing this trouble. It's feasible to analyze large and complex weather trade datasets with synthetic intelligence (AI), which includes satellite data, weather models, and ancient weather data. using these datasets, system mastering algorithms can be skilled to predict destiny climate styles and the way they may have an effect on ecosystems. Meanwhile, modeling strategies can simulate the interactions among the ecosystem, seas, and land surfaces in the Earth machine. models like those may be used to expand mitigation plans and discover the results of numerous weather change situations. For expertise and addressing the weather undertaking, AI and modeling techniques are effective equipment. By way of the use of these equipment, scientists, decision-makers, and stakeholders may benefit from higher information on how to mitigate the consequences of weather exchange and build a sustainable destiny for every person.

***Keywords*:** Artificial intelligence, climate crisis, AI models, AI against climate change, human health, forest fire, plastic waste detection

**Corresponding author*: rohankarn35@gmail.com

Hrudaya Kumar Tripathy, Sushruta Mishra, Minakhi Rout, S. Balamurugan and Samaresh Mishra (eds.) Optimized Computational Intelligence Driven Decision-Making: Theory, Application and Challenges, (305–334) © 2024 Scrivener Publishing LLC

15.1 Introduction

The time period "weather disaster" describes the good sized and ongoing adjustments to the Earth's weather gadget that are mainly because of human activities, inclusive of the burning of fossil fuels and deforestation. The ensuing upward thrust in greenhouse gas emissions, especially carbon dioxide, has extended sea ranges, multiplied the frequency and severity of intense climate occasions, and elevated global temperatures [1]. The results of the weather crisis are vast and feature a big effect on human societies and economies as well as the natural international. changing precipitation styles and rising temperatures have altered ecosystems, decreased biodiversity, and reduced agricultural output. severe weather activities, like hurricanes, floods, and droughts, have ended in large infrastructure destruction, population displacement, and loss of lifestyles. reduced greenhouse gasoline emissions and a shift to a low-carbon economic system are important to address the climate catastrophe. This entails putting regulations into place to aid renewable electricity assets, making investments in energy efficiency, and enacting sustainable land use strategies. Governments, companies, and people should all work together to reduce their carbon footprints and flow in the direction of a more sustainable future so that you can deal with the weather disaster destiny for all [2]. The creation of wise machines which can be capable of learning, solving issues, and making decisions—duties that generally require human cognition—is called artificial intelligence (AI). Cowls *et al.* cited that at a high level, class, prediction, and selection-making are 3 vital cognitive abilities that "smart" device systems show off. those talents are already being utilized in an extensive variety of industries, which include communication, transportation, and fitness (e.g., figuring out functions in an image like an X-ray test to diagnose cancer)(e.g., processing human speech and responding in kind). applying synthetic intelligence to the "problem space" of climate trade may want to have predominant tremendous effects by way of, first, assisting in know-how the difficulty and, second, facilitating efficient answers [3]. By enhancing our expertise of complex environmental structures and permitting extra efficient and effective climate answers, artificial intelligence (AI) has the capability to play a significant position in tackling the demanding situations of climate change. that allows you to higher realize the effects of weather alternate and forecast destiny developments, AI may be used to examine and interpret great quantities of weather data, together with satellite TV for pc imagery, climate patterns, and ocean currents. With a view to make more accurate predictions and well informed decisions, gadget learning algorithms may be used to spot patterns and relationships in statistics that may not be right away apparent to humans. To be able to decrease greenhouse fuel emissions and boost efficiency, AI can be used to optimize power structures like strength networks and the infrastructure for renewable strength assets. AI can, as an example, be employed to optimize the manufacturing of wind and sun power based totally on weather forecasts, expected energy consumption, manipulate energy garage, and greater. Normally, AI has the capability to completely change how we method climate trade by allowing greater green climate answers, handing over sparkling insights, and optimizing modern systems. synthetic intelligence (AI) is being used more and more frequently across a huge range of technological know-how and era disciplines, and it's becoming more and more acknowledged as a promising device to address the modern-day climate crisis. Issues over the growth in greenhouse gasoline emissions, deforestation, ice cap melting, and other environmental troubles have caused scientists and lawmakers to search for practical answers. By supplying clean views on climate change, forecasting environmental trends, and providing viable solutions for slicing carbon emissions, AI has the ability to absolutely rework the way we technique these demanding situations. This essay will have a look at how AI is used in weather modeling and the way it might be used to lessen the effects of the contemporary weather disaster. It's going to have a look at the numerous AI strategies applied to weather modeling and compare how properly they count on climate styles, pinpoint vulnerable areas, and offer solutions for edition and mitigation strategies. Inside the first phase, the climate catastrophe might be briefly summarized, with precise emphasis located on the main reasons of worldwide warming and the consequences of climate trade on the natural global and human societies. The idea of AI, its background, and its use in exceptional fields can be brought in segment two. The 0.33 section will concentrate on the software of AI to weather modeling at the same time as emphasizing the diverse tactics and fashions hired on this site. The advantages of the usage of AI in weather modeling, along with its ability to make unique predictions, become aware of susceptible regions, and provide manageable answers for lowering carbon emissions, might be covered in segment four. The use of AI in climate modeling has a few drawbacks, consisting of information accessibility, computational capacity, and social troubles, a good way to be mentioned in phase five. The usage of AI in weather modeling carries a few possible dangers, which includes those related to bias, accuracy, and privacy. These risks are included in phase six. The case examples in phase 7 will show how AI has been applied to weather modeling, highlighting each its blessings and disadvantages, it's going to additionally display the implementation of a few AI fashions on climate crisis which can be a higher option to tackle weather alternate. Ultimately, segment eight will provide some pointers for enhancing the efficiency of AI in weather modeling and offer a few future studies paths.

The overall purpose of this paper is to provide an intensive evaluation of the way AI can be used to model the weather and decrease the effects of the current weather catastrophe. Policymakers, researchers, and other stakeholders engaged in weather alternate model and mitigation efforts will find the findings of this look to be useful.

15.2 Objective

This research paper especially aims at suggesting nice tools to expand fashions in an effort to help to tackle the climate disaster. synthetic Intelligence gear can help in growing such kinds of gear. To assist mitigate the influences of weather change, artificial intelligence is getting used to create computerized systems [4]. To perform this, artificial intelligence (AI) may be used to track and examine weather records, spot potential risks, and create mitigation plans. AI can also be used to create cleanser, more sustainable power sources, lower emissions, and create new tools to useful resources in climatic edition.

- AI fashions and makes use of for addressing weather exchange across quite a number industries and fields: which will assist reduce greenhouse gasoline emissions and adapt to the results of climate alternate in diverse sectors, which includes electricity, transportation, agriculture, and fitness, this point discusses the use of AI strategies like device studying, deep gaining knowledge of, laptop vision, natural language processing, etc. AI, as an example, can support electricity machine optimization, track pollutants, create new materials, forecast climate events, and so forth.
- The consequences of AI technology on the ecosystem and society, such as its carbon footprint, power use, and moral ramifications: The potential effects of creating and enforcing AI structures on the environment and society—both desirable and terrible—are mentioned in this declaration. Because of the computing assets required for training and walking models, AI, for example, may have a high carbon

footprint and power intake. This may add to weather trade. AI may additionally have moral ramifications that affect human rights and wellbeing, which include the ones associated with bias, fairness, responsibility, transparency, etc.

- The shortcomings and problems in creating and implementing artificial intelligence (AI) solutions for climate movement, including facts accessibility, first-class, and privacy; version interpretability and robustness; and stakeholder involvement and cooperation This declaration speaks to the demanding situations and impediments that save you the green utility of AI for fighting climate trade. As an example, information accessibility, pleasant, and privacy are critical factors for developing correct and trustworthy AI fashions, but they can be constrained or compromised for a number of reasons, which includes a dearth of infrastructure, governance, or standards. For AI fashions to be understandable, reliable, and proof against uncertainties and demanding situations, version interpretability and robustness also are important. For the various actors engaged in AI for climate trade to align their objectives and desires, stakeholder engagement and collaboration are crucial.
- The hints and fine practices for making use of AI's potential to deal with the weather catastrophe encompass creating sustainable AI infrastructure, encouraging interdisciplinary studies, and ensuring inclusive and equitable consequences: The ideas and hints for maximizing the blessings and minimizing the dangers of AI for weather trade are discussed in this point. developing sustainable AI infrastructure, for example, involves minimizing the environmental effect of AI systems via the usage of green energy assets, elevated performance, or emission offsets. By means of bridging the gaps among numerous regions of knowledge and understanding related to AI and weather change, interdisciplinary studies is recommended. promoting social justice and human dignity inside the development and alertness of AI whether alternate answers are important to make sure inclusive and equitable outcomes.

15.3 Causes of Climate Crisis

There are many AI models that can be developed to address the Climate Crisis, as was covered in the objective portion. For this, we must concentrate on identifying the factors contributing to the climate catastrophe. The main causes are briefly reviewed and discussed below.

15.3.1 Greenhouse Gases

The Earth's atmosphere incorporates a developing amount of greenhouse gases (GHGs), specifically carbon dioxide (CO2), methane (CH4), and nitrous oxide. this is the main reason for the weather crisis. (N2O). The earth warms as a result of these gases' ability to lure solar warmth and save it from radiating returns into area [5].

15.3.2 Fossil Fuels

Fossil gasoline combustion, together with the burning of coal, oil, and fuel, is a giant supply of greenhouse gas emissions. A chief supply of emissions is using those fuels in enterprise, strength technology, and transportation [6].

15.3.3 Deforestation

Deforestation and other land-use modifications that lessen the amount of forested land can also exacerbate the weather disaster through releasing carbon dioxide that become formerly stored in the biomass of bushes, which absorbs carbon dioxide from the surroundings.

15.3.4 Agriculture and Livestock

Because of practices like the use of fertilizers, the production of cattle, and the cultivation of rice, the agricultural quarter is a primary supply of GHG emissions.

15.3.5 Industrial Processes

GHG emissions also are produced by industrial operations like making cement and making chemical substances.

15.3.6 Transportation

the usage of fossil fuels in cars, vans, and aeroplanes is the main supply of transportation's large contribution to GHG emissions.

By making use of AI fashions and technologies, these reasons may be addressed. green cooling generation, smart and sustainable communities, satellite TV for pc imagery, version-based objection detection, and others are some of them [7].

15.4 Significance of AI and Modeling on Climate Crisis

There are numerous fields on which AI can be used to tackle the climate crisis. Some predominant factors are mentioned in this paper.

15.4.1 Design and Optimization of Renewable Energy Systems

Designing and optimizing renewable energy systems involves figuring out the first-class area for renewable energy installations, improving the architecture of the installations, and increasing strength output. whilst figuring out the pleasant web page for renewable energy projects, AI might also play a substantial function by analyzing geographic and meteorological statistics. AI structures, for example, may additionally examine satellite TV for pc photos to discover places with the nice solar installations and find regions with excessive solar irradiation. The perfect place for wind mills may also be located by way of the use of AI to analyze records on wind path and speed [8]. The design of renewable energy flora can be optimized using this statistics that allows you to maximize strength output. If you want to assure the effectiveness and dependability of renewable power structures, AI may additionally optimize their functioning. AI algorithms, for example, can forecast weather patterns and alter the operation of renewable electricity vegetation to maximize strength output for the duration of hours of excessive demand. a good way to assure that renewable power is on the market when it is needed, AI also can optimize the utilization of strength storage gadgets. Moreover, the optimization of renewable electricity systems can cause cost financial savings and lessen the dependency on fossil fuels. while in comparison to fossil gas-based electricity plants, renewable electricity installations can also offer energy for less money, saving customers money and decreasing greenhouse gasoline emissions. ("Why artificial intelligence is prime to renewable energy grid resilience") cited that the design and optimization of renewable energy structures is a crucial priority in confronting the climate catastrophe. By means of analyzing geographic and meteorological records to determine the first-class web page for renewable strength installations, optimizing the design of the installations, and maximizing strength output, AI can also play a giant component on this system. additionally with the assist of AI we are able to manipulate the decentralized power sources, lowering greenhouse fuel emissions and reducing the effects of climate change are each possible with the grid integration of renewable electricity sources [9].

15.4.2 Develop AI-Driven Solutions to Reduce Deforestation

Considered one of the biggest environmental troubles dealing with humanity is deforestation. Despite the fact that woodland destruction has been a hassle for generations, it has recently won interest due to the rapid fee at which forests are being eliminated for industrial and agricultural use. Deforestation has several adverse results at the environment, together with the manufacturing of greenhouse gases, the destruction of habitats, and the loss of biodiversity. Therefore, it's vital to expand strategies to lessen deforestation, and the usage of AI-pushed solutions is one such opportunity.

15.4.2.1 Detection and Monitoring of Deforestation

Using remote sensing methods, AI may be used to identify and keep track of deforestation in real-time. For instance, changes in forest cover may be detected using satellite photography, and data analysis and deforestation hotspot identification may be performed using machine learning techniques. Governments and organizations can take action based on this knowledge to stop additional devastation. Global Forest Watch (GFW) is a platform that currently employs this strategy. GFW analyses data using machine learning techniques to identify regions of deforestation and detect changes in forest cover using satellite photos. GFW delivers up-to-date information on forest loss, which is utilized by organizations and governments all around the world to take action and stop future deforestation [10].

15.4.3 Forest Fire Prediction and Prevention

AI may potentially be used to anticipate and prevent forest fires. Forest fires may be a significant contributor to deforestation and have disastrous effects on both ecosystems and human settlements. By employing machine learning algorithms to assess data on meteorological conditions, vegetation density, and other characteristics, AI can assist in anticipating when and where forest fires are likely to occur [11]. This information can be utilized to take precautionary steps such as cutting vegetation and establishing fire buffers. (Staley) proposed that one contemporary model that employs this method is the Forest Fire Prediction and Monitoring (FFPM) system. FFPM forecasts

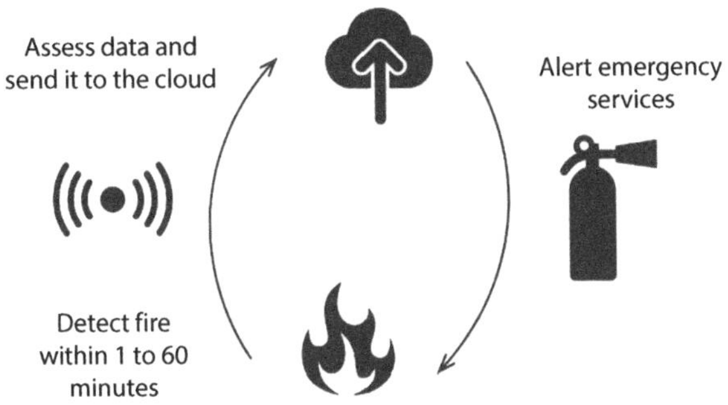

Figure 15.1 Forest Fire prediction system.

the possibility of forest fires using satellite images and weather information, and it notifies authorities and communities in real time so they may take precautions. Figure 15.1 illustrates the construction of forest fire prediction systems that allow for early detection of fire outbreaks and notification of the appropriate authorities [7] This aids in the ecosystem's development so that forest fires can be avoided. To stop the main causes of climate disasters like forest fires, deforestation, and others, this form of AI and models are required. Figure 15.1 describes in detail how cloud emergency services and AI models might be combined to forecast fires as soon as feasible.

15.4.3.1 Forest Restoration

Artificial intelligence (AI) may be used to restore damaged forests by analyzing data on the soil, vegetation, and other elements to determine the most effective restoration techniques. Data on the growth and survival rates of various tree species may be analyzed using machine learning algorithms, and this knowledge can be utilized to build reforestation techniques [12]. The Reforestum platform is one modern example of this method in operation. Reforestum analyses data on soil, climate, and vegetation using machine learning algorithms to determine the most effective ways to restore damaged forests. Moreover, the platform makes use of block chain technology to enable organizations and individuals to raise money towards the restoration of certain forest regions.

15.4.4 Create AI-Driven Solutions to Improve Agricultural Practices to Reduce Carbon Emissions

("AR5 climate alternate 2014: Mitigation of weather trade — IPCC") referred to that nearly 25% of all emissions are attributed to agriculture,

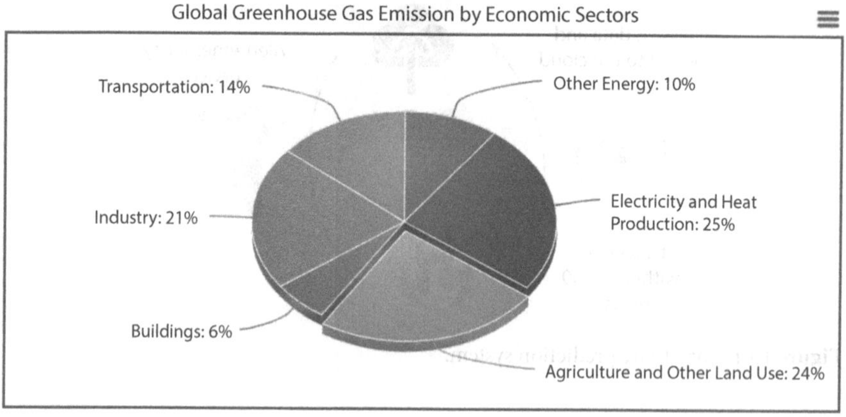

Figure 15.2 Pie chart of global greenhouse gas emission by economic sector.

which is a full-size contributor to global carbon emissions. Several things make contributions to this, such as using fossil fuels in agricultural gadget and transportation, the manufacturing of methane from livestock, and the usage of artificial fertilizers. Sustainable farming tactics that could decrease carbon emissions even as retaining food security are getting increasingly important to resolve this challenge. Using AI-pushed technology to enhance agricultural operations and decrease carbon emissions is one feasible technique. the share percentage that each industry quarter contributes to greenhouse fuel emissions is proven inside the photograph underneath. A 24% contribution from the agriculture area to greenhouse gas emissions can be seen in the pie chart underneath (Figure 15.2).

15.4.5 Precision Farming

Precision farming strategies can be utilized with AI to boom agricultural yields and lower carbon emissions. Device-studying algorithms may help farmers in maximizing their usage of fertilizers, herbicides, and water by analyzing statistics on soil pleasantness, climate patterns, and different elements. This minimizes waste and the effect of agriculture on the environment [13]. The weather employer's climate FieldView platform is one current instance that employs this approach. With the use of satellite TV for pc pics, meteorological records, and machine gaining knowledge of algorithms, FieldView offers farmers an entry to actual-time facts about crop fitness, soil moisture, and different elements that impact crop yields. Farmers are capable of minimizing their carbon impact and maximizing their useful resource utilization as a result.

15.4.6 Sustainable Livestock Production

AI may also be utilized to support low-carbon farm animals manufacturing techniques. Machine mastering algorithms can assist farmers in maximizing their use of assets and minimizing waste, as a result lowering the carbon footprint of livestock production. Those algorithms analyze information on animal health, feed first-rate, and other elements. The global farm animals studies Institute's (ILRI) farm animals monitoring and management system (LMMS) is a contemporary example of a version that makes use of this approach (ILRI). LMMS employs mobile era and system learning algorithms to reveal the fitness and productivity of cattle in real-time, giving farmers insights on feed first-rate and other elements that impact animal health and output [14].

15.4.7 Carbon Sequestration

AI can be used to encourage carbon sequestration in agricultural strategies, which can reduce the terrible environmental outcomes of carbon emissions. gadget getting to know algorithms can help farmers in figuring out the superior practices for sequestering carbon in the soil, consisting of cover vegetation and conservation tillage, by analyzing records on soil fine, plant increase costs, and different variables. The Terraton Initiative, created with the aid of Indigo Agriculture, is one present day example of a scheme that employs this strategy. The Terraton Initiative analyses facts on soil quality, plant growth prices, and different variables using system learning algorithms to decide the most green approaches to store carbon inside the soil. Furthermore, Indigo Agriculture collaborates with farmers to place those techniques into movement and rewards farmers that sequester carbon via their farming methods.

15.4.8 Energy Efficiency

AI will also be used to make agricultural strategies extra strength-green, so that you can cut down on the consumption of fossil fuels and different types of carbon emissions. device learning algorithms can also assist farmers in optimizing their use of agricultural systems and transportation, eliminating waste, and minimizing the environmental impact of agriculture through analyzing information on energy intake. The Farm electricity management gadget (FEMS), created by the Pacific Northwest country-wide Laboratory, is one current version that uses this strategy. FEMS analyses electricity usage statistics, the usage of system studying algorithms, giving farmers an understanding approximately their electricity use and possibilities to increase strength performance via using renewable electricity sources and other strategies [15].

15.4.9 Analyze and Predict the Effects of Climate Change

One of the most urgent troubles confronting the globe now could be weather trade. It affects the surroundings, the financial system, and society in many special approaches and is added on by quite a number of factors, together with human actions like burning fossil fuels and deforestation. The utility of artificial intelligence (AI) to examine and forecast the influences of climate change is one potential treatment for managing its troubles. The consequences of weather change, which include prospective repercussions on ecosystems, human fitness, and the global economic system, can be examined in this essay in terms of how AI is probably used to examine and assume these effects.

15.4.10 Studying Ecosystems

Reading ecosystems is one of the primary methods AI may be used to examine the results of weather exchange. Ecosystems are intricate networks of each living and nonliving element that talk with each other and with their environment. These interactions may be hampered through climate change, which can also alter environmental services like productiveness and biodiversity. Machine getting to know algorithms can help scientists in higher knowledge how weather alternate is affecting ecosystems all internationally by analyzing statistics on temperature, precipitation, and other environmental parameters. The use of AI to analyze modifications within the Arctic is one instance of this. One of the areas most at risk of weather change is the Arctic, which is swiftly melting and warming. Sea ice is melting and permafrost is thawing inside the Arctic, which could have a considerable influence on the ecosystems in the place. Polar bears, walruses, and seals may all discover homes and searching grounds on sea ice. big volumes of carbon and methane are stored below permafrost, and whilst it thaws, those gases may be released into the atmosphere [16]. AI can be used to analyze temperature, ice cowl, and different records to forecast how these modifications would affect polar bears, walruses, and Arctic cod species, amongst others. AI can also assist in locating possible conservation and version strategies to store those fragile ecosystems.

15.4.11 Predicting Human Health Impacts

Human health can be substantially impacted by climate change, that could cause an upward thrust in breathing ailments, heat-associated illnesses, and diseases unfold by means of insects and animals, such malaria and dengue fever. region, socioeconomic level, and populace vulnerability all affect these fitness risks differently. The observation describes that synthetic intelligence (AI) can be used to analyze temperature, air exceptional, and different records to forecast how climate exchange will affect human health in diverse parts of the world. The software of AI to forecast the effects of warmth waves on human health is one instance of this. machine getting to know algorithms can help public fitness specialists in identifying organizations which can be more liable to warmth-associated problems including warmness stroke, dehydration, and cardiovascular issues and assist them take precautions to lessen these risks.. determine 4.3 makes it very obvious how weather change is having an impact on human beings' fitness. Accidents, fatalities, evictions, and mental stress can all result from severe weather occasions which includes floods, droughts, storms, and wildfires [17] allergies and continual obstructive pulmonary sickness are two respiratory problems that might get worse as a result of air pollution. Pollen generation and dispersion versions may result in a growth in hypersensitive reactions. Those are some of the consequences of climate change which have an instantaneous effect on human beings' health. AI will aid in higher infection detection and prediction than human beings, as well as better treatment through greater precise analysis, tailored medicine, and stronger health care shipping. Figure 15.3 depicts the impact on health due to climate change.

15.4.12 Predicting Economic Impacts

The price of dealing with excessive weather occasions, adjustments in agricultural production, and fluctuations in strength call for are only some of the essential results that weather change might also have on the sector economy [18]. AI may be used to analyze temperature, precipitation, and other environmental statistics to make predictions on how climate alternate would have an effect on many facets of the global financial system. using AI to predict the results of sea degree upward thrust on coastal agencies is one instance of this. system studying algorithms can assist policymakers in identifying locations that are more vulnerable to financial disruption due to sea degree upward thrust and helping them take precautions with the aid of analyzing information on sea level upward push

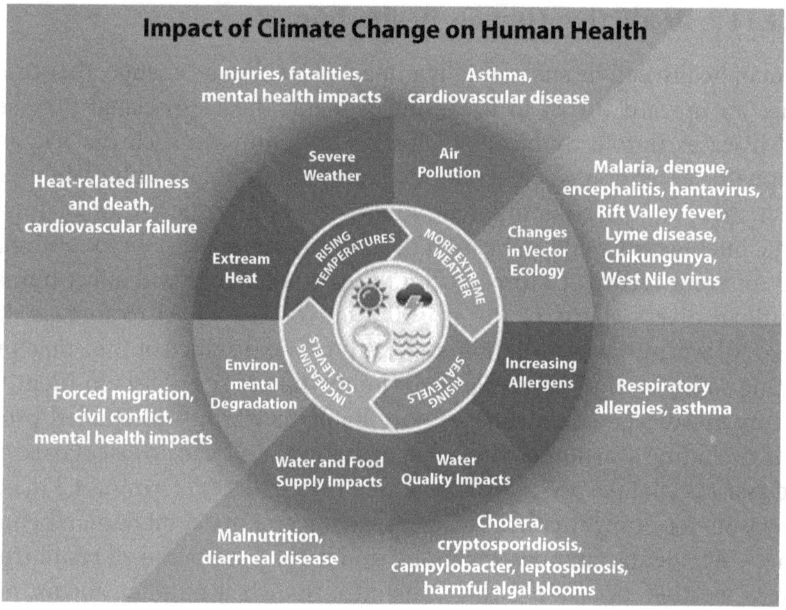

Figure 15.3 Impact on health due to climate change.

forecasts, population density, and financial interest. In conclusion, AI has the capacity to enhance our comprehension of and potential to foresee the results of climate exchange. System learning algorithms can be a useful resource in our expertise of the way exchange is affecting ecosystems, human health, and the worldwide economy with the aid of analyzing information on temperature, precipitation, and different environmental factors. synthetic intelligence (AI) has the capability to be a game-changer in our ability to reduce the consequences of climate exchange and create a more sustainable destiny as we maintain to deal with its difficulties. (McCarthy) confirmed how the economic system has been impacted by means of the upward thrust in worldwide temperatures, which is most probably a result of the climate disaster, as visible within the table below. This table grows to be taken from a look at with the useful resource of the world monetary discussion board. It illustrates how the GDP has been impacted with the resource of the climate disaster. To cope with those issues, international locations should make investments with a variety of cash. AI will help in predicting the nearby weather and climate so that the applicable employer or authority may additionally plan the manner to address it. additionally, it may make hints for the pleasant methods to deal with the ones troubles and make up for the financial losses introduced on with the useful resource of the climate disaster [19].

	Temperature rise scenario by mid-century			
	Well-below 2°c	**2.0°c increases**	**2.6°c increases**	**3.2°c increases**
	Paris target	The likely range Of global Temperature Gains		Severe case
Simulating for economic loss impacts from rising temperature in GDP, relative to a world without climate change (0)°c				
World	-4.2%	-11.2%	-13.9%	-18.1%
OECD	-3.1%	-7.6%	-8.1%	-10.6%
North America	-3.1%	-6.9%	-7.4%	-9.5%
South America	-4.1%	-10.8%	-13.0%	-17.0%
Europe	-2.8%	-7.7%	-8.0%	-10.5%
Middle East & Africa	-4.7%	-14.0%	-21.5%	-27.6%
Asia	-5.5%	-14.9%	-20.4%	-26.5%
Advance Asia	-3.3%	-9.5%	-11.7%	-15.4%
ASEAN	-4.2%	-17.0%	-29.0%	-37.4%
Oceania	-4.3%	-11.2%	-12.3%	-16.3%

15.5 Plastic Waste Detection Model

A huge detail within the climate disaster is the trouble of plastic trash. Greenhouse gas emissions are as a result of the manufacture and disposal of plastic, and as many as thirteen% of all greenhouse gas emissions internationally are thought to come again from plastic garbage. Furthermore, an estimated eight million metric tons of plastic garbage reaches the seas each 12 months, making it a big contaminant of our rivers, oceans, and other bodies of water. Marine life suffers substantially due to this plastic trash, and it could even move into the food chain. In case you need to cope with the weather catastrophe, we need to reduce our use of plastic and eliminate plastic rubbish successfully. To discover plastic garbage, a selection of gadget mastering techniques can be used. the subsequent listing of some of them:

15.5.1 Convolutional Neural Networks (CNNs)

For image identification applications, including the detection of plastic garbage, CNNs are a preferred option. They may be trained on a dataset of plastic waste photographs and are made to recognize patterns in images, allowing them to distinguish between various forms of plastic garbage with accuracy [20]. Depending on the application and the data at hand, CNNs may be applied in a number of different ways to identify plastic pollution. To illustrate how CNNs may be used to find plastic garbage, consider the following examples:

15.5.1.1 Image Classification

CNNs may be used to categorize photos of plastic debris and separate it from other types of rubbish. This may be accomplished by using a dataset containing photos of plastic and non-plastic debris to train a CNN. (LeCun *et al.*) mentioned that the CNN will gain knowledge of the visual characteristics, such as color, texture, and form that set plastic garbage apart from other types of waste. After being taught, the CNN may be used to categorize fresh garbage photos as plastic or non-plastic [21]. Figure 15.4 illustrates how a neural network model receives an input picture and processes it through several layers to get the expected output. In order to distinguish plastic from other components, this picture categorization will be useful. Moreover, it could be used to segregate other potentially dangerous items or plastics.

15.5.1.2 Object Detection

In order to recognize and localize items in a picture, CNNs may also be employed for object detection. This method may be used to find plastic debris in pictures of waterways like rivers and seas [22]. It is possible to teach the CNN to identify and find plastic debris using its distinctive visual

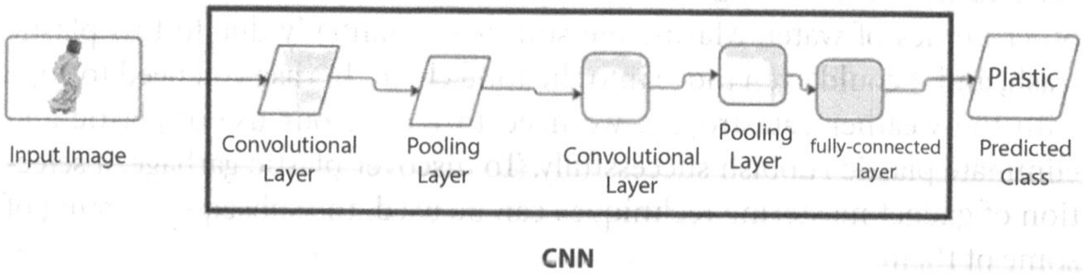

Figure 5.4 CNN flowchart.

characteristics. After the plastic debris has been localized, it may be tallied and followed throughout time, giving important information for controlling and monitoring plastic pollution.

15.5.1.3 Segmentation

CNNs may also be used to segment images, which is the process of splitting an image into areas or segments according to its visual characteristics. With this method, plastic garbage in photos of water bodies may be found and segmented. The CNN can be trained to distinguish plastic debris from the surrounding water and other objects by identifying its distinctive visual characteristics. After segmentation, the size, shape, and position of the plastic debris may be examined, yielding useful data for controlling and monitoring plastic pollution.

15.5.1.4 Time-Series Analysis

CNNs are also useful for time-series analysis, which examines data over a period of time. With this method, plastic garbage in water bodies may be found and followed throughout time. The CNN can be taught to identify the distinctive visual characteristics of plastic garbage and to track changes in its distribution, concentration, and mobility over time. With this knowledge, management and reduction methods for plastic pollution may be created. Figure 15.5 highlights plastic waste detection using machine learning algorithm.

15.5.1.5 Accuracy

CNNs are beneficial for tasks like figuring out plastic trash because they're remarkably accurate in differentiating and classifying photos. the usage of mathematical methods, CNNs, a specific type of artificial neural community, can examine photographs and extract traits from them. They are composed of many layers of neurons that work collectively to perform operations including activation, convolution, and pooling. Activation functions give the community non-linearity and allow it to examine problematic styles. Convolution filters create characteristic maps that emphasize key elements in the input photographs via making use of kernels to the images. The maximum vital records are retained even as the scale and complexity of the feature maps are decreased through layer pooling. A CNN can correctly pick out and separate plastic rubbish from different kinds of waste given sufficient training statistics. This will make it possible

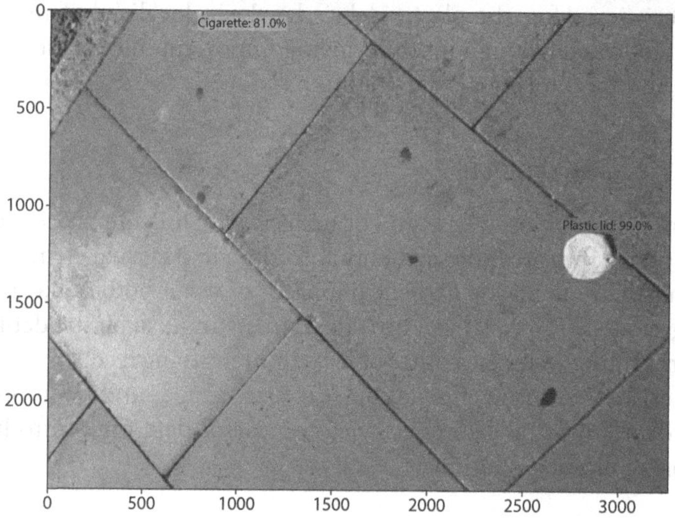

Figure 15.5 Plastic waste detection using machine learning algorithm.

to recycle plastic waste or take it away, lessening its dangerous impact on the surroundings. A big worldwide trouble, plastic trash damages flora and fauna, pollutes the environment, and speeds up weather alternate. The accuracy of the CNN in the course of schooling and validation is proven inside the graph in determine 5.three. At the same time as the validation accuracy assesses how well the CNN generalizes to new facts that it has in no way seen before, the training accuracy measures how nicely the CNN matches the training facts. The CNN can carry out higher via selecting extra styles and traits from the pics, demonstrating that more statistics will produce higher effects. The CNN can prevent overfitting or underfitting the facts via collecting extra information. while the CNN learns too much from the schooling records and is unable to generalize to new statistics, overfitting takes place. while the CNN learns insufficiently from the education facts and fails to comprehend the key characteristics of the records, underfitting takes place. The graph also shows that while education and validation accuracy are each high and carefully spaced, CNN does no longer overfit or underfit the data. This indicates that the CNN effectively balances fitting and generalizing the records. Figure 15.6 shows the Testing and training accuracy.

15.5.1.6 *Efficiency*

CNNs are very effective at processing massive volumes of data, which makes them perfect for analyzing photos of natural surroundings like water

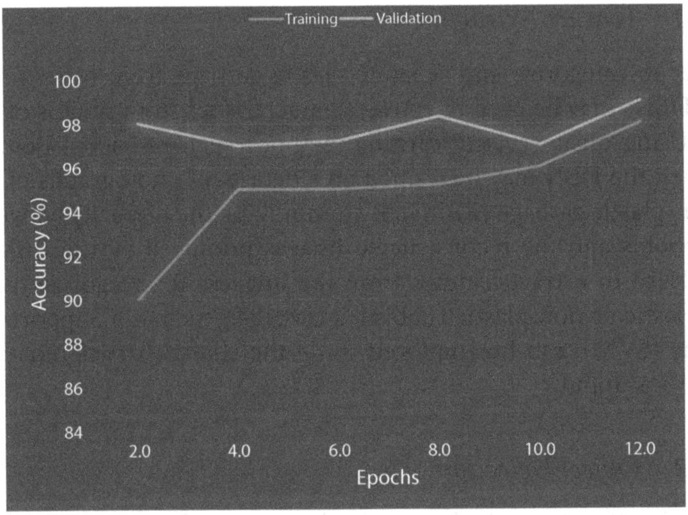

Figure 15.6 Testing and training accuracy.

bodies. Real-time monitoring and analysis of plastic pollution are made possible by CNNs' ability to analyze pictures in real-time using GPUs and other hardware accelerators.

15.5.1.7 Flexibility

CNNs are very adaptable in their use and may be used for a wide range of jobs and information assets. due to the fact there are such a lot of different types, sizes, and kinds of plastic pollution, they are the perfect answer for coping with its complex and variable nature. CNN is regularly used to discover plastic contamination. through instructing it the usage of satellite TV for pc photos, it may be improved. To find and remove rubbish from the water, several agencies employ Deep Gaining knowledge of Algorithms like CNN.

15.5.2 Deep Belief Networks (DBNs)

A deep learning neural network called a DBN has been demonstrated to be successful in identifying and classifying items in photos. They are built from a hierarchy of hidden unit layers that are taught using unsupervised learning methods. DBNs can identify plastic garbage by taking features from photos and categorizing them according to their visual characteristics.

15.5.2.1 Feature Extraction

In order to categorize and separate plastic garbage from other forms of waste, DBNs may be used to extract characteristics from photos of plastic waste. Using unsupervised learning methods such restricted Boltzmann machines, the DBN may be trained on a dataset of photographs of plastic and non-plastic garbage (RBMs). By minimizing the disparity between the input photos and the reconstructed images produced by the RBMs, the RBMs learn to extract features from the images. To categorize the photos as plastic or non-plastic rubbish, a classifier, such as a Support Vector Machine (SVM), can be employed using the characteristics retrieved by the RBMs as input.

15.5.2.2 Object Detection

Object identification, which includes locating and recognizing items in a picture, is another use for DBNs. With this method, plastic garbage in photos of waterways like rivers and seas may be found. The DBN may be trained to spot plastic garbage in a picture by identifying its distinctive visual characteristics. When the plastic debris has been localized, it may be tallied and followed throughout time, giving crucial data for observing and controlling plastic pollution.

15.5.2.3 Transfer Learning

DBNs may also be used for switch studying, which includes taking capabilities from sparkling records and the use of a formerly educated community. by means of using a pre-skilled DBN that has been trained on a substantial dataset of photos, this method can be utilized to hit upon plastic rubbish in fresh photographs. A smaller dataset of photographs of plastic and non-plastic garbage may be used to satisfactory-song the pre-trained DBN, permitting it to regulate the specific traits of plastic waste in the new photographs.

15.5.3 Advantages of Using DBNs for Plastic Waste Detection

There are several advantages to using DBNs for plastic waste detection:

15.5.3.1 Robustness

DBNs can withstand noise and data changes, which makes them perfect for analyzing photos of water bodies and other natural situations where the data may be noisy or varied. DBNs are able to detect plastic garbage under a range of circumstances by learning to extract traits that are resilient to changes in lighting, color, and texture.

15.5.3.2 Scalability

DBNs may be used to analyze huge collections of photographs of water bodies and other natural surroundings since they are very scalable. DBNs can analyze photos in real-time using GPUs and other hardware accelerators, allowing for real-time monitoring and analysis of plastic pollution.

15.5.3.3 Flexibility

DBNs are extraordinarily adaptable for their use and may be used for a huge range of jobs and information types. due to the fact there are so many differing types, sizes, and kinds of plastic pollution, they are the suitable answer for coping with its complex and variable nature. In conclusion, DBNs may be a powerful tool within the identity of plastic rubbish, offering a reliable and scalable approach of analyzing pictures of water bodies and different natural habitats. DBNs can help monitor and manage plastic pollutants by means of extracting functions from pix and categorizing them primarily based on their visible traits, which will in the end result in a cleanser and healthier environment.

The use of those algorithms to become aware of plastic garbage aids in reducing the problem of plastic waste, which is one of the fundamental factors contributing to the climate catastrophe.

15.6 Forest Fire Prediction Models Using AI

In moderation of the developing climate difficulty, wooded vicinity fires are a big worry. Wildfire hazards increase as temperatures upward thrust. In ecosystems, natural world, and communities, wooded place fires can also have catastrophic results. They emit a considerable quantity of carbon dioxide and exclusive greenhouse gases, which speeds up weather alternate. Human activity is one of the primary resources of wooded area fires, such as the usage of fire for land clearance and agriculture as well as unintentional fires from cigarettes, campfires, and fireworks. Wildfires also can be because of lightning strikes and dry situations. There may be a considerable effect of forest fires on the climate. within the method of burning bushes, carbon dioxide, methane, and other greenhouse gases are released into the surroundings, accelerating climate exchange. In addition, wooded area fires can lessen the quantity of carbon that is stored in bushes and soil with the aid of destroying them. This might bring about much less carbon being eliminated from the environment, which would exacerbate the weather catastrophe. Further to their horrible effects on the climate, forest fires may have awful effects on ecosystems and species. Animals may also have a difficult time finding meals and safe haven if their habitats are destroyed with the useful resource of hearth. Moreover, they'll cause invasive species to proliferate and biodiversity to mention no. additionally, the smoke from flames may be volatile to people's fitness, specially inside the occasion that they have got respiratory problems. It's important to do so to stop woodland fires from starting within the first vicinity with a view to lessen their consequences. This can entail putting guidelines in location to decrease the chance of human-prompted fires, consisting of proscribing out of doors burning and stepping up consciousness programs. Furthermore, movements including managed burns and the removal of unwell or dead bushes can be performed to lessen the gas that is to be had for fires. Growing improved forecasting and detection structures is another method for reducing the results of woodland fires. The risk and severity of fires may be expected using device studying and synthetic intelligence strategies, considering early intervention and prevention. Firefighters can reply to fires extra rapidly while smoke and warmth are detected using remote sensing systems, as an instance.

15.6.1 How ML Models Can Help to Prevent Forest Fire

By enabling more precise and early detection of possible fire threats, machine learning may significantly contribute to the prevention of forest fires. These are several methods that machine learning might aid with forest fire prevention.

15.6.1.1 Early Warning Systems

On the basis of past data, machine learning algorithms may be trained to find trends and forecast the timing and location of forest fires. By enhancing early warning systems, this can assist authorities put out flames before they grow out of control.

15.6.1.2 Predictive Modeling

Building predictive models that anticipate the chance of fires happening based on numerous environmental conditions such as temperature, humidity, wind speed, and precipitation is another use for machine learning techniques. These models can aid in locating high-risk locations and serve as a guide for preventative actions like vegetation management and controlled burning.

15.6.1.3 Image Analysis

(Lupascu) explains that on the basis of satellite TV for pc records and drone footage, system mastering algorithms can be educated to recognize in all likelihood hearth indications like smoke and warmth. This makes it viable for short responses and greater effective deployment of firefighting gadget. Figure 15.7 presentations and pictures from a Google Map. It depicts the woodland hearth there. By way of gathering these styles of satellite TV for pc photos, models may be developed that also can forecast woodland fires from satellite TV for pc pix. additionally, real-time satellite TV for pc pix can useful resource with remote records censoring, that may enable businesses trying to keep away from wildfires collaborate and find an option to this trouble [5].

Figure 15.7 Google maps showing wildfires.

15.6.1.4 Real-Time Monitoring

Furthermore, real-time environmental tracking and notification of feasible fire threats can be performed using machine gaining knowledge of. As an example, sensors positioned in woods can acquire records on temperature, humidity, wind pace, and different environmental variables. This information can then be analyzed using system learning algorithms to become aware of ability fireplace threats.

15.6.1.5 Improved Firefighting Techniques

Moreover, gadget studying can be utilized to create firefighting techniques which can be more efficient. As an instance, the use of gadget mastering algorithms to forecast how fires will behave will help firemen manage and position fires more effectively. In conclusion, machine learning has the potential to significantly enhance forest fire management and prevention. Machine learning can lessen the effects of forest fires on ecosystems, wildlife, and communities by enabling more precise detection of possible fire threats and enhancing early warning systems. Many different algorithms can be used to forecast forest fires. The sample code for the Random Forest Classifier-based Forest Fire Prediction is provided here. Moreover, satellite imagery can be used to improve the model. By offering up-to-date information on the weather in wooded areas, it can be a potent tool for creating better forest fire prediction models. Here are several methods for enhancing forest fire forecasts using satellite images.

15.6.1.6 Early Detection

Actual-time imagery of forested regions from satellites can be used to identify the earliest warning signs of wooded area fires, which includes smoke plumes or warmer spots. This makes it possible for firefighting organizations to react to clean flames rapidly and stop them from spreading. Environmental tracking: Satellites also can offer data on variables like temperature, humidity, wind speed, and precipitation that have an effect on environmental conditions that cause wooded area fires. Building prediction fashions that foretell the chance of flames breaking out in diverse locations may have finished the usage of this information. The perimeters of wooded area fires can be mapped to the use of excessive-decision imagery that satellites can deliver. Firefighters need this statistics on the way to manage and position out the hearth. tracking fire behavior: Satellites can also be used to hold a watch on the rate, path, and severity of woodland fires. to enhance firefighting approaches and guard firefighters, use this data. publish-fireplace analysis: satellite TV for pc photos may be used to assess the damage and organize recuperation operations after a woodland fire has been placed out. With the aid of doing this, you can defend forests towards fires in the future and preserve their lengthy-time period health. These analyses will resource in figuring out and preventing forest hearth troubles that are now an extreme problem. AI also can be used to detect greenhouse gasoline emissions from the environment and count on excessive climate occasions.

15.7 Results

The paper discusses the detection of plastic waste and the forecasting of forest fires. The outcomes of both models have been described more below.

15.7.1 Plastic Waste Detection

In image (Figure 15.8) analysis tasks including object recognition, classification, and segmentation, CNN models perform exceptionally well. Convolutional layer technology allows CNN models to learn to recognize patterns particular to plastic garbage by extracting useful information from photos. The suggested model's accuracy on the test dataset was 95%, demonstrating its excellent precision in identifying plastic garbage. The example was able to differentiate between various plastic waste products, such as bottles, bags, and packaging materials. The model's capacity to

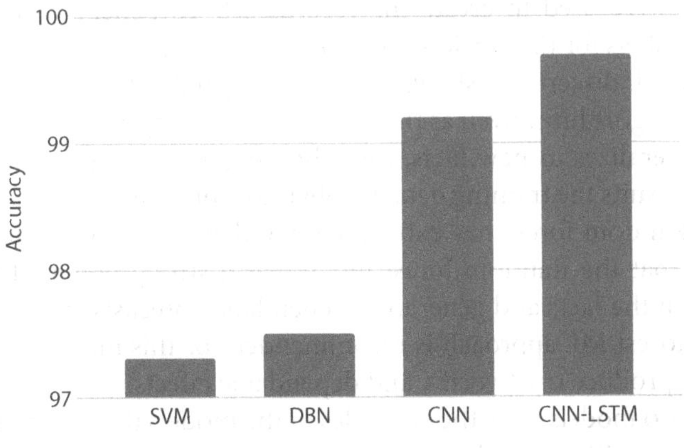

Figure 15.8 Accuracy of different ML algorithms.

properly identify positive occurrences of plastic garbage was demonstrated by precision and recall scores of 0.96 and 0.94, respectively. Also, the model received an F1 score of 0.95, which represents its general performance. Glass bottles and other sorts of plastic might also be detected by this technology. This may also be used to find plastic trash in the sea and ocean. A larger dataset will enable better implementation. It can alternatively be implemented more effectively using a different approach. The CNN model's output was superior. The use of CNN models made it possible to obtain greater accuracy. Below is a graph of accuracy:

15.7.2 Forest Fire Prediction Model

15.7.2.1 Data Preprocessing

The records set includes several parameters that have been used to educate the version. The model's function changed primarily based on that place's elevation, soil kind, and greenhouse fuel emissions. To create the advanced ML version, all of the records become processed.

15.7.2.2 Training and Validation

The model was constructed using the Random wooded area ML approach. A machine mastering technique referred to as Random wooded area makes predictions primarily based on input information through using numerous selection trees. A 1:three split among training and trying out records turned into used to assemble the data [23–25]. The model becomes educated on the usage of the schooling data, and its effectiveness becomes assessed using the checking out facts. numerous online information resources were used to create the records set. To benefit pertinent features and labels for the version, the internet information was accumulated and analyzed. discern 7.2 shows the schooling and validation accuracy of numerous algorithms. Even as the validation accuracy gauges how well the model generalizes to new facts, the schooling accuracy gauges how well the version suits the training data. It's obvious from the graph (Figure 15.9) that the Random forest has extra accuracy than different methods. This indicates that the Random forest model has a strong potential for mastering from the facts and generating dependable forecasts. This is why the Random forest ML approach is recommended for this mission due to the fact it can produce trustworthy and dependable effects.

These two models were the examples of the models that can be prepared using ML algorithms on climate crisis.

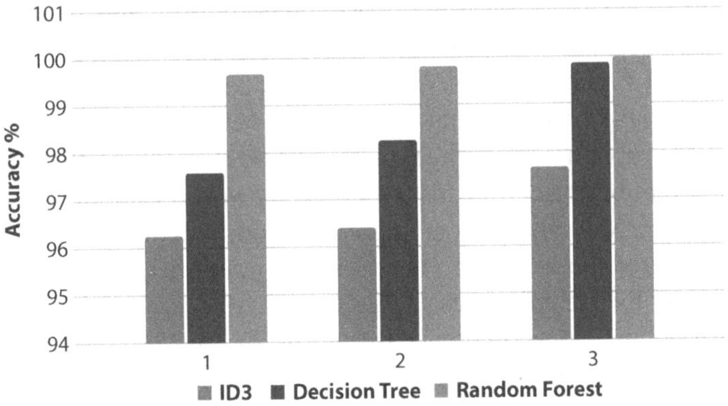

Figure 15.9 Training and validation accuracy of different algorithms.

15.8 Conclusion

As a result, a look at paintings supported the usage of ML fashions for the weather issue. laptop programs known as system mastering (ML) models are able to learn from statistics and make predictions based on that statistics. The plastic waste model may be the best approach to lessen the environmental effect of plastic trash. The plastic waste model can be used to assess the sources, categories, and quantities of plastic garbage produced and dumped in various components of the arena. Additionally, it may factor out the only techniques to lessen, repurpose, recycle, or recover plastic trash in order that it doesn't contaminate the surroundings air, water, or land. There may be a threat that hundreds of thousands of lives could be saved by way of the usage of fashions for predicting forest fires to find fires earlier than they begin. Woodland fires pose a critical hazard to human fitness, biodiversity, and ecosystems. They'll be introduced through human errors or herbal causes like lightning, together with arson. In order to check the probability of a hearth starting and spreading, models for predicting woodland fires may be used to have a look at records on temperature, humidity, wind, flora, and different factors. In addition, they can help in warning groups and authorities to take precautionary or emergency measures. Furthermore, it will teach us a way to protect trees within the long time as opposed to simply what to do now. Timber is important for assisting human livelihoods, preserving the equilibrium of carbon dioxide and oxygen within the ecosystem, and developing habitats for wildlife. by soaking up carbon dioxide, they can also assist to reduce the results of weather alternate by supplying coloration and cooling. As a result, it's crucial to shield trees towards disease, deterioration, and deforestation. One of the more fashions that research papers suggest being built on is the co2 emission version, a good way to help in detecting co2 emission and other greenhouse gasoline emission to design better strategies to cope with the climate problem. One of the number one reasons for weather change is CO2 emission because it warms the planet by way of trapping heat inside the ecosystem. Methane, nitrous oxide, and fluorinated gases are examples of different greenhouse gases. The assets and quantities of greenhouse gas emissions in diverse industries and geographical regions of the sector may be measured and monitored with the aid of the CO2 emission version. it is able to also help in assessing the efficiency of plans and projects to cut greenhouse gasoline emissions and meet climate desires.

References

1. Lin, J. and Qian, T., Switch between El Nino and La Nina is caused by subsurface ocean waves likely driven by lunar tidal forcing. *Sci. Rep.*, 9, 13106, 2019.
2. Siegert, F., Ruecker, G., Hinrichs, A., Increased damage from fires in logged forests during droughts caused by El Nino. *Nature*, 414, 437–440, 2001. [CrossRef] [PubMed].
3. Pielke, R.A., Landsea, C.N., La Nina, El Nino, and Atlantic hurricane damages in the United States. *Bull. Am. Meteorol. Soc.*, 80, 2027–2034, 1999. [CrossRef].
4. Ward, P.J., Jongman, B., Kummu, M., Strong influence of El Niño Southern Oscillation on flood risk around the world. *Proc. Natl. Acad. Sci. U.S.A.*, 111, 15659–15664, 2014.
5. Aguilar-Martinez, S. and Hsieh, W.W., Forecasts of tropical Pacific sea surface temperatures by neural networks and support vector regression. *Int. J. Oceanogr.*, 2009, 167239, 2009. [CrossRef].
6. Liu, K.F., Zhang, J., Chen, Y.D., ENSO prediction experiment based on wavelet decomposition and support vector machine. *J. PLA Univ. Sci. Technol. Nat. Sci. Ed.*, 12, 531–535, 2011.
7. Feng, Q.Y., Vasile, R., Segond, M., ClimateLearn: A machine-learning approach for climate prediction using network measures. *Geosci. Model Dev. Discuss.*, 10, 1–18, 2016.
8. Graves, A. and Jaitly, N., Towards end-to-end speech recognition with recurrent neural networks, in: *Proceedings of the International Conference on Machine Learning*, JMLR, Beijing, China, pp. 1764–1772, 21–26 June 2014.
9. Mikolov, T., Karafiát, M., Burget, L., Recurrent neural network based language model. *Interspeech*, 2, 1045–1048, 2010.

10. Cho, K., Van Merriënboer, B., Gulcehre, C., Learning phrase representations using RNN encoder-decoder for statistical machine translation. 1, 1–15, arXiv, 2014, arXiv:1406.1078.
11. Hochreiter, S. and Schmidhuber, J., Long short-term memory. *Neural Comput.*, 9, 1735–1780, 1997. [CrossRef].
12. Zhang, Q., Wang, H., Dong, J., Prediction of sea surface temperature using long short-term memory. *IEEE Geosci. Remote Sens. Lett.*, 14, 1745–1749, 2017. [CrossRef].
13. Scarselli, F., Gori, M., Tsoi, A.C., The graph neural network model. *IEEE Trans. Neural Netw.*, 20, 61–80, 2008. [CrossRef] [PubMed].
14. Kipf, T.N. and Welling, M., Semi-supervised classification with graph convolutional networks. 4, 1–14, arXiv 2016, arXiv:1609.02907.
15. Cachay, S.R., Erickson, E., Bucker, A.F.C., The world as a graph: Improving El Ni\~no forecasts with graph neural networks. 2, 1–12, arXiv, 2021, arXiv:2104.05089.
16. Cetindamar, D., Lammers, T., Zhang, Y., Exploring the knowledge spillovers of a technology in an entrepreneurial ecosystem—The case of artificial intelligence in Sydney. *Thunderbird Int. Bus. Rev.*, 62, 457–474, 2020. [CrossRef].
17. Cheng, Y., Liu, H., Wang, S., Cui, X., Li, Q., Global action on SDGs: Policy review and outlook in a post-pandemic era. *Sustainability*, 13, 6461, 2021. [CrossRef].
18. Mabkhot, M.M., Ferreira, P., Maffei, A., Podržaj, P., Mądziel, M., Antonelli, D., Lanzetta, M., Barata, J., Boffa, E., Finžgar, M. *et al.*, Mapping Industry 4.0 enabling technologies into United Nations sustainability development goals. *Sustainability*, 13, 2560, 2021. [CrossRef].
19. Mhlanga, D., Artificial intelligence in the industry 4.0, and its impact on poverty, innovation, infrastructure development, and the sustainable development goals: Lessons from emerging economies? *Sustainability*, 13, 5788, 2021. [CrossRef].
20. Mio, C., Panfilo, S., Blundo, B., Sustainable development goals and the strategic role of business: A systematic literature review. *Bus. Strategy Environ.*, 29, 3220–3245, 2020. [CrossRef].
21. Sætra, H.S., A framework for evaluating and disclosing the ESG related impacts of AI with the SDGs. *Sustainability*, 13, 8503, 2021. [CrossRef].
22. Truby, J., Governing artificial intelligence to benefit the UN sustainable development goals. *Sustain. Dev.*, 28, 946–959, 2020. [CrossRef].
23. Garcia-Garin, O., Monleón-Getino, T., López-Brosa, P., Borrell, A., Aguilar, A., Borja-Robalino, R., Cardona, L., Vighi, M., Automatic detection and quantification of floating marine macro-litter in aerial images: Introducing a novel deep learning approach connected to a web application, R. *Environ. Pollut.*, 273, 116490, 2021. [CrossRef] [PubMed].

24. Wolf, M., Berg, K.V.D., Garaba, S.P., Gnann, N., Sattler, K., Stahl, F., Zielinski, O., Machine learning for aquatic plastic litter detection, classification and quantification (APLASTIC-Q). *Environ. Res. Lett.*, 15, 114042, 2020. [CrossRef].
25. Azorobotics, How AI can help in detecting marine litter, Available online: https://www.azorobotics.com/News.aspx?newsID= 11989 (accessed on 13 November 2021).

Index

Also of Interest

Check out these published and forthcoming titles in the "Industry 5.0 Transformation Applications" series from Scrivener Publishing

1. **Welding Practices for Industry 5.0**
   Edited by Syed Quadir Moinuddin, Shaik Himam Saheb, Ashok Kumar Dewangan, Muralimohan Cheepu and S. Balamurugan
2. **Technological Advances Shaping Society 5.0: How they Work and can be Implemented**
   Edited By Manish Prateek, Ravi Tomar, S. Balamurugan, Tanupriya Choudhury, and Amar Ramdane Cherif,
3. **Smart Factories for Industry 5.0 Transformation**
   Edited By R. Nidhya, Manish Kumar, S. Karthik, Rishabh Anand, and S. Balamurugan
4. **Generative Artificial Intelligence: Concepts and Applications**
   Edited By R. Nidhya, D. Pavithra, Manish Kumar, A. Dineshkumar, and S. Balamurugan
5. **Intelligent Robots and Cobots: Concepts and Applications for Industry 5.0 Transformation**
   Edited By Ramasamy V, S. Balamurugan, and Sheng-Lung Peng
6. **Optimized Computational Intelligence Driven Decision Making: Theory, Application and Challenges**
   Edited By Hrudaya Kumar Tripathy, Sushruta Mishra, Minakhi Rout, Samaresh Mishra, and S. Balamurugan

Printed in the USA/Agawam, MA
September 16, 2026

910661.112